Isotopes and Radiation in Agricultural Sciences

Isotopes and Radiation in Agricultural Sciences

Volume 1

Soil–Plant–Water Relationships

Edited by

M. F. L'ANNUNZIATA

International Atomic Energy Agency
Vienna, Austria

and

J. O. LEGG

Department of Agronomy
University of Arkansas, Fayetteville, USA

1984

ACADEMIC PRESS

(*Harcourt Brace Jovanovich, Publishers*)

London Orlando San Diego San Francisco New York
Toronto Montreal Sydney Tokyo São Paulo

ACADEMIC PRESS INC. (LONDON) LTD
24–28 Oval Road
London NW1 7DX

US Edition published by
ACADEMIC PRESS INC.
(Harcourt Brace Jovanovich, Inc.)
Orlando, Florida 32887

British Library Cataloguing in Publication Data

Isotopes and radiation in agricultural sciences.
Vol. 1: Soil-plant-water relationships
1. Radioisotopes in agriculture
I. L'Annunziata, M.F. II. Legg, J.O.
630'.2'41388 S598.5

ISBN 0-12-436601-5
LCCCN 83 72401

Filmset by Eta Services (Typesetters) Ltd., Beccles, Suffolk
Printed in Great Britain by Thomson Litho Ltd., East Kilbride, Scotland

Contributors

D. K. Cassel *Department of Soil Science, North Carolina State University, Raleigh, North Carolina 27650, USA*

R. Fiedler *International Atomic Energy Agency, Wagramerstrasse 5, PO Box 200, A-1400 Vienna, Austria*

J. N. Ladd *CSIRO Division of Soils, Private Bag No. 2, Glen Osmond, South Australia 5064, Australia*

M. F. L'Annunziata *International Atomic Energy Agency, Wagramerstrasse 5, PO Box 200, A-1400 Vienna, Austria*

J. O. Legg *Department of Agronomy, University of Arkansas, Fayetteville, Arkansas 72701, USA*

J. K. Martin *CSIRO Division of Soils, Private Bag No. 2, Glen Osmond, South Australia 5064, Australia*

R. G. Menzel *USDA-SEA-ARS, Southern Plains Watershed and Water Quality Laboratory, PO Box 1430, 801 Wilson Street, Durant, Oklahoma 74701, USA*

D. R. Nielsen *Department of Land, Air and Water Resources, University of California, Davis, California 95616, USA*

S. J. Smith *USDA-SEA-ARS, Southern Plains Watershed and Water Quality Laboratory, PO Box 1430, 801 Wilson Street, Durant, Oklahoma 74701, USA*

D. C. Wolf *Department of Agronomy, University of Arkansas, Fayetteville, Arkansas 72701, USA*

Foreword

by Hans Blix

Nuclear techniques in agricultural research have played a major role in increasing world food production to the level at which it is today. A few areas in which these techniques, largely based on the use of isotopes and radiation, have attained prominence are insect pest control, food preservation, plant nutrition, animal health and production, crop management and pesticide residue studies. Nuclear methods have contributed in no small way to our understanding of biological processes and, when translated into practice, have made a subsequent impact in various fields of agriculture.

According to a recent FAO report, the rate at which food production increased in the developing countries during the past decade was just sufficient to meet the increased demand for food resulting from population growth. While there have been remarkable advances in the agricultural sciences in recent years, the need for increasing global food production is still critical and will become more so as time goes on. If further advances are to be made in this respect, our need for nuclear techniques will become even greater.

I am pleased that this book, comprising contributions from international experts in various agricultural disciplines, is now available. The information it contains will doubtless facilitate the application of nuclear methods in meeting the challenges facing agriculture and, as a consequence, contribute to the improvement of human health the world over.

April 1983

HANS BLIX
Director General
International Atomic Energy Agency

Preface

Almost three decades have passed since the publication, in 1955, of C. L. Comar's pioneering book "Radioisotopes in Biology and Agriculture". At the time of its publication, research methods utilizing radioisotopes as tracers were in the very early stages of development. Since then, however, so many advancements have been made that the theory and applications of isotopes and radiation in various branches of agriculture have become distinct sciences in themselves. In view of this fact, it would be fatuous for any one person to attempt to author a book of any depth concerning the application of isotopes and radiation in all of the agricultural sciences. Our only alternative was to seek out contributors specialized in various disciplines in order to produce a treatise that would represent the "state of the art" on applications of nuclear techniques in agricultural research.

Although we cannot claim to have been exhaustive in our coverage of the agricultural sciences, we have tried to include those fields whose development has depended to a great extent on the use of isotope tracers and nuclear radiation. The outcome of this endeavour is presented as two volumes. Chapters related directly to studies with soils, soil water and the improvement of crop production through soil and water management are found in the first volume subtitled "Soil–Plant–Water Relationships". The second volume, subtitled "Animals, Plants, Food and the Environment", covers the isotope tracer techniques used in studies on the biochemistry and physiology of living systems important to agriculture, the preservation of food and the environment.

Emphasis in Volume 1 has been placed on nuclear techniques used in the study of soil–plant–water relationships; through the use of these techniques, it has been possible to achieve significant increases in crop production. It was also felt necessary to include in the first volume chapters on soil organic matter and soil microbiology. Radiocarbon has played a key role as a tracer in studies on the dynamics of soil carbon—in particular, the formation and degradation of organic matter affecting the physical and chemical properties of soils. Our knowledge of the role of microorganisms in soil and plant biochemical processes as agents of biological dinitrogen fixation, which affects soil fertility and plant growth, is largely a result of the use of isotope tracers. The tracer techniques used to study these processes are discussed in this volume.

As an aid to researchers and teachers in the agricultural sciences and for the sake of completeness, we have included a chapter on the detection and measurement of radionuclides used as tracers in the agricultural and biological sciences. A separate chapter on the measurement of the stable isotope ^{15}N was prepared because of the ever increasing use of this nuclide as a tracer in agricultural research, particularly in studies of biological dinitrogen fixation and fertilizer use efficiency. These chapters have been included in Volume 1; nevertheless, the theory and techniques presented in these two chapters are directly relevant to all fields of agricultural science.

Practical applications of isotope tracers and radiation in studies on the health and reproduction of animals of economic importance could not be ignored. Radioimmunoassay and the preparation of radiation-attenuated vaccines are but two of the many techniques covered in Volume 2 that have been effectively utilized to improve animal health and increase animal production.

Isotope tracer and radiation techniques employed to study and control insect pests, which affect crop production and animal health, are also dealt with in the second volume, including the sterile-insect-technique which has been used to control and/or eradicate populations of insect pests.

We have included in Volume 2 chapters on isotope techniques used to study biochemical and physiological transformations in microorganisms, plants and animals; our knowledge of anabolic and catabolic reactions in living systems has been acquired, for the most part, through isotope tracer studies. Classical work on plant photosynthesis and cholesterol biosynthesis is presented to illustrate remarkable advances made in elucidating biosynthetic pathways with the aid of ^{14}C as a carbon isotope tracer. We have also tried to show in this volume how nuclear magnetic resonance spectrometry, used in conjunction with the stable isotope ^{13}C, has totally revolutionized and simplified studies on the biosynthesis of carbon compounds.

While the objective of most agricultural research is to improve the quantity and quality of food, great losses of agricultural produce may occur in storage. Nuclear irradiation is used to reduce storage losses of foods of many kinds, and much research is in progress aimed at broadening the applicability of this method. Accordingly, the current situation and techniques used in irradiation preservation research are dealt with in Volume 2.

Agricultural pesticides have been shown to be effective in controlling microbial, insect and weed infestations detrimental to the overall yield and quality of crops; however, the presence of pesticide residues in crops has given rise to concern due to the possible adverse effects of such residues on human health and the environment. A chapter in Volume 2 is devoted to isotope tracer techniques employed in investigating the fate of such residues in crops and the environment.

A field that cannot be overlooked due to its significant role in plant production is mutation breeding. We have not covered crop mutation breeding in this book principally because the use of nuclear radiation to induce mutation in plant genetic material represents only one of many techniques employed in typical plant breeding programmes. Furthermore, the use of chemical mutagens, which lies outside the scope of isotope and radiation techniques, is, in many cases, a more effective means of inducing mutation. For those interested in this field, the text "Induced Mutations—A Tool in Plant Research", published by the International Atomic Energy Agency in Vienna, is recommended. It contains the proceedings of an FAO/IAEA symposium, held in Vienna during 9–13 March 1981, on the use of induced mutations for crop plant improvement.

While it was our primary intention to prepare a treatise for the research investigator and student in the agricultural sciences, we feel that these volumes should also be of value to anyone working or studying in the field of biochemistry and the life sciences. The isotope tracer techniques employed in studies of microbial, plant and animal systems in agriculture are common to other fields where biochemical processes are investigated. Likewise, the chapters on detection and measurement of radionuclides and stable isotopes in Volume 1 should be of interest to anyone who uses or plans to use isotopes as tracers in fields such as medicine, hydrology, industry, biology and chemistry, to mention only a few.

We are most grateful to the authors for their co-operation and devotion to this project. Their contributions were not only submitted on time but they have covered their fields of expertise meritoriously, which we believe fulfils the objectives of these volumes. Sincere thanks are extended to C. Vélez-Ocón, S. Srisukh, P. Schultze-Kraft and S. B. Hammond and others at the International Atomic Energy Agency for having supported this endeavour. Furthermore, we gratefully acknowledge the assistance of the following persons who reviewed some of the material in this book: E. H. Belcher, K. Buchtela, D. J. W. Bullock, M. Davis, J. L. Gringorten, K. S. Kumarasinghe, P. H. Makin, J. P. Martin, E. A. Paul, G. Wagner and R. W. Weaver. Above all, we thank our wives, Reyna and Effie, for their help, understanding and unrelenting support.

January 1984

M.F.L'A.
J.O.L.

Contents

4. Soil Microbiology

D. C. WOLF and J. O. LEGG

5. The Detection and Measurement of Radionuclides

MICHAEL F. L'ANNUNZIATA

6. The Measurement of ^{15}N

RUDOLF FIEDLER

Contents of Volume 2

Symbols, Abbreviations and Acronyms

Å	angstrom
AA	atomic absorption
ACP	acyl carrier protein
ADC	analogue-to-digital converter
AES	automatic external standardization
α	alpha particle
AMP	adenosine monophosphate
ANDA	7-amino-1,3-naphthalenedisulphonic acid
APS	adenosine phosphosulphate
ATP	adenosine triphosphate
ATPase	adenosine triphosphase
β^-	negatron, negative beta particle
B_{12}	vitamin B_{12}
β^+	positron, positive beta particle
BK	background
Bq	becquerel, 1 disintegration per second (dps)
BP	before present
c	cycles
C	coulomb
°C	degrees Celsius
CAC	Codex Alimentarius Commission
CAM	crassulacean acid metabolism
Ci	curie, 3.7×10^{10} dps, 3.7×10^{10} Bq
CIPC	chlorpropham, isopropyl-3-chlorophenyl-carbamate
cm	centimetre
CoA	coenzyme A
cpm	counts per minute
CRT	cathode ray tube
C3	three-carbon photosynthetic pathway or plant
C4	four-carbon photosynthetic pathway or plant
D	deuterium
D_{10}	dose lethal to 90% of a microbial population
2,4-D	2,4-dichlorophenoxyacetic acid
DAHP	3-deoxy-D-arabino-heptulosonate-7-phosphate
DBCP	1,2-dibromo-3-chloropropane
d.c.	direct current
DDT	dichlorodiphenyltrichloroethane
DFP	diisopropylfluorophosphate

DHAP	dihydroxyacetone phosphate
DHQ	dehydroquinate
DNA	deoxyribonucleic acid
D_2O	heavy water, deuterium oxide
dpm	disintegrations per minute
dps	disintegrations per second, Bq
E	counting efficiency
EC	electron capture
EDB	ethylene dibromide
EDTA	ethylenediaminetetraacetic acid
E_{max}	maximum energy
ESCR	external standard channels ratio
ESP	inverse relative pulse height
eV	electron volt
FAO	Food and Agriculture Organization of the United Nations
FDA	United States Food and Drug Administration
FDNP	1-fluoro-2,4-dinitrophenol
f.f.	forest floor
g	gramme
γ	gamma ray
GA_3	gibberellic acid, gibberellin A_3
GC	gas chromatography
GM	Geiger–Müller
GOGAT	glutamine oxoglutarate amino transferase
GS	glutamine synthetase
Gy	gray, 1 J kg^{-1}, 100 rad
h	hour
ha	hectare
HPLC	high performance liquid chromatography
HTO	tritiated water
IAA	indole acetic acid
IAEA	International Atomic Energy Agency
IFIP	International Food Irradiation Project
in	inch
i.r.	infra red
i.v.	intravenous
J	joule
JH	juvenile hormone
K	kelvin
kCi	kilocurie
keV	thousand electron volts
kG	kilogauss
kg	kilogramme
kGy	kilogray
kJ	kilojoule
K_m	Michaelis constant

kR	kiloroentgen
kW	kilowatt
l	litre
LD_{50}	dose lethal to 50% of the animals tested
LED	light-emitting diode
M	molar solution
m	metre
mbar	millibar
MCi	megacurie
mCi	millicurie
Mc	megacycles
Me	methyl, CH_3
m/e	mass-to-charge ratio
MeV	million electron volts
mg	milligramme
MHz	megahertz
min	minute
MIT	mineralization-immobilization turnover
ml	millilitre
mM	millimolar
mm	millimetre
mosm	milliosmole
MRR	mark-release-recapture
MS	mass spectrometry
MW	molecular weight
μCi	microcurie
μg	microgramme
μl	microlitre
μm	micrometre
μs	microsecond
N	normal solution
n	index of refraction
NAA	neutron activation analysis
NAD	nicotinamide adenine dinucleotide
NADH	reduced nicotinamide adenine dinucleotide
NADPH	reduced nicotinamide adenine dinucleotide phosphate, TPNH
NAG	*N*-acetylglucosamine
NAM	*N*-acetylmuramic acid
nm	nanometre
NMR	nuclear magnetic resonance
NOE	nuclear Overhauser enhancement
ns	nanosecond
ORNL	Oak Ridge National Laboratory
osm	osmole
PAPS	3′-phosphoadenosine-5′-phosphosulphate
P_i	inorganic phosphate

pCi	picocurie
PCR	photosynthetic carbon reduction
PEP	phosphoenol pyruvate
PEPC	phosphoenol pyruvate carboxylase
PGA	3-phosphoglyceric acid
PH	pulse height
PM	photomultiplier
POPOP	2,2′-*p*-phenylene-*bis*-5-phenyloxazole
PP_i	pyrophosphate
ppm	parts per million
PPO	2,5-diphenyloxazole
PPP	pentose pyrophosphate pathway
PVC	polyvinylchloride
R	roentgen, 2.58×10^{-4} C kg^{-1}
rad	100 erg g^{-1}, 10^{-2} J kg^{-1}
RBC	red blood cells
RuBP	ribulose-1,5-*bis*-phosphate
RuBPC	ribulose-*bis*-phosphate carboxylase
rem	10^{-2} sievert
r.h.	relative humidity
RNA	ribonucleic acid
rRNA	ribosomal ribonucleic acid
tRNA	transfer ribonucleic acid
RPH	relative pulse height
Rubisco	ribulose-*bis*-phosphate carboxylase
RWC	relative water content
S	sedimentation coefficient
s	second
SA	specific activity
SCR	sample channels ratio
SIE	spectral index of the external standard
SIS	spectral index of the sample
SIT	sterile insect technique
S/N	signal-to-noise
Sv	sievert, 1 J kg^{-1}, 100 rem
T	tritium; tesla = 10^4 gauss
t	tonne
$t_{1/2}$	half-life
TCA	tricarboxylic acid
THO	tritiated water
TLC	thin layer chromatography
TMS	tetramethylsilane
TPNH	reduced triphosphopyridine nucleotide, NADPH
TU	tritium unit = 1 atom ^{3}H/^{18}10 atoms ^{1}H
UDP	uridine diphosphate
UDPG	uridine diphosphate glucose

USDA	United States Department of Agriculture
u.v.	ultraviolet
V	volt
V_{max}	maximum velocity
W	watt
WHO	World Health Organization
w:v	weight-to-volume
y	year

1

Soil Fertility and Plant Nutrition

R. G. Menzel and S. J. Smith

I. Introduction

Soon after their discovery, isotope tracers were introduced into investigations on soil fertility and plant nutrition. Tracers have provided a standard for the measurement of the quantities of soil nutrients that are available to plants. They have provided information on the movement and fate of fertilizer materials in the soil–plant system that could not be obtained by other means. Isotope tracers have facilitated studies of the distribution of plant roots, especially the actively absorbing roots that are most important in plant nutrition. Finally, they have permitted better understanding of soil erosion processes which deplete soil fertility.

In this chapter, we will discuss the applications of isotopic and related techniques, including autoradiography, radiation absorption, radiation scattering and activation analysis. We will point out the unique information that can be obtained with isotopes and radiation techniques. The advantages and disadvantages of these techniques will be discussed in relation to other methods of obtaining similar information. After a general review of current techniques, specific applications will be described in more detail.

II. Isotope tracers

The first extensive use of isotope tracers in plant nutrition studies was made in the 1940s. The stable isotope of nitrogen, ^{15}N, was used by Norman and Werkman (1943) to determine the amount of nitrogen taken up by soybeans from plant materials decomposing in the soil. The radioactive isotope of phosphorus, ^{32}P, was used to study the utilization of phosphatic fertilizers by various crops in a series of greenhouse (Dean *et al.*, 1948) and field experiments (Nelson *et al.*, 1948). These experiments established patterns for tracer studies that have been repeated many times with other plant nutrients and fertilizer materials. Essentially, a tracer was incorporated into a fertilizer material, or other nutrient source of interest, and the amount of tracer taken up by the plant was measured directly.

An ideal tracer should behave exactly as the material being traced, and should have no effect on the behavior of that material in the system under study. Because their atomic weights differ, isotopes have slightly different reaction rates (Melander, 1960; Edwards, 1975). These differences can be neglected in most tracer applications. Furthermore, detection of many isotope tracers is so sensitive that the amount of material added in tracer experiments can be very small compared with the amounts already present. Thus, the system is disturbed very little.

When an isotope tracer is added to a soil–plant system, it moves into or exchanges with various components of the system. If its radiation is sufficiently penetrating, a radioactive isotope can be traced without destructive sampling as it arrives in various parts of a plant. More generally, samples are removed from the plant or soil, and different fractions are examined for the presence of tracer. Quantitative measurements of process kinetics or equilibrium distributions require an understanding of the concepts of specific activity and isotope dilution.

Specific activity is usually expressed as the amount of radioactivity per gram of element. The chemical and physical form of the element must also be defined. A pure radioactive isotope is "carrier-free", and has the highest possible specific activity for that isotope. Some carrier (stable isotopes of the same element) is applied with most radioactive isotopes. The atom percent abundance of stable isotope tracers is analogous to the specific activity of radioactive tracers.

Isotope dilution is the reduction in specific activity that occurs as a tracer isotope exchanges with other isotopes in an experimental system. If the tracer equilibrates with a well defined compartment, the amount of material in the compartment can be measured by determining specific activity before and after equilibration. In this case, the initial specific activity is $S_1 = X^*/X$, and the final specific activity is $S_2 = X^*/(X + A)$, where X is the amount of added element and A is the amount of equilibrating element in the compartment. Since the amount of tracer isotope, X^*, does not change,

$$S_1 X = S_2(X + A) \quad \text{and} \quad A = \frac{S_1 X}{S_2} - X. \tag{1.1}$$

Soil–plant systems consist of many compartments, and a tracer exchanges with each at characteristic rates. By determining the rate of change in specific activity in different compartments, the mechanisms and kinetics of chemical reactions in soil and nutrient uptake by plants have been determined. Many examples will be discussed in this chapter.

Detection of isotope tracers requires special instrumentation which is more complex and costly for stable than for radioactive isotopes. Stable isotopes are usually detected by mass spectrometry, although optical emission is becoming popular for the analysis of ^{15}N (Chapter 6 in this volume). Radioactive isotopes are usually detected by ionization or scintillation counting techniques, depending on the types and energies of the radiations (Chapter 5 in this volume). The radioactive isotopes most used in soil fertility and plant nutrition studies emit beta and/or gamma radiation. Beta radiation penetrates only a few millimeters of water, soil or plant material, and hence is detected inefficiently in large samples. The more penetrating gamma radiation can be measured through several centimeters of sample material. Unlike beta radiation, gamma radiation occurs at characteristic energies, allowing isotopes to be identified by gamma-ray spectroscopy.

Suitable isotopes, either stable or radioactive, exist for most elements of interest in soil fertility and plant nutrition. A list of isotopes with potential for use as tracers is given in Table 1.1. Some typical applications are indicated with selected literature references for many of the isotopes. In general, stable isotopes have the advantages of no radiation hazard and no loss of tracer by radioactive decay, while radioactive isotopes have the advantages of high sensitivity and low cost.

Radiation hazards must be controlled in all uses of radioactive isotopes or radiation techniques. Books by L'Annunziata (1979) and Vose (1980) discuss the hazards and appropriate protective measures in relation to agricultural research. Personnel handling the radioactive material must take precautions

Table 1.1 Isotope tracers of interest in soil fertility and plant nutrition studies, with some examples of their use

(a) Stable tracers

Isotope (approx. price,[a] if available)	Natural abundance (%)	Typical application	References
Boron-10	19.7	Foliar absorption, neutron activation	Martini and Thellier (1980)
Nitrogen-14 ($200/kg, 99.99 % enriched)	99.634	Single season fertilizer utilization (suitable for use on large field plots)	Hauck and Bremner (1976), Edwards (1978), Vose (1980)
Nitrogen-15 ($100/g, 99 % enriched)	0.366	Nitrogen transformations in soils	Hauck and Bystrom (1970), Hauck and Bremner (1976), Vose (1980)
		Biological N fixation	Fried and Broeshart (1975), Bremner (1977)
		Long-term fertilizer utilization (cost limits field use to small plots)	Westerman and Kurtz (1972) Chichester and Smith (1978)
Magnesium-26[b]	11.29		
Potassium-41[b]	6.77		
Copper-65[b]	30.9		

[a] Prices are given in US dollars as of 1 January 1983.
[b] These nuclides are of potential interest because of the absence of long-lived radioactive tracers for magnesium and copper, and of the low commercial availability of radioactive ^{40}K. ^{65}Cu has been used in animal nutrition studies (Harvey, 1978; Buckley *et al.*, 1982).

(b) Radioactive tracers

Isotope (approx. price,[a] if available)	Half-life	Radiation characteristics		Typical applications	References
		Type	Energy[b]		
Nitrogen-13	10.0 min	β^+	1.20 (100)	Biological N fixation Denitrification (Very short-term studies)	Dazzo and Hubbell (1974) Tiedje (1978)
		γ	0.511 (from annihilation radiation)		
Magnesium-28	21.2 h	β^-	0.46 (100)	Movement in plants	Schimansky (1977)
		γ	0.031 (100), 0.400 (37.8) 0.95 (37.8) 1.35 (56.8)		
Phosphorus-32 ($50/mCi)	14.3 day	β^-	1.71 (100)	Exchangeable P in soils	Jose and Krishnamoorthy (1972)
				Availability of P fertilizers	Dean *et al.* (1948), Nelson *et al.* (1948)
				Root distribution in soil	Goodman and Collison (1981)
				Residual fertilizer availability	Spratt *et al.* (1980)
Phosphorus-33 ($200/mCi)	24.4 day	β^-	0.248 (100)	Autoradiography of roots Diffusion in soil Double labeling (with ^{32}P)	Claassen *et al.* (1981) Bhat and Nye (1973) Christians *et al.* (1981), Broeshart and Nethsinghe (1972)

Table 1.1—*cont.*

Isotope (approx. price,[a] if available)	Half-life	Radiation characteristics		Typical applications	References
		Type	Energy[b]		
Sulfur-35 ($50/mCi)	87.9 day	β^-	0.167 (100)	Uptake from atmosphere Availability from soils Cycling in pastures	Furrer (1967) Vimal (1972) Till and May (1971)
Chlorine-36 ($1000/mCi)	3.08×10^5 y	β^- β^+	0.714 (98.1) 0.115 (1.9)	Solute movement in soil	Paetzold and Scott (1978)
Potassium-40	1.26×10^9 y	β^-	1.314 (89.3)	Exchangeable K in soils	Graham and Fox (1971) Fardeau *et al.* (1979)
		γ	1.46 (10.6), from EC (10.6)		
Potassium-42 ($100/mCi)	12.36 h	β^-	3.52 (82) 1.97 (18)	Exchangeable K in soils Ion uptake mechanisms	Graham and Silva (1979) Bouat (1969)
		γ	1.524 (18)		
Potassium-43	22.4 h	β^-	1.82 (1.3), 1.2 (3.5), 0.83 (87), 0.46 (8)	Exchangeable K in soils	Fardeau *et al.* (1979)
		γ	0.373 (85), 0.39 (11), 0.59 (13), 0.619 (80)		
Calcium-45 ($50/mCi)	165 day	β^-	0.252 (100)	Exchangeable Ca in soil Autoradiography of roots Ion uptake mechanisms Movement in plants	Blume and Smith (1954) Barber and Ozanne (1970) Malzer and Barber (1976) Schimansky (1977)

Manganese-54 ($200/mCi)	303 day	γ	0.835 (100)	Availability from soils	Salcedo and Ellis (1979)
				Complexing in soil solution	Geering *et al.* (1969)
Iron-59 ($100/mCi)	45.6 day	β^-	0.475 (51.2)	Fertility erosion	Wooldridge (1965)
			0.273 (48.5)	Diffusion in soils	O'Connor *et al.* (1971)
		γ	1.095 (56.5)	Plant nutrition	Chatterjee and Agarwala (1979)
			1.292 (43.2)	Availability from soils	Wallace *et al.* (1976)
Copper-64 ($100/mCi)	12.8 h	β^-	0.573 (39.6)	Complexing in soil solution	Hodgson *et al.* (1966)
		β^+	0.656 (19.3)		
		γ	1.34 (0.6), from EC (41.1)		
Copper-67	58.5 h	β^-	0.577 (20)		
			0.484 (28)		
			0.395 (51)		
			0.091 (6.2)		
		γ	0.092 (35)		
			0.184 (45)		
Zinc-65 ($100/mCi)	245 day	β^+	0.327 (1.46)	Availability from soils	Tiller and Wasserman (1972)
				Complexing in soil solution	Hodgson *et al.* (1966)
		γ	1.115 (50.7), from EC (98.5)	Fertilizer efficiency	Amer *et al.* (1980)
Rubidium-86 ($100/mCi)	18.66 day	β^-	1.78 (91.2)	Ion uptake mechanisms	Epstein (1972)
			0.71 (8.8)	Substitute tracer for K	Mackie and Fried (1955)
		γ	1.078 (8.8)		
Strontium-85 ($100/mCi)	64.0 day	γ	0.514 (100)	Cation exchange capacity of soil	Francis and Grigal (1971)
				Ion uptake mechanisms	Malzer and Barber (1976)
Strontium-89 ($200/mCi)	52.7 day	β^-	1.463 (100)	Substitute tracer for Ca	Russell and Newbould (1969)

Table 1.1—*cont.*

Isotope (approx. price,[a] if available)	Half-life	Radiation characteristics		Typical applications	References
		Type	Energy[b]		
Molybdenum-99 ($200/mCi)	66.7 h	β^-	1.23 (82)	Plant nutrition	Chatterjee and Agarwala (1979)
			0.45 (17)		
		γ	0.041 (1)		
			0.181 (7)		
			0.366 (1.8)		
			0.740 (15)		
			0.780 (4)		
Cesium-134 ($200/mCi)	2.046 y	β^-	0.662 (71)	Fertility erosion	Goss *et al.* (1973)
			0.089 (28)		
		γ	0.57 (14)		
			0.605 (98)		
			0.796 (88)		
			1.038, 1.168, 1.365		
Cesium-137 ($50/mCi)	30.0 y	β^-	1.176 (6)	Fertility erosion	Ritchie and McHenry (1975)
			0.514 (94)		
		γ	0.662 (85)		

[a] Prices are given in US dollars as of 1 January 1983.

[b] Units of energy are million electron volts and represent E_{max} for β^- and β^+ radiations. The values within parentheses alongside each radiation energy value represent the percentage abundances of the emissions. Only radiations emitted with appreciable abundance are listed. The abbreviation EC indicates decay by electron capture.

against overexposure. Experimental areas must be posted with appropriate warnings. Assess to these areas, whether in laboratory, greenhouse or field plots, may have to be limited until the radioactivity has decayed to a level comparable with that of natural background. Radioactive wastes must be properly disposed of. None of these precautions are necessary with the use of stable isotopes.

The decay of radioactive isotopes limits the period of time that they can be easily detected. One-half of any given amount of a radioactive isotope decays in a time that is characteristic for that isotope, and is known as its half-life (Table 1.1). One-fourth of the isotope remains after two half-lives, one-eighth after three half-lives, etc. Since the initial level of radioactivity must be low enough to avoid biological damage to the experimental system, the period for detection is usually not more than 10 half-lives. All aspects of a soil fertility or plant nutrition study, including, if necessary, fertilizer equilibration with soil, growth and harvest of plants, and analysis of soil and plant materials, must be completed within this period. Thus, the most useful radioactive isotopes for field studies are those with half-lives between 10 and 100 days. With shorter-lived isotopes, it is difficult to complete a meaningful experiment. With longer-lived ones, it may be difficult to limit access to the experimental area for the required time. With stable isotopes, there are no limitations on the time for a study except for physical losses of the isotope from the system and dilution of the isotope to the point where its isotopic ratio is indistinguishable from that of the general environment.

The sensitivity of tracer studies with most radioactive isotopes is very high because each radiation event represents the disintegration of a single atom. Approximately 100% of the disintegrations are detected with modern counting techniques. With ^{32}P, for example, the specific activity of the pure isotope is 3×10^{11} μCi g^{-1}. One microcurie of radioactivity is defined as 2.2×10^6 disintegrations per minute (dpm), hence 1 g of ^{32}P yields 6.6×10^{17} dpm. Since a count rate of 1000 counts per minute (cpm) allows quick and accurate assay, less than 10^{-14} g of ^{32}P is required in the sample when it is counted. A 1 g sample is normally convenient for counting. In such a sample, the ^{32}P may be diluted with 10^{14} times its weight of inert material without interfering with its determination.

With stable isotopes, the sensitivity of detection is relatively low because of analytical limitations and natural variations in isotope abundance (Hauck and Bremner, 1976). In order to distinguish a sample of isotopically enriched nitrogen from environmental nitrogen, the abundance of ^{15}N in the sample must lie outside the range of natural variations (approximately 0.366 $\pm$ 0.004%). Thus, the permissible dilution of ^{15}N in an experiment is limited to ~100/0.004 or ~25 000 times its weight, and that of ^{14}N (^{15}N-depleted tracer) is even less.

The cost of radioactive isotopes is generally less than that of stable isotopes. Because large amounts of most radioactive isotopes can be produced in nuclear reactors, their cost is an insignificant part of the cost of experiments in soil fertility and plant nutrition. The cost of stable isotopes is usually also insignificant except for use in field experiments.

In view of their widespread use in soil fertility and plant nutrition studies, we summarize here the relevant properties of nitrogen and phosphorus isotopes.

A. Nitrogen isotopes

The nitrogen isotopes utilized in agronomic tracer studies are the stable ^{15}N and ^{14}N, and the radioactive ^{13}N. Materials enriched in ^{15}N have been available since 1936 (Edwards, 1978) and, by far, the preponderant attention has focused on this isotope. Its use has allowed marked progress in determining the behavior and fate of nitrogen in soil–plant studies. ^{15}N has contributed to an understanding of essentially all the nitrogen processes involved, including mineralization and immobilization, denitrification, leaching, biological fixation, mineral fixation, plant uptake and volatilization. In fact, the literature on ^{15}N in soil–plant studies is now so vast that adequate treatment in a general review is not possible. One bibliography book exists (Hauck and Bystrom, 1970), and a revision is in preparation with approximately 1500 additional references (R. D. Hauck, personal communication, 1982). Discussion and references to much of the latest work can be found in reviews by Hauck and Bremner (1976), Edwards (1978) and Vose (1980).

The ^{15}N enrichment used in soil–plant studies must be considered in relation to analytical and instrument limitations (Hauck and Bremner, 1976; Edwards, 1978). Some double collector mass spectrometers have reported precisions of $<0.05\%$ relative error (Edwards, 1978). In practice, analytical and instrument errors with most single collector mass spectrometers require about 0.15% enrichment in ^{15}N for routine work (Vose, 1980). In other words, for the isotope dilution to be determined with reasonable accuracy, more than 0.5% of the nitrogen atoms in the analytical sample should be ^{15}N (0.366% natural abundance + 0.15% enrichment or 0.516% ^{15}N).

Actual ^{15}N enrichments in soil–plant studies have varied from about 1% to nearly 100%, but are usually of the order of 5–10% for most vegetative tests (Allison, 1973). To determine field fertilizer nitrogen efficiency over just one season by plant analysis, the 1% enrichment may be adequate (Rennie and Fried, 1971). This is provided, of course, that the crop takes up a substantial portion of the fertilizer. For those studies involving nitrogen cycling (e.g. mineralization–immobilization, leaching, etc.) and nitrogen disposition (e.g. balance sheet approach), much higher atom percent excess ^{15}N must be used to provide for the enormous dilution effect. Depending on the length and conditions of the experiment, from 10% to nearly 100% atom excess ^{15}N may be desired.

The use of ^{15}N-depleted (i.e. essentially 100 atom percent ^{14}N) fertilizers in soil–plant studies was introduced by Edwards and Hauck (1968). Since the materials are now available in large amounts as a by-product from ^{15}N enrichment, considerable interest lies in their potential for large-scale field experiments. Practical considerations on the use of ^{15}N-depleted materials center around analytical sensitivity and nitrogen dilution encountered in the experiment. Generally, the use of ^{15}N-depleted materials is restricted to single season studies of plant uptake of the material, or studies involving the movement of applied nitrogen from soils to water (Hauck and Bremner, 1976). On the other hand, J. J. Meisinger and J. O. Legg (personal communication, 1982) were able to determine a nitrogen fertilizer balance after 3 years of ^{15}N depleted fertilization and cropping in long-term, no-till corn experiments. This was because the immobilized ^{15}N-depleted materials built up in the soil, which had a moderately low (0.1%) initial nitrogen content, to a level that could be readily detected by mass spectrometry.

Variations in natural isotope abundance of nitrogen have been used in some nitrogen cycling and environmental studies (Vose, 1980). However, the natural variations are small, usually within ± 10 ΔN-15 units. (One ΔN-15 unit equals ± 0.00037 atom percent ^{15}N), and may change during nitrogen mineralization (Bremner and Tabatabai, 1973). Consequently, the analytical and procedural problems are considerable. To date, the approach has not received general acceptance.

Use of the radioactive ^{13}N isotope has been reported in a few studies involving short-term biological nitrogen fixation (Dazzo and Hubbell, 1974; Hauck and Bremner, 1976) and denitrification (Gersberg *et al.*, 1976; Tiedje, 1978; Hollocher *et al.*, 1980). The ^{13}N technique is extremely sensitive and highly accurate, but the short half-life (10.0 min) limits its use to experiments which can be completed in a few hours (Hauck and Bremner, 1976). Moreover, the tracer is very costly, requiring generation of the ^{13}N gas in a cyclotron and purification to remove other radioactive isotopes produced in the cyclotron target chamber. Consequently, ^{13}N is impractical for routine use in agronomic studies.

B. Phosphorus isotopes

Two radioactive isotopes of phosphorus, ^{32}P and ^{33}P, are suitable for agronomic studies. The first readily available isotope was ^{32}P, and it has commonly been used as a tracer for phosphorus. The higher beta energy of

^{32}P is more easily detected than that of ^{33}P by ionization counting techniques that were available in the 1950s. However, detection efficiency is high for both isotopes with modern liquid scintillation counters. The lower energy of ^{33}P presents less radiation hazard, and its longer half-life allows significantly longer plant growth periods. Also, better resolution can be obtained in autoradiography with ^{33}P than with ^{32}P (L'Annunziata, 1979; Claassen *et al.*, 1981). The availability of two suitable isotopes makes it possible to use double labeling techniques in phosphorus studies (Broeshart and Nethsinghe, 1972; Christians *et al.*, 1981). The technique used to assay ^{33}P and ^{32}P as a double label is discussed in Chapter 5 in this volume.

III. Autoradiography

The technique of autoradiography is based on the darkening of photographic film by radiation. The autoradiograph is a picture of the distribution of radionuclides in an object. A review of the theory and practice of autoradiography is given by L'Annunziata (1979). Autoradiography has been used to study nutrient mobility in soils (Bhat and Nye, 1973), nutrient accumulation or depletion at root surfaces (Walker and Barber, 1961), and the distribution of various elements in plants (Tiller, 1979).

While the visual impression of radionuclide distribution may be very informative, its resolution is sometimes poor, and it is difficult to obtain quantitative results. Resolution in an autoradiograph depends on the range of the radiation, and on the distance between the object and the photographic film. Low energy beta emitters, such as ^{33}P, ^{35}S and ^{45}Ca, are best for autoradiography because the range of their radiations in photographic film is only a few tenths of a millimeter. On the other hand, the low energy means that only the surface of the object will contribute to the photographic image. Close contact between the object and the film improves resolution. When resolution is satisfactory, densitometry can be used for quantitative interpretation of autoradiographs (L'Annunziata, 1979; Claassen *et al.*, 1981). Where satisfactory resolution for autoradiography cannot be obtained, the object may be sectioned with a microtome and the distribution of radionuclides determined by counting each section.

IV. Radiation absorption and scattering

Absorption and scattering of neutrons in soil has become widely used for measuring soil moisture content (Nielsen *et al.*, 1978). Changes in soil moisture content may indicate root distribution, which is closely related to the uptake of nutrients by plants. While many investigators have used the neutron moisture meter to study soil moisture changes in relation to crop growth and efficient use of soil moisture, the technique has not been accepted to study root development and distribution. Soil moisture depletion rates are greatly affected by soil moisture content as well as rooting density (Taylor and Klepper, 1973). Thus, the interpretation of root distribution is complicated. Use of the neutron moisture meter in relation to soil moisture is discussed in Chapter 2 in this volume.

A technique of neutron radiography has been developed (Willatt *et al.*, 1978) for nondestructive studies of root growth in a small container. Since roots represent local concentrations of water in the soil, they attenuate thermal neutrons more than does the surrounding soil. The neutrons passing through the soil activate a metal foil, producing a radioactive image of the root system which can be transferred to photographic film. Repeated measurements on the same root system are possible. However, access to an atomic reactor is required.

V. Activation analysis

Activation analysis is well suited for determining trace concentrations of many elements in plants and soils. Stable atoms in the analytical sample are converted to radioactive atoms by irradiation in a flux of neutrons or charged particles. The technique requires a source of radiation, and a system for detecting the radioactivity induced in the sample (Lyon, 1964; De Soete *et al.*, 1972). With the use of high resolution γ-ray detectors, it is possible to determine many elements simultaneously. By proper choice of irradiation conditions and detection methods, it is often possible to avoid interferences in determining the elements of interest.

Applications of activation analysis in agriculture and soil science have been reviewed by Campbell and Bewick (1978). The technique offers a practical method for determining concentrations of elements such as Co, Cu, Mn, Mo, Se and Zn in the soil solution, which has rarely if ever been carried out by other techniques (Bowen, 1967).

It is possible to use activation analysis as a sensitive method for detecting a stable isotope tracer. Knaus (1981) studied the distribution of Mo and La in a stream microcosm using the stable isotopes ^{98}Mo and ^{139}La and the short-lived radioactive isotopes ^{99}Mo and ^{140}La as tracers. Long after the radioactive isotopes had decayed below detectable levels, the stable isotopes could be detected by activation analysis.

One of the major advantages of activation analysis is that many of the usual sources of contamination in trace analyses can be avoided. Samples can be irradiated with a minimum of physical preparation, without digestion or addition of reagents until after the irradiation.

VI. Soil nutrients in relation to plant nutrition

Only a portion of the total nutrient content in soils is available for uptake by plants. The available portion of some nutrients may change greatly with changes in soil conditions such as acidity or alkalinity, oxidation or reduction, and wetting or drying. Isotopic techniques have been used extensively to measure the amounts of exchangeable nutrients in soils, their rates of exchange and their mobility and availability to plants, and to investigate mechanisms of nutrient uptake by plants.

A. Exchangeable nutrients

The amounts of exchangeable nutrients can be measured by observing changes in the isotope ratio of a tracer added to a soil suspension. Whereas conventional methods for measuring exchangeable nutrients require their displacement with concentrated salt solutions, the isotopic method permits equilibration with a solution whose composition approximates that of a normal soil solution. This may give a more realistic measure of exchangeable nutrients. For example, Blume and Smith (1954) measured exchangeable calcium by adding a solution containing known quantities of ^{45}Ca and calcium to a soil. The solution contained 100 mg l^{-1} of calcium as $Ca(NO_3)_2$ and 800 μCi ^{45}Ca per gram of calcium. Exactly 50 ml of the solution was added to 25 g of soil. After 3 h of equilibration on a shaker, an aliquot of the solution was separated from the soil and the specific activity of the calcium was determined (in microcuries of ^{45}Ca per gram of calcium). The amount of exchangeable calcium in the soil was calculated from the isotope dilution formula. In Blume and Smith's (1954) notation,

$$X = \frac{AC}{B} - C, \tag{1.2}$$

where X is the exchangeable calcium, A is the initial specific activity of the solution, B is the final specific activity of the solution, and C is the amount of calcium added in the solution.

In the procedure used by Blume and Smith, any soluble calcium in the soil is included in the determination of exchangeable calcium. They also found that a portion of the lime in calcareous soils exchanged with the radioisotope. However, the amount that was isotopically exchangeable was less than the amount of lime that dissolved in the ammonium acetate solution normally used to extract exchangeable cations from soils. For a wide variety of soils, the isotopic exchange procedure gave consistent and seemingly realistic values for the amount of soil calcium that existed in equilibrium with the soil solution.

Analogous procedures have been used to determine amounts of other exchangeable cations in the soil. Major nutrient cations that have been studied, in addition to calcium, are magnesium and potassium (Graham and Silva, 1979). The isotopically exchangeable amounts corresponded very closely to the amounts determined with usual ion exchange techniques. Minor nutrient cations that have been studied include manganese, zinc, iron, cobalt and copper. Difficulty arises with soils that are alkaline or contain large quantities of manganese oxides. In these soils, isotopic exchange of the minor nutrients is slow and erratic (Tiller, 1979), and the amount of nutrient in solution is often too small to determine accurately. Consistent results can be obtained using an extractant such as diethylenetriaminepentacetic acid (DTPA) with the equilibrating radioisotope solution (Lopez and Graham, 1972). However, the results seem to reflect merely the amounts of minor nutrients dissolved by the extractant.

Many investigators have studied the isotopic exchange of phosphate anions in soils. Results obtained prior to 1972 were discussed and summarized by Jose and Krishnamoorthy (1972). An initial rapid exchange lasting for a few hours is followed by slow exchange for many days. To determine the readily exchangeable or surface phosphate, it is necessary to choose a time when the rapid exchange is nearly complete, but the slow exchange is negligible. Different investigators have chosen equilibration times ranging from 1 to 14 days for this purpose (Jose and Krishnamoorthy, 1972). The amount of exchangeable phosphate is not affected by adding moderate amounts of phosphate in the equilibration solution if the equilibration time is longer than 2 days. The addition of phosphate increases the amount of phosphate in solution after equilibration, and thus facilitates its accurate determination. Ryden and Syers (1977) showed, with several New Zealand topsoils and subsoils, that the amount of phosphate isotopically exchangeable in 30 min corresponded closely to the amount that could be removed in two 1-h extractions with distilled water (1 g soil:40 ml water). This quantity, said to represent "labile" phosphate, is of interest in plant nutrition and in environmental studies of phosphorus in runoff water.

Radioisotope techniques can also be used to determine cation exchange capacity. The exchange capacity must first be saturated with a suitable cation. After washing to remove excess saturating cation, the amount of adsorbed cation can be measured directly, or determined by isotope dilution as outlined above for calcium. Advantages of the method are that it is adaptable to micro techniques and is rapid. In one procedure (Francis and

Grigal, 1971), samples ranging from 50 mg clay to 3 g soil were saturated with 0.1 or 1.0 N $SrCl_2$ solution containing a known specific activity of ^{85}Sr. Successive saturating and washing solutions were separated from the sample by centrifuging and decanting from a 50 ml centrifuge tube. Radioactivity of the ^{85}Sr retained by the clay or soil sample was measured in the same tube, avoiding any need to transfer the sample.

B. Mobility of nutrients

The mobility of nutrients affects directly the volume of soil from which plants derive nutrients. Nutrient movement occurs via the soil solution, and its composition is thus a primary factor in nutrient movement. The sensitivity of isotopic techniques allows the determination of important trace constituents in the soil solution. Movement of solutes in the soil can be observed directly with isotope tracers, and this has improved our understanding of the processes involved.

1. *Soil solution*

The composition of soil solution as it exists under field moisture conditions is difficult to determine. The most satisfactory method for obtaining a sample of soil solution for analysis is to displace it with a dilute electrolyte, such as saturated $CaSO_4$ solution (Adams, 1974). Solutions can be displaced only from relatively moist soils, and there is some risk that the solution composition may be changed by displacing even the few milliliters required for analysis of most nutrient elements.

Neutron activation analysis has sufficient sensitivity to determine several critical plant nutrients using less than 1 ml of soil solution. As shown in Table 1.2, the sensitivity is adequate for Na, K, Mg, Ca, S, Mn, Cu and Cl. Although several laboratories have developed the neutron activation technique for analysis of soil solution (Cho and Axmann, 1965; Zmijewska and Minczewski, 1965), no systematic investigations have been reported using this technique. Contributing factors in the slow application of neutron activation techniques are the common availability of atomic absorption instruments with high sensitivity for many of the plant nutrients (Table 1.2).

Hinkley (1979) used stable isotope dilution for analysis of K, Rb, Cs, Ca, Sr and Ba in only 0.3 g of soil solution. The soil solution was sampled by pressing a stack of specially cleaned absorbent paper against a moist soil surface for a few seconds. A known amount of isotopically enriched tracer for each of the above elements was equilibrated with the sample. The resultant isotope ratios were measured on a solid source, thermal emission mass spectrometer. The actual metal masses that were determined ranged from 0.02×10^{-9} g for Cs to 160×10^{-9} g for K.

Table 1.2 Expected composition of soil solutions in relation to sensitivity of neutron activation analysis (NAA) and atomic absorption analysis (AA)

Element	Expected concentration in soil solution[a] ($\mu g\ ml^{-1}$)	Sensitivity of NAA[b] (μg)	Sensitivity of AA[b] ($\mu g\ ml^{-1}$)
Na	2–400	0.0004	0.003
K	5–100	0.002	0.01
Mg	10–200	0.04	0.003
Ca	100–700	0.09	0.03
P	0.01–1	0.02	—
S	10–100	2	—
Cl	5–300	0.002	—
Cu	0.004–0.04	0.0002	0.04
Zn	0.002–0.2	0.02	0.01
Mn	0.01–4	0.000004	0.02

[a] Adams (1974); Hodgson *et al.* (1966) (Cu, Zn); Geering *et al.* (1969) (Mn).
[b] For routine analysis under interference free conditions, Reeves and Brooks (1978).

Radioactive tracers were instrumental in demonstrating the importance of organic complexes of Cu, Zn and Mn in solutions from a wide variety of soils (Hodgson *et al.*, 1966; Geering *et al.*, 1969). In the presence of a complexing agent, distribution of the metals between water and organic solvent, or between water and ion-exchange resin, was changed. Total concentrations of the dissolved metals were determined by atomic absorption spectrophotometry. The radioactive tracer made it possible to determine the distribution of metals between phases even when the concentrations were too low to determine by atomic absorption. The portions of complexed metals in the soil solutions were approximately 90–99% for Cu, 50–60% for Zn and 85–98% for Mn.

2. *Leaching*

Movement of nutrients in the soil can be conveniently followed using radioactive tracers. A typical study involves placing the tracer on the surface of a soil column, then leaching the column with water. The rate of movement is determined by analyzing successive samples of leachate, or by analyzing segments of the column after leaching (L'Annunziata and Fuller, 1968; Fuller and L'Annunziata, 1969). Radioisotopes that emit gamma or strong beta radiations can be determined by counting leachate or soil samples without any chemical separations. In one example of this method, D'Souza and Mistry (1979) showed that ^{59}Fe, ^{58}Co, ^{54}Mn and ^{65}Zn were completely immobile when leached with water in a black soil and a laterite soil of India.

However, leaching with 10^{-2} M ethylenediaminetetraacetic acid (EDTA) induced a rapid breakthrough of all four radionuclides in the leachate. A field application of the method was made by Riekerk (1971). He applied ^{32}P, ^{45}Ca and ^{86}Rb (as a tracer for potassium) on the surface of a forest soil. Leachate from natural rainfall was collected with tension lysimeters placed beneath 1–2 cm of forest floor and beneath the A horizon at 20 cm depth. Riekerk found that the annual amounts leached were as follows:

	P	K	Ca
Percentage leached through forest floor (f.f.)	21.2	9.1	1.7
Percentage of f.f. leached through A horizon	0.07	0.11	3.4

3. *Diffusion*

Diffusion may become important in nutrient movement when flow of the soil solution ceases, or when strong concentration gradients are imposed by nutrient uptake, by precipitation reactions, or by dissolution of fertilizer granules. These phenomena have been studied very effectively with isotope tracers (Nye, 1979). Self-diffusion, that which occurs with no concentration gradients, can only be studied with isotope tracers.

The technique for self-diffusion studies was described by Graham-Bryce (1965). Two similar cylindrical sections of soil are prepared in the same way except that one contains a uniformly distributed radiotracer of the ion being studied. When the two sections are placed in contact, the tracer ions diffuse from the labeled to the unlabeled soil. After an appropriate time, the quantities of radioactivity in the two sections are determined. The coefficient of diffusion (D) is calculated from a simplified solution of Fick's law: $Q/Q_\infty = 2\sqrt{Dt}/L$, where Q is the quantity of radioactivity that has diffused across the boundary in time t, Q_∞ is the quantity that would diffuse in infinite time, i.e. half the total quantity of radioactivity present, and L is the length of one soil section. The simplified solution is satisfactory for values of Q/Q_∞ less than 0.5.

The coefficients of self-diffusion in soil were approximately 1/100 of the values in aqueous solutions (Graham-Bryce, 1965). In homoionic soils, monovalent ions diffuse faster than divalent ions, and in soils containing both monovalent and divalent ions, diffusion coefficients for monovalent ions are reduced. The diffusion coefficients are also greatly reduced as soils become drier, and this is probably the most important factor affecting diffusion in the field. Concentration of the diffusing ion in the soil solution is also significant. Lewis and Quirk (1965) found that additions of soluble phosphate to a sandy loam soil increased the self-diffusion coefficient for phosphate.

C. Nutrient uptake by plants

Plants take up nutrients from a film of soil solution immediately surrounding the plant roots. Concentrations of nutrients in the roots and in the adjacent soil may be higher or lower than those in the mass of soil away from the roots. Concentrations in soil next to the root increase when nutrients arrive faster than they are taken up, and decrease when nutrients arrive slower than they are taken up. Likewise, concentrations inside the root may be very low if the nutrient is rapidly translocated to plant tops, and concentrations may be high if the nutrient translocation is slow.

Such concentration gradients have been shown graphically for many nutrients by autoradiography. Improvements in technique now permit quantitative measurements of nutrient concentrations at plant roots where uptake is occurring. Experiments with ^{86}Rb and ^{45}Ca (Barber, 1978) and with ^{33}P (Hendriks *et al.*, 1981) have been described. With ^{86}Rb, which diffused relatively rapidly and similarly to potassium, the depletion zone extended into the soil several millimeters from the root, and concentrations varied markedly within the root, being much higher near root tips than away from them. With ^{45}Ca, mass flow of the soil solution led to an accumulation of ^{45}Ca for several millimeters next to the root, but its concentration within the root was very low. With ^{33}P, the depletion zone extended to about 2 mm from the root.

Claassen *et al.* (1981) considered the problem of correcting densitometric measurements for "cross-fire", blurring of the radiographic image by beta particles that traveled across the radioactivity gradient. The cross-fire distance is greater for higher energy beta particles. If the cross-fire distance is greater than the width of the depletion zone, an accurate correction cannot be made. For this reason, ^{32}P was not satisfactory for studying phosphate concentrations near plant roots, but with the lower energy beta particles of ^{33}P the concentrations could be measured within $\pm 10\%$.

The mechanisms of nutrient uptake through plant roots have been classified as "passive" and "active" (Epstein, 1972). As opposed to "passive" uptake, "active" uptake depends on plant metabolism, is largely unidirectional rather than readily reversible, and is highly selective. Isotope tracers aided greatly in characterizing the two mechanisms of uptake, since the fraction of nutrients that were reversibly taken up could easily be determined by isotope dilution measurements. At the same time, the great sensitivity of radioactive tracers permitted accurate measurement of very small amounts of "active" uptake.

The sensitivity of the radioactive tracer technique has perhaps been most useful in studying the kinetics of nutrient uptake at low nutrient concentrations. In a typical experiment (Epstein, 1972), barley seedlings are grown in 0.2 mM $CaSO_4$ for 5 days. The roots are excised and 1.00 g samples enclosed in cheesecloth "teabags". The samples can then be handled with ease by means of a thread attached to the "teabag" and transferred into and out of experimental solutions with precise timing. The ion under investigation is usually radioactively labeled, and routine experimental periods are 10 or 20 min. At the end of the absorption period, the sample is transferred to an unlabeled solution of the same ion to remove the reversibly adsorbed tracer ion, leaving for final radioassay only that fraction that has been taken up by metabolically active processes. With this technique, significant accumulation can be shown within a minute.

Foliar uptake of nutrients has also been studied with isotope tracers. Experiments completely analogous to those with excised roots have been done with discs cut from leaf tissue (Epstein, 1972). The same mechanisms of uptake have been identified. In one field study, Vasilas *et al.* (1980) determined nitrogen recovery from ^{15}N-labeled urea applied to soybean foliage, and its distribution through the plant.

Isotope tracers are valuable in distinguishing between atmospheric and soil sources of nutrients. This is done by labeling one of the sources. Porter *et al.* (1972) exposed corn seedlings to ^{15}N-labeled ammonia in the atmosphere within a sealed enclosure. The enclosure was sealed for 24 h, and the 30 day old plants took up 30–45% of the labeled ammonia at concentrations ranging from 1 to 20 ppm. It is simpler experimentally to label the rooting medium rather than the atmosphere because a sealed enclosure is not necessary in the former case. Solution cultures or potted soils have been labeled in greenhouse studies to determine the amounts of sulfur (Jensen, 1963; Furrer, 1967) and nitrogen (Matsumaru *et al.* 1981) that were derived from the atmosphere. If the rooting medium is not uniformly labeled, it is necessary to compare isotopic ratios in control and treated plants. For example, Matsumaru *et al.* (1981) used the following isotope dilution formula:

$$\text{N from atmospheric NO}_2 = \text{total N in NO}_2\text{-treated plant}$$
$$\times \left(1 - \frac{^{15}\text{N atom \% excess in NO}_2\text{ treated plant}}{^{15}\text{N atom \% excess in control plant}}\right). \quad (1.3)$$

Exactly analogous procedures are suitable for determining amounts of biologically fixed nitrogen. Vose (1980) states, "The use of ^{15}N labeled atmospheres to determine actual N_2-fixation must now be regarded as the primary standard against which other methods are judged." While the ^{15}N-labeled atmosphere provides the only direct measure of nitrogen fixation, it is not widely used because of the experimental difficulties. When ^{15}N-labeled fertilizer is added to the soil, the isotope dilution equation (1.3) given above may be used, with appropriate modifications, to measure nitrogen fixation.

In this case, the control plant must be grown under conditions that preclude nitrogen fixation. A non-legume or a non-nodulated legume may serve as the control plant. The rate of fertilizer added should not exceed 30 kg of nitrogen per hectare in order that nitrogen fixation will not be reduced by the fertilizer itself. Ham (1977) obtained good agreement between the ^{15}N isotope dilution method and other methods for determining nitrogen fixation with soybeans.

D. Measures of nutrient availability

Isotope dilution forms the basis for a widely used technique to evaluate nutrient availability from soils and fertilizers. The technique has become the reference standard against which are compared the chemical extraction procedures for determining availability of various nutrients (Vose, 1980). Depending upon the system used for equilibration of the isotope tracer, these measures of availability have been designated as *E*, *L* and *A* values (Fried, 1964). The values may be computed using equation (1.1):

$$E, L \quad \text{or} \quad A = \frac{S_1 X}{S_2} - X.$$

The *E* value is determined by equilibration of a soil suspension. This technique measures exchangeable ions (see Section VI.A). The equilibrations are usually for time periods of a few days or less, and use a relatively high ratio of solution to soil. In equation (1.1), S_1 is the specific activity of the added tracer, X is the amount of added carrier, and S_2 is the specific activity of tracer in the equilibrium solution.

The *L* value (Larsen, 1952) is determined by growing a plant in the equilibrated soil. Hence, the equilibration is normally for a period of weeks at moisture and temperature conditions that are suitable for plant growth. The measurement also reflects any effect that the plant has on the equilibrating system. In equation (1.1), S_1 and X are the same as for the *E* value, but S_2 is the specific activity of the tracer in the plant.

The *A* value measures amounts of available nutrients in terms of a reference source, usually a fertilizer material, that is taken as the standard of availability (Fried and Dean, 1952). Thus, S_1 and X are the specific activity and amount, respectively, of fertilizer element added, and S_2 is the same as for the *L* value. It is assumed that a plant will take up nutrient from different sources in the soil–plant system in proportion to the amounts available from each source. Ideally, the reference source would not interact with the soil and would maintain the same standard of availability for long periods and in different types of soil. Perhaps this ideal is never attained, although the use of granular materials and banded placement of them reduce the degree of interaction. Even so, many useful comparisons of nutrient availability from soils and other sources have been obtained.

The *A* value is most often used to measure amounts of available nutrient in different soils or fertilizer materials. It can also be used to measure indirectly the availability of nutrients from sources that cannot be labeled conveniently. Such sources include rock phosphate (Fried, 1953; Ensminger and Pearson, 1957) and residues from organic materials, or from previous applications of fertilizer (Spratt *et al.*, 1980). Comparisons are usually made in a greenhouse or growth chamber so that test conditions are similar for all soils. However, the *A* value concept is also useful for studying effects of varying field conditions on nutrient availability (Vose, 1980).

Early work (Shaw *et al.*, 1954; Fried, 1957) demonstrated relative constancy of *A* values with different rates of labeled fertilizer, addition of other nutrients, and different plant species. A number of later experiments (Terman and Khasawneh, 1968; Tiller and Wasserman, 1972) have shown that *A* values may increase with increasing rates of labeled phosphate or zinc fertilizers. However, Stanford *et al.* (1973) found that *A* values for 39 soils did not vary with rates of ^{15}N-labeled fertilizer providing 100 or 200 mg of N per kilogram of soil. The occasional increases have not been satisfactorily explained. Possibly, fertilization sometimes increases root activity and thereby enhances uptake of soil nutrient more than that of the labeled fertilizer. One of the distinct advantages of the *A* value determination is that it need not be affected by marked changes in the uptake of soil and fertilizer nutrients (Smith and Legg, 1971). The non-labeled "difference" method (Dean, 1954) of distinguishing uptake of soil and fertilizer nutrients usually assumes a constant amount of soil nutrient uptake and a constant percentage of fertilizer nutrient uptake. These conditions are more restrictive than the conditions for a constant *A* value.

VII. Fertilizer disposition

Increasing fertilizer costs and the necessity for minimizing environmental pollution have given added impetus to fertilizer disposition studies. While non-labeled rate-of-fertilizer experiments may have some utility for determining fertilizer uptake by plants (Terman and Brown, 1968), tracer fertilizers provide the only definitive means for determining both the behavior and fate of applied nutrients. A basic assumption is that the isotopically labeled element behaves in an identical manner to the non-labeled element both physically and chemically, and the plant cannot distinguish between labels. Also, in the case of the stable nitrogen isotopes, natural nitrogen is assumed to have a constant isotopic composition. For most ^{15}N enrichment studies, these assumptions generally hold. Nitrogen tracer experiments, reviewed by Allison (1966) and Vose (1980), indicate that plant recoveries of nitrogen fertilizer under single-season field conditions range generally between 20 and 70%. Higher recoveries may be obtained in pot experiments (Allison, 1973; Stanford *et al.*, 1973), where the fertilizer is confined to a relatively small volume. On the other hand, plants rarely recover more than 20% of the applied phosphorus in a single season (Black, 1968). On soils high in phosphorus, Nelson *et al.* (1948) found cotton absorbed approximately 2.5% of the applied ^{32}P-tagged fertilizer, while potatoes absorbed approximately 10%.

The most appropriate application of labeled fertilizer depends upon the supply of nutrients in the soil, behavior of the fertilizer, nature of the crop, environmental conditions, and other factors. In field experiments, especially, factors influencing the efficient utilization of the fertilizer include placement, time of application and fertilizer source. Since phosphorus is relatively immobile in soil, its placement, as a band or broadcast, is much more critical than that of nitrogen to achieve acceptable efficiency.

Assuming usual isotope dilution principles, the percentage of the nutrient in the plant derived from labeled fertilizer may be determined as follows:

$$\text{Plant nutrient derived from fertilizer (\%)} = \frac{\text{Atom \% excess (or specific activity) in plant sample}}{\text{Atom \% excess (or specific activity) in fertilizer}} \times 100. \quad (1.4)$$

Since the readily available radioactive isotope of potassium, ^{42}K, has a half-life of only 12.4 h, it has very limited use as a fertilizer tracer in nutrient uptake studies. Moreover, the long-lived ^{40}K that occurs naturally is far too expensive to be used. Consequently, ^{86}Rb has received considerable attention as a substitute fertilizer tracer for potassium. Collander (1941) was one of the first to show that rubidium could partially substitute for potassium in plants. The affinity of rubidium and potassium for soil clays is approximately the same (Grim, 1953), and both use the same biological carrier when taken up by plant roots (Epstein, 1972). Results involving use of ^{86}Rb as a substitute fertilizer tracer for potassium in plant growth studies are somewhat mixed (Menzel, 1954; Mackie and Fried, 1955; Menzel and Heald, 1955; Fried *et al.*, 1959), especially when using soils with high potassium fixation capacities. Use of ^{86}Rb as a substitute fertilizer tracer for potassium appears to be best suited to "short- or medium-term" nutrient solution studies, where variations due to soil, and/or reproductive and vegetative plant parts, will be avoided. Under field conditions, reaction rate differences between the two cations, and the extra radiation protection needed for the energetic gamma emissions of ^{86}Rb, are added complicating factors (Vose, 1980).

Strontium is similar to calcium in its soil chemistry and plant uptake. Because the weak beta radiation of ^{45}Ca was much more difficult to measure than the more energetic radiations of ^{85}Sr or ^{89}Sr before the advent of modern liquid scintillation spectrometers, the strontium isotopes have sometimes been used as substitute tracers for calcium (Russell and Newbould, 1969).

The distribution function, k, is an index of similarity in the behavior of two elements toward plant uptake. For example, the relationship may be stated as follows:

$$\frac{^{85}\text{Sr in plant}}{\text{Calcium in plant}} = k \times \frac{^{85}\text{Sr in nutrient solution (or soil)}}{\text{Calcium in nutrient solution (or soil)}}. \quad (1.5)$$

A value of unity indicates that behavior of the two elements is identical. A distribution function greater than unity indicates that ^{85}Sr is preferentially absorbed. Likewise a distribution function less than unity indicates that calcium is preferentially absorbed by the plant.

Residual effectiveness of some labeled radioactive fertilizers is difficult to determine due to relatively short half-life. With ^{32}P, for instance, the 14.3 day half-life precludes its use for more than single season studies. A similar problem exists with use of the potassium tracer substitute, ^{86}Rb. However, with the stable ^{15}N isotope tracer, residual effectiveness can be determined simply by growing a new crop on the previously fertilized soil and employing the isotope dilution formula. With nitrogen, residual tracer studies indicate labeled ^{15}N uptake by the plant the second year of cropping is about one-tenth that of the first year (Legg and Allison, 1967; Westerman and Kurtz, 1972; Chichester and Smith, 1978; Kissel and Smith, 1978).

Of additional interest in many cases is the availability of the residual fertilizer relative to that of the associated soil nutrient. In this case, an availability ratio (Broadbent and Nakashima, 1967) can be obtained as a measure of residual fertilizer availability with time. In 1982, Smith *et al.* used the technique as a laboratory tool for predicting the potential for relative leaching of fertilizer-derived nitrate from soils. The ratio may be calculated as

$$\text{Fertilizer availability ratio} = \frac{\text{Residual fertilizer in plant/Total nutrient in plant}}{\text{Residual fertilizer in soil/Total nutrient in soil}}. \quad (1.6)$$

A ratio of unity means that the residual and indigenous soil nutrients are identically available, whereas a ratio greater than unity means that the residual fertilizer nutrient is more available, and vice versa.

The balance sheet approach offers the most complete way for determining the disposition of applied labeled fertilizer. Again, due to the relatively short half-lives of ^{32}P and ^{86}Rb, most labeled fertilizer balance sheet approaches have centered on the stable isotope ^{15}N. As noted by Legg and Meisinger (1982), there is no single best way for determining labeled nitrogen balance sheets. The proper experimental method depends upon the resources available, the objectives of the study, and the major nitrogen transformations (i.e. mineralization, denitrification, immobilization, leaching, etc.) of the system investigated. Due to costs, sampling problems and spatial variability, most ^{15}N field studies have been confined to small plots, generally $<1\ m^2$. The sampling and spatial variability problems can be more critical with nitrogen than with phosphorus or potassium, due to nitrogen transformations related to soil water content and movement. Small plots allow all the soil to be removed for sampling, thus avoiding complications due to coring and spatial variability. However, such procedure does entail destruction of the plot. Moreover, small plots also tend to restrict fertilizer disposition studies to fibrous rooted crops.

A general example of stable isotope usage to determine nitrogen fertilizer disposition may be illustrated from the results of Smith *et al.* (1982). Summarized in Table 1.3 are single-season results for a 3-year field plot study on the fate of fertilizer ^{15}N nitrate associated with sorghum-sudangrass production in the Southern Plains of the USA. The study was conducted in open-ended field lysimeters (30 cm × 30 cm), and all soil was removed for sampling at the end of the experiment. Included are plant data for individual harvests, stubble and roots, and soils data for organic and inorganic nitrogen. The fertilizer (27% ^{15}N enrichment) was applied at an annual rate of 112 kg N per hectare to eight area soils representing five soil orders. On a seasonal average, slightly more than half of the fertilizer was taken up by the plants (tops plus roots), and about a third was incorporated into the soil organic matter. Very little (<5%) remained in an inorganic leachable form, and about 7% was not recovered. The unrecovered nitrogen is considered to represent denitrification, since profile sampling during and after the cropping season showed no evidence of fertilizer nitrogen leaching below the root zone. Any losses from runoff can be discounted due to the border around the plots. It should be noted that, in this study, the non-isotopic difference method was not found satisfactory for estimating fertilizer nitrogen efficiency (Smith *et al.*, 1982). Therefore, the isotope was necessary for determining fertilizer nitrogen efficiency, as well as determining the fate of the residual nitrogen.

Table 1.3 Disposition of ^{15}N-labeled fertilizer applied to sorghum-sudangrass in the Southern Plains (Smith *et al.*, 1982)

	Percentage of applied ^{15}N-labeled fertilizer[a]			
	1977	1978	1979	Average
Plant uptake				
June harvest	16 ± 10	42 ± 10	39 ± 20	32 ± 18
July harvest	22 ± 8	(no harvest)	25 ± 14	24 ± 11
September harvest	10 ± 7	4 ± 2	4 ± 2	6 ± 5
Stubble/roots	4 ± 1	5 ± 2	6 ± 1	5 ± 2
Plant total	52 ± 5	50 ± 11	74 ± 12	59 ± 15
Soil residual (0–180 cm depth)				
Organic N	27 ± 11	36 ± 11	25 ± 9	29 ± 11
Inorganic N	6 ± 10	6 ± 12	0.1 ± 0.1	4 ± 9
Soil total	33 ± 10	42 ± 11	25 ± 7	33 ± 9
Recovered plant + soil	86 ± 7	92 ± 11	99 ± 8	93 ± 9
Not recovered	15	8	1	7

[a] 112 kg of nitrogen per hectare was applied as KNO_3 (27% ^{15}N enrichment) each May to new plots. The ± values represent single standard deviations for eight soils. Columns may not sum exactly, due to rounding of decimals.

VIII. Plant root distribution

There are three types of study that utilize tracers or radiation techniques to determine the distribution of plant roots. Each gives slightly different information about the root system. The three types are (i) uptake of tracers from soil; (ii) translocation of tracer into roots; and (iii) root system radiography.

The uptake of tracers from different locations in the soil is used to determine the depth and lateral extension of actively absorbing roots. This knowledge is important for optimizing fertilizer placement and timing (Vose, 1980), and cannot be obtained by any other technique (Russell and Newbould, 1969). For measuring the relative uptake from different locations, the isotope dilution of the tracer at each location must be determined. This may be done by the *E* value method. It is necessary that the tracer equilibrate with the same volume of soil at each location, and that the distribution of roots not be affected by the placement of the tracer. However, useful qualitative information can often be obtained without these conditions. The technique has been used very extensively to study the effects of various soil factors (e.g. moisture, cultivation and fertilization) and plant factors (e.g. age and competition) on the absorbing activity of root systems. Some examples include studies on cultivated crops (Hall *et al.*, 1953; Russell and Newbould, 1969), mixed grassland communities (Boggie *et al.*, 1958; Goodman and Collison, 1981), and tree crops (Broeshart and Nethsinghe, 1972).

Some useful refinements have been made to facilitate the safe handling and exact placement of the tracers. The following method was devised by the IAEA for extensive studies on tree crops (Vose, 1980). Glass ampules are filled with 4 ml of a solution containing 300 μCi of ^{32}P and 4 mg of phosphorus, and sealed by machine with suitable shielding. The ampules are transported in a shielded and cushioned container. In the field, up to 16 holes are prepared in a ring around each tree at a specified depth and distance from the trunk. One ampule is placed in each hole and crushed with a rod. At regular intervals after application, tree leaves are sampled for ^{32}P assay. If the trees and leaves are similar physiologically, the treatments resulting in the highest activity in the leaf samples correspond with the location of highest root activity.

Various isotopic tracers may be used, and may not necessarily give the same results. Russell and Newbould (1969) found that the relative ability of different parts of the ryegrass root system to absorb phosphorus and calcium changed with time. However, Broeshart and Nethsinghe (1972) found no difference in depth of nutrient absorption by apple trees using ^{15}N and ^{32}P tracers. The most commonly used tracer is ^{32}P because of its rapid uptake by plants, relative immobility in soils, convenient half-life for field studies and ease of purchase.

Translocation of mobile ions from plant stems or leaves can be used to determine the location of living roots in the soil, and the extent of translocation through rhizomes and other connections between root systems of different plants. Suitable tracers are ^{32}P, ^{86}Rb and ^{42}K. They may be injected into the stem with a microsyringe (Rennie and Halstead, 1965; Russell, 1977) or imbibed through leaves (Halm *et al.*, 1972). The results depend markedly on the age and metabolic activity of the roots, but depend less on current soil conditions than when the tracer is placed in the soil. Distribution of the tracer is determined by removing soil cores or (with gamma ray-emitting tracers) by lowering a radiation detector down tubes in the soil. Core sampling is more precise, but also more laborious, than the tube method.

Living roots can be observed nondestructively in a soil environment by neutron radiography (Willatt *et al.*,1978). This technique (Section IV) allows measurement of root growth in soil blocks of up to 5 cm thickness. Growth rates of roots smaller than 0.33 mm diameter could not be measured at the currently available resolution. Previous observations of root growth in soil have been made at the glass/soil interface of a glass-sided box, where growing conditions are probably different from those in bulk soil.

IX. Fertility erosion

Erosion constitutes a major source of soil productivity loss. This is because most plant nutrients are present in the surface soil, associated with either the mineral or organic fraction. By labeling the soil with an appropriate tracer, both the extent and source of soil loss can often be determined. Nuclides that have been applied for this purpose include the gamma ray emitters ^{59}Fe, ^{46}Sc, ^{110m}Ag, ^{198}Au, ^{134}Cs and ^{51}Cr (Wooldridge, 1965; Coutts *et al.*, 1968; McHenry, 1969; Goss *et al.*, 1973). Gamma ray emitters are generally recommended for field erosion studies. They are more readily detected at distances and in thick overburden. Also, differential gamma spectrometry allows simultaneously detecting more than one gamma emitter. The double labeling technique allows detection of the concurrent movement of two sizes of particles or the contribution from two sources (McHenry, 1969).

Techniques which are available for labeling sand, silt and clay for use in erosion studies (McHenry, 1969) include the following: (i) neutron activation of natural or simulated sand grains; (ii) surface binding of radioactivity to sand or silt particles by chemical or physical means; (iii) adsorption labeling of radioactivity to fine silt and clay.

To date, most tracer studies for determining erosion have provided qualitative rather than quantitative information. Although volume estimates of sand and silt transport can be made, quantitative measurements require additional refinement.

Nuclear weapons testing has provided a convenient fallout label for use in erosion studies (Ritchie and McHenry, 1975). The fallout radionuclides have been dispersed by wind and water around the globe, labeling the land surface with recognizable and specific tracers. Use has been made of ^{90}Sr, ^{239}Pu, ^{240}Pu and, especially, ^{137}Cs in fallout (Frere *et al.*, 1963; McHenry *et al.*, 1973; Muller *et al.*, 1978). ^{90}Sr, a beta emitter, and ^{239}Pu and ^{240}Pu, both alpha emitters, require extraction and concentration from large volumes of soil before radiological analysis. ^{137}Cs, which occurs in a ratio of 1.6:1.0 compared with ^{90}Sr, has a strong gamma emission of 0.662 MeV and is much easier to determine. It has a convenient half-life of 30 years, and is strongly held by various soil components, particularly vermiculite and illite, tending to remain on or near the soil surface (Schultz *et al.*, 1960; Ritchie and McHenry, 1975). Radioactive fallout peaked in 1963, with a smaller peak occurring in 1959 (Hardy, 1974). Measurement of ^{137}Cs in field studies has been enhanced greatly by the development of scintillation counters in combination with differential spectrometers and multichannel pulse height analyzers. Consequently, a powerful tool presently exists for identifying and measuring low levels of ^{137}Cs in environmental samples (McHenry *et al.*, 1973).

Studies by Ritchie and McHenry (1975) showed that fallout ^{137}Cs loss from watersheds correlated logarithmically with soil loss, as calculated by the universal soil loss equation and measured from small runoff plots. The distribution of ^{137}Cs in vertical profiles of sediment deposited in impoundments or flood plains correlated with yearly atmospheric fallout rates.

In 1981, Brown *et al.* estimated soil erosion and deposition rates within an agricultural watershed using fallout ^{137}Cs by two different approaches. The first approach was to measure volumes of accumulated sediment in deposition zones, assuming that movement of ^{137}Cs by leaching and biological activity was negligible. Therefore, occurrence of ^{137}Cs deeper than the plowing depth indicated a deposition zone. The second approach was to estimate weights of transported soil from differences in ^{137}Cs concentrations between deposition zones and upland sites. The deposition rate was obtained by dividing the estimated accumulation of ^{137}Cs (in picocuries per square meter) by the weight concentration of ^{137}Cs (in picocuries per gram) in the deposition zone topsoil. The erosion rate from upland sites was obtained similarly by dividing the estimated depletion of ^{137}Cs (in picocuries per square meter) by the weight concentration of ^{137}Cs (in picocuries per gram) in the upland topsoil. Both approaches also required an estimate of the watershed delivery ratio, the fraction of total erosion that leaves the watershed without being deposited, in order to estimate the soil erosion rate. The erosion rates estimated by the two approaches were internally consistent, but depended critically on the delivery ratio.

Applied ^{32}P has also been used in soil nutrient/transport studies (Kazo' and Gruber, 1962, Ahuja *et al.*, 1981). In the latter reference ^{32}P was used as a tracer to determine the depth of interaction between surface soil and runoff in the transport of phosphorus. By placing ^{32}P at different depths in soil boxes (4% slope), the zone of interaction was found to be very thin, ranging from 1 to 3 mm. For three soil types the depths of interaction were related to the degree of soil aggregation. As the degree of aggregation increased, the depth of interaction decreased, and consequently the transport of phosphorus and soil material decreased.

X. Evaluation and prospects

In the 40 years since their introduction, tracer and radiation techniques have been widely applied in soil fertility and plant nutrition research. There is still an active interest in some of the earliest applications, and some new approaches to old problems are made possible by advances in technology.

Isotope tracers give direct information on fertilizer use efficiency, which has increased in importance because of increasing fertilizer costs and environmental concerns arising in the 1970s. New fertilizer materials and new practices for crop, soil and residue management require continued testing. The quantitative information provided by stable and radioactive tracers insures their continued and expanded use in soil–plant research. Moreover, availability of large quantities of ^{15}N-depleted material now allows field-scale tracer experiments with nitrogen. Stable isotopes of other elements, including ^{10}B, ^{26}Mg, ^{41}K and ^{65}Cu are available in small amounts for use in laboratory or greenhouse experiments. The experiments can be performed without the hazards or time limitations associated with radioactive isotopes. Isotope ratios can be determined by mass spectrometry or activation analysis. These possibilities are of great interest for these important plant nutrients for which no suitable tracer has previously been available.

Potential uses of mass spectrometry and activation analysis have been somewhat limited by the cost and unavailability of suitable instrumentation. However, various instrumentation options are now available for nitrogen isotope ratio analysis. For less exacting work, relatively inexpensive and simple ^{15}N emission spectrometers are available. For precise work, the ion ratio mass spectrometers can be used. In addition, highly sensitive, automated mass spectrometry has been introduced (McInteer and Montoya, 1981), allowing a single instrument to analyze hundreds of samples per day. Such development presents the likelihood of commercial, centralized analysis of nitrogen isotopes, thus freeing individual laboratories of the burden of maintaining and servicing costly instrumentation. Finally, the possibility remains that increased usage of nitrogen isotopes may stimulate the commercial introduction of a fairly simple, relatively low-cost mass spectrometer designed specifically for nitrogen isotope ratio analysis.

References

Adams, F. (1974). *In* "The Plant Root and its Environment" (E. W. Carson, ed.), pp. 441–481, University Press of Virginia, Charlottesville, Va.

Ahuja, L. R., Sharpley, A. N., Yamamoto, M. and Menzel, R. G. (1981). *Water Resources Res.* **17**, 969–974.

Allison, F. E. (1966). *Adv. Agron.* **18**, 219–258.

Allison, F. E. (1973). "Soil Organic Matter and its Role in Crop Production", Elsevier, Amsterdam.

Amer, F., Rezk, A. I. and Khalid, H. M. (1980). *Soil Sci. Soc. Am. J.* **44**, 1025–1030.
Barber, S. A. (1978). *Trans. 11th Int. Congr. Soil Sci., Edmonton* **3**, 45–60.
Barber, S. A. and Ozanne, P. G. (1970). *Soil Sci. Soc. Am. Proc.* **34**, 635–637.
Bhat, K. K. S. and Nye, P. H. (1973). *Plant Soil* **38**, 161–175.
Black, C. A. (1968). "Soil Plant Relationships", John Wiley and Sons, New York.
Blume, J. M. and Smith D. (1954). *Soil Sci.* **77**, 9–17.
Boggie, R., Hunter, R. F. and Knight, A. H. (1958). *J. Ecol.* **46**, 621–632.
Bouat, M. (1969). *Ann. Agron.* **20**, 89–104.
Bowen, H. J. M. (1967). *In* "Nuclear Activation Techniques in the Life Sciences", pp. 287–299, IAEA, Vienna.
Bremner, J. M. (1977). *In* "Biological Nitrogen Fixation in Farming Systems of the Tropics" (A. Ayanaba and P. J. Dart, eds), pp. 335–352, John Wiley and Sons, New York.
Bremner, J. M. and Tabatabai, A. (1973). *J. Environ. Qual.* **2**, 363–365.
Broadbent, F. E. and Nakashima, T. (1967). *Soil Sci. Soc. Am. Proc.* **31**, 648–652.
Broeshart, H. and Nethsinghe, D. A. (1972). *In* "Isotopes and Radiation in Soil–Plant Relationships Including Forestry", pp. 453–463, IAEA, Vienna.
Brown, R. B., Kling, G. F. and Cutshall, N. H. (1981). *Soil Sci. Soc. Am. J.* **45**, 1191–1197.
Buckley, W. T., Huckin, S. N., Budac, J. J. and Eigendorf, G. K. (1982). *Anal. Chem.* **54**, 504–510.
Campbell, J. A. and Bewick, M. W. M. (1978). "Neutron Activation Analysis – A Review of the Method and its Present and Potential Uses in Agriculture and Soil Science", Special Publication No. 7, Commonwealth Bureau of Soils, Commonwealth Agricultural Bureaux, Farnham Royal, England.
Chatterjee, C. and Agarwala, S. C. (1979). *J. Nucl. Agric. Biol.* **8**, 21–23.
Chichester, F. W. and Smith, S. J. (1978). *J. Environ. Qual.* **7**, 227–233.
Cho, C. M. and Axmann, H. (1965). *In* "Plant Nutrient Supply and Movement", pp. 125–130, IAEA Technical Report Series no. 48, IAEA, Vienna.
Christians, N. E., Karnok, K. J. and Logan, T. J. (1981). *Commun. Soil Sci. Plant Anal.* **12**, 765–774.
Claassen, N., Hendriks, L. and Jungk, A. (1981). *Zeit. Pflanzenernahr. Bodenk.* **144**, 306–316.
Collander, R. (1941). *Plant Physiol.* **16**, 691–720.
Coutts, J. R. H., Kandil, M. F., Nowland, J. L. and Tinsley, J. (1968). *J. Soil Sci.* **19**, 311–324.
Dazzo, F. B. and Hubbell, D. H. (1974). *Proc. Soil Crop Sci. Soc. Fla* **34**, 71–79.
Dean, L. A. (1954). *Soil Sci. Soc. Am. Proc.* **18**, 462–466.
Dean, L. A., Nelson, W. L., MacKenzie, A. J., Armiger, W. H. and Hill, W. L. (1948). *Soil Sci. Soc. Am. Proc.* **12**, 107–112.
De Soete, W., Gijbels, A. and Hoste, J. (1972). "Neutron Activation Analysis", John Wiley and Sons, New York.
D'Souza, T. J. and Mistry, K. B. (1979). *In* "Isotopes and Radiation in Research on Soil–Plant Relationships", pp. 407–425, IAEA, Vienna.
Edwards, A. P. (1975). *In* "Isotope Ratios as Pollutant Source and Behaviour Indicators", pp. 455–468, IAEA, Vienna.
Edwards, A. P. (1978). "A Guide to the Use of ^{14}N and ^{15}N in Environmental Research", Special Report no. 78-18, US Army Corps of Engineers, Hanover, NH.
Edwards, A. P. and Hauck, R. D. (1968). *Agron. Abst.* **1968**, 101–102.
Ensminger, L. E. and Pearson, R. W. (1957). *Soil Sci. Soc. Am. Proc.* **21**, 80–84.

Epstein, E. (1972). "Mineral Nutrition of Plants: Principles and Perspectives", John Wiley and Sons, New York.
Fardeau, J. C., Hetier, J. M. and Jappe, J. (1979). *Compt. Rend. Acad. Sci. Paris D*, **288**, 1039–1042.
Francis, C. W. and Grigal, D. F. (1971). *Soil Sci.* **112**, 17–21.
Frere, M. H., Menzel, R. G., Larson, K. H., Overstreet, R. and Reitemeier, R. F. (1963). "The Behavior of Radioactive Fallout in Soils and Plants", Publication no. 1092, National Academy of Sciences–Nature Research Council, Washington, D.C.
Fried, M. (1953). *Soil Sci. Soc. Am. Proc.* **17**, 357–359.
Fried, M. (1957). *In* "Atomic Energy and Agriculture" (C. L. Comar, ed.), pp. 1–18, American Association for the Advancement of Science, Washington, D. C.
Fried, M. (1964). *Trans. 8th Int. Congr. Soil Sci. Bucharest* **4**, 29–39.
Fried, M. and Broeshart, H. (1975). *Plant Soil* **43**, 707–711.
Fried, M. and Dean, L. A. (1952). *Soil Sci.* **73**, 263–271.
Fried, M., Hawkes, G. and Mackie, W. Z. (1959). *Soil Sci. Soc. Am. Proc.* **23**, 360–362.
Fuller, W. H. and L'Annunziata, M. F. (1969). *Soil Sci.* **107**, 223–230.
Furrer, O. J. (1967). *In* "Isotopes in Plant Nutrition and Physiology", pp. 403–407, IAEA, Vienna.
Geering, H. R., Hodgson, J. R. and Sdano, C. (1969). *Soil Sci. Soc. Am. Proc.* **33**, 81–85.
Gersberg, R., Krohn, K., Peek, N. and Goldman, C. R. (1976). *Science* **192**, 1229–1231.
Goodman, P. J. and Collison, M. (1981). *Ann. Appl. Bot.* **98**, 499–506.
Goss, D. W., Smith, S. J., Stewart, B. A. and Jones, O. R. (1973). *Water Resources Res.* **9**, 668–675.
Graham-Bryce, I. J. (1965). *In* "Plant Nutrient Supply and Movement", pp. 42–56, IAEA Technical Report Series no. 48, IAEA, Vienna.
Graham, E. R. and Fox, R. L. (1971). *Soil Sci.* **111**, 318–322.
Graham, E. R. and Silva, C. G. (1979). *Soil Sci.* **128**, 17–22.
Grim, R. E. (1953). "Clay Mineralogy", McGraw-Hill, New York.
Hall, N. S., Chandler, W. F., Van Bavel, C. H. M., Reid, P. H. and Anderson, J. H. (1953). "A Tracer Technique to Measure Growth and Activity of Plant Root Systems", University of North Carolina Agricultural Experiment Station Technical Bulletin no. 101, University of North Carolina, Raleigh, N.C.
Halm, B. J., Stewart, J. W. B. and Halstead, R. L. (1972). *In* "Isotopes and Radiation in Soil–Plant Relationships Including Forestry", pp. 571–586, IAEA, Vienna.
Ham, G. E. (1977). *In* "Biological Nitrogen Fixation in Farming Systems of the Tropics" (A. Ayanaba and P. J. Dart, eds), pp. 325–334, John Wiley and Sons, New York.
Hardy, E. P., Jr. (1974). Health and Safety Laboratory Report no. HASL-286, New York.
Harvey, B. R. (1978). *Anal. Chem.* **50**, 1866–1870.
Hauck, R. D. and Bremner, J. M. (1976). *Adv. Agron.* **28**, 219–266.
Hauck, R. D. and Bystrom, M. (1970). "^{15}N, A Selected Bibliography for Agricultural Scientists," Iowa State University Press, Ames, Iowa.
Hendriks, L., Claassen, N. and Jungk, A. (1981). *Zeit. Pflanzenernahr. Bodenk.* **144**, 486–499.
Hinkley, T. (1979). *Nature (Lond.)* **277**, 444–446.
Hodgson, J. R., Lindsay, W. L. and Trierweiler, J. R. (1966). *Soil Sci. Soc. Am. Proc.* **30**, 723–726.

Hollocher, T. C., Garber, E., Cooper, A. J. L. and Reiman, R. E. (1980). *J. Biol. Chem.* **255**, 5027–5030.
Jensen, J. (1963). *Soil Sci.* **95**, 63–68.
Jose, A. I. and Krishnamoorthy, K. K. (1972). *Soils Fertilizers* **35**, 620–627.
Kazo', B. and Gruber, L. (1962). *In* International Association for Scientific Hydrology, Publication no. 59, pp. 62–66, IASH, New York.
Kissel, D. E. and Smith, S. J. (1978). *Soil Sci. Soc. Am. J.* **42**, 77–80.
Knaus, R. M. (1981). *Environ. Sci. Tech.* **15**, 809–812.
L'Annunziata, M. F. (1979). "Radiotracers in Agricultural Chemistry", Academic Press, New York and London.
L'Annunziata, M. F. and Fuller, W. H. (1968). *Soil Sci.* **105,** 311–319.
Larsen, S. (1952). *Plant Soil* **4**, 1–10.
Legg, J. O. and Allison, F. E. (1967). *Soil Sci. Soc. Am. Proc.* **31**, 403–406.
Legg, J. O. and Meisinger, J. J. (1982). *In* "Nitrogen in Agricultural Soils" (F. J. Stevenson, ed.), pp. 503–566, American Society of Agronomy, Madison, Wisc.
Lewis, D. G. and Quirk, J. P. (1965). *In* "Plant Nutrient Supply and Movement", pp. 71–77, IAEA Technical Report Series no. 48, IAEA, Vienna.
Lopez, P. L. and Graham, E. R. (1972). *Soil Sci.* **114**, 295.
Lyon, W. S. (ed.) (1964). "Guide to Activation Analysis", D. Van Nostrand, Princeton.
McHenry, J. R. (1969). *Isotopes Radiat. Tech.* **6**, 280–287.
McHenry, J. R., Ritchie, J. C. and Gill, A. C. (1973). *Water Resources Res.* **9**, 676–686.
McIntеer, B. B. and Montoya, J. G. (1981). *In* "Recent Developments in Mass Spectrometry in Biochemistry, Medicine and Environmental Research", Vol. VII (A. Frigerio, ed.), pp. 343–349, Elsevier, Amsterdam.
Mackie, W. Z. and Fried, M. (1955). *Soil Sci.* **80**, 308–312.
Malzer, G. L. and Barber, S. A. (1976). *Soil Sci. Soc. Am. J.* **40**, 727–731.
Martini, F. and Thellier, M. (1980). *Planta* **150**, 197–205.
Matsumaru, T. T., Yoneyama, T., Tosuka, T. and Matsuoka, Y. (1981). *Soil Sci. Plant Nutr.* **27**, 255–261.
Melander, L. (1960). "Isotopic Effects on Reaction Rates", Ronald Press, New York.
Menzel, R. G. (1954). *Soil Sci.* **77**, 419–425.
Menzel, R. G. and Heald, W. R. (1955). *Soil Sci.* **80**, 287–293.
Muller, R. N., Sprugel, D. G. and Kohn, B. (1978). *J. Environ. Qual.* **7**, 171–174.
Nelson, W. L., Krantz, B. A., Colwell, W. E., Woltz, W. G., Hawkins, Arthur, Dean, L. A., MacKenzie, A. J. and Rubins, E. J. (1948). *Soil Sci. Soc. Am. Proc.* **12**, 113–118.
Nielsen, D. R., Walker, H. O. and Corey, J. C. (1978). *Trans. 11th Int. Congr. Soil Sci., Edmonton* **3**, 32–44.
Norman, A. G. and Werkman, C. H. (1943). *Agron. J.* **35**, 1023–1025.
Nye, P. H. (1979). *Adv. Agron.* **31**, 225–272.
O'Connor, G. A., Lindsay, W. L. and Olsen, S. R. (1971). *Soil Sci. Soc. Am. Proc.* **35**, 407–410.
Paetzold, R. F. and Scott, H. O. (1978). *Soil Sci. Soc. Am. J.* **42**, 874–877.
Porter, L. K., Viets, F. G. and Hutchinson, G. L. (1972). *Science* **175**, 759–761.
Reeves, R. D. and Brooks, R. R. (1978). "Trace Element Analysis of Geological Materials", John Wiley and Sons, New York.
Rennie, D. A. and Fried, M. (1971). *In* "Proceedings of the International Symposium on Soil Fertility Evaluation, New Delhi", Vol. I, pp. 639–656, FAO/IAEA, Vienna.

Rennie, D. A. and Halstead, E. H. (1965). *In* "Isotopes and Radiation in Soil–Plant Nutrition Studies", pp. 489–504, IAEA, Vienna.
Riekerk, H. (1971). *Soil Sci. Soc. Am. Proc.* **35**, 350–356.
Ritchie, J. C. and McHenry, J. R. (1975). *J. Soil Water Cons.* **30**, 283–286.
Russell, R. S. (1977). "Plant Root Systems", McGraw-Hill, Maidenhead.
Russell, R. S. and Newbould, P. (1969). *In* "Root Growth" (W. J. Whittington, ed.), pp. 148–169, Plenum Press, New York.
Ryden, J. C. and Syers, J. K. (1977). *Soil Sci.* **123**, 353–361.
Salcedo, I. H. and Ellis, B. G. (1979). *Soil Sci.* **127**, 227–234.
Schimansky, C. (1977). *Mitteilungen Klosterneuberg* **27**, 181–184.
Schultz, R. K., Overstreet, R. and Barshad, I. (1960). *Soil Sci.* **80**, 16–27.
Shaw, E., Menzel, R. G. and Dean, L. A. (1954). *Soil Sci.* **77**, 205–214.
Smith, S. J., Dillow, D. W. and Young, L. B. (1982). *J. Environ. Qual.* **11**, 341–344.
Smith, S. J. and Legg, J. O. (1971). *Soil Sci.* **112**, 373–375.
Spratt, E. D., Warder, F. G., Bailey, L. D. and Read, D. W. L. (1980). *Soil Sci. Soc. Am. J.* **44**, 1200–1204.
Stanford, G., Legg, J. O. and Smith, S. J. (1973). *Plant Soil* **39**, 113–124.
Taylor, H. M. and Klepper, Betty (1973). *Agron. J.* **65**, 965–968.
Terman, G. L. and Brown, M. A. (1968). *Plant Soil* **29**, 48–65.
Terman, G. L. and Khasawneh, F. E. (1968). *Soil Sci.* **105**, 346–354.
Tiedje, J. M. (1978). *In* "Microbiology" (D. Schlessinger, ed.), pp. 362–366, American Society for Microbiology, Washington, D.C.
Till, A. R. and May, P. F. (1971). *Aust. J. Agric. Res.* **22**, 391–400.
Tiller, K. G. (1979). *In* "Isotopes and Radiation in Research on Soil–Plant Relationships", pp. 359–372, IAEA, Vienna.
Tiller, K. G. and Wasserman, P. (1972). *In* "Isotopes and Radiation in Soil–Plant Relationships Including Forestry", pp. 517–529, IAEA, Vienna.
Vasilas, B. L., Legg, J. O. and Wolf, D. C. (1980). *Agron. J.* **72**, 271–275.
Vimal, O. P. (1972). *Indian J. Agron.* **17**, 54–57.
Vose, P. B. (1980). "Introduction to Nuclear Techniques in Agronomy and Plant Biology", Pergamon Press, Oxford.
Walker, J. M. and Barber, S. A. (1961). *Science* **133**, 881–882.
Wallace, A., Patel, P. M. and Mueller, R. T. (1976). *Commun. Soil Sci. Plant Anal.* **7**, 65–75.
Westerman, R. L. and Kurtz, L. T. (1972). *Soil Sci. Soc. Am. Proc.* **36**, 91–94.
Willatt, S. T., Struss, R. G. and Taylor, H. M. (1978). *Agron. J.* **70**, 581–586.
Wooldridge, D. D. (1965). *Soil Sci. Soc. Am. Proc.* **29**, 469–472.
Zmijewska, W. and Minczewski, J. (1965). *In* "Plant Nutrient Supply and Movement", pp. 110–122, IAEA Technical Report Series no. 48, IAEA, Vienna.

2

Soil Water Management

D. R. Nielsen and D. K. Cassel

I. Introduction

The technology to monitor and manage soil water resources is of prime importance in agriculture, hydrology and civil engineering. That technology was enhanced immensely by the development of lightweight, robust, portable, neutron moisture gauges. Before that development, soil water contents were estimated by indirect methods or by laborious, destructive sampling techniques that were inefficient and time-consuming. As a result, the neutron moisture gauge, now considered a reliable instrument, is used routinely for both research and practical investigations in agriculture, hydrology and civil as well as industrial engineering. Its major use in agriculture is aimed at the conservation of water and increased crop production. Research on soil–water–plant relations, irrigation and drainage, evapotranspiration and soil moisture storage in relation to agronomic alternatives relies heavily on the use of the neutron moisture gauge. In hydrology, it is used primarily to ascertain changes of water retention within consolidated and unconsolidated strata of the Earth's mantle as well as the density distribution within glaciers and snowfields. In civil engineering, the neutron moisture gauge is used in construction involving the preparation of soil foundations for roads, airfields, buildings and dams. Industrial applications allow accurate water content determinations of concrete, sand and gravel mixes, sinter and coke. All of the above uses of the neutron moisture gauge take advantage of its simplicity to operate and the fact that measurements are nearly instantaneous.

Soil water management decisions have also been improved with information derived from the interaction of gamma radiation with soil systems. The transmission of gamma radiation through soil or scattering within soil varies with soil properties, including bulk density. The bulk density also depends upon soil water content. Hence, both laboratory and field methods relying upon instruments to measure gamma radiation transmission and attenuation in soils have improved soil water management technology, especially when used conjunctively with neutron methods. In agriculture, hydrology and civil and industrial engineering, measurements of bulk density may be just as important as those of soil water. If the bulk density is known, fundamental relations between soil particle density, water content on a mass basis and water content on a volume basis are explicit. Changes in bulk density ascertained in the field allow assessments of soil compaction and related soil properties of both agronomic and engineering interest.

Decisions regarding soil water management hinge not only on the assessment of soil water content and bulk density, but upon the movement of dissolved constituents with the soil water – fertilizers, pesticides, and other inorganic and organic materials which undergo physical, chemical and microbiological reactions within soil profiles. Both radioactive and stable isotopes are used in agriculture and hydrology to identify pathways, reactions and retention times of the constituents in soils and groundwater aquifers. These isotopic tracer methods, based upon dilution and kinetics of reaction in closed and open systems, are especially useful where fertilizers and pesticides for agricultural production and solid and liquid wastes from municipalities and commercial industries are applied to the land. With tracers providing estimates of the relative movements of solutes and water, application rates can be more easily optimized for agricultural production as well as to reduce the pollution potential of surface and groundwaters.

Unlike analyses made in the laboratory, identification of the movement and retention of soil water must be made over relatively large land areas treated as a management unit in agriculture, hydrology, etc. And hence, the principles of radiation techniques for soil water management are complicated by spatial and temporal variations of soil properties. In addition to the deterministic procedures for measuring the interaction of the radiation and the soil system, it is of equal importance to ascertain the number and spacing of observations that should be taken to represent the management unit. And because soil water, as a result of infiltration, drainage, redistribution and evapotranspiration, is constantly changing, the question of when to take observations must also be considered.

II. Neutron moisture gauges

The principle of the neutron moisture gauge was indicated by the early work of Chadwick (1932), who discovered neutrons through their interaction with materials rich in hydrogen. He noticed that when alpha particles collided with a piece of beryllium in the presence of paraffin, protons were detected. In the absence of paraffin, protons were not detected. From this and later experiments, it was found that neutrons are uncharged particles of mass slightly larger than that of a proton, and that they interact with soil and other materials primarily in two ways – scattering and capture. Scattering occurs when a neutron collides with a nucleus of an atom and transfers some or all of its kinetic energy to the nucleus. The scattering process continues until the kinetic energy of the neutron approaches that of the surrounding atoms. Capture (or absorption) occurs when a neutron enters the nucleus of the atom and ends its existence through a nuclear reaction. Depending upon the kinetic energy of the neutron as well as the kind of atom penetrated, an atom of a different element will be formed with the emission of a proton or an alpha particle, as well as gamma radiation. The utility of a neutron moisture gauge is based upon the physical laws governing neutron scattering and absorption.

A. Principle of method

Neutrons emitted from sources used in neutron moisture gauges have energies ranging from near zero to 10 MeV and are classified as fast neutrons. Neutron scattering reduces their energies to those in thermal equilibrium with the constituent atoms of the soil, and such neutrons are classified as slow neutrons. Slow neutrons are further classified into thermal neutrons (energies up to about 0.5 eV) and epithermal neutrons (from 0.5 to about 2 eV).

When a source of fast neutrons is placed in a soil, the neutrons collide with atomic nuclei of the soil and are scattered in all directions. With each collision, a neutron not only loses some of its energy but changes its direction of movement. The average energy loss by a fast neutron is greater for collisions with atoms of low atomic weight than for collisions with heavier atoms. Hydrogen, a component of water, is the most effective neutron scatterer in the soil. Hence, the greater the soil water content, the greater is the opportunity for fast neutrons to be reduced to slow neutrons in the vicinity of their source. Once neutrons have slowed to thermal energies, they diffuse through the soil very much like one gas diffuses into another, except that their lifetime and spatial distribution are limited by capture (absorption). A source of fast neutrons placed in a moist soil immediately becomes surrounded by a cloud of thermal neutrons. The density of this cloud represents an equilibrium between the rate of emission of fast neutrons, the rate of their scattering or thermalization and the rate of their capture. A slow neutron detector insensitive to fast neutrons placed in the vicinity of the fast neutron source allows a relationship to be established between the soil water content and the number of slow neutrons in the cloud.

After Belcher *et al.* (1950), Spinks *et al.* (1951) and Gardner and Kirkham (1952) applied the above principles to the detection of soil water, and Underwood *et al.* (1954) and Stone *et al.* (1955) developed portable, self-contained equipment to measure soil water in the field, various forms of commercial equipment soon became available. Today, neutron moisture gauges contain a source of fast neutrons, a detector for slow neutrons, an instrument to determine the count rate from the detector and a suitable standard to verify the performance of the equipment. A schematic diagram of a neutron moisture gauge is given in Fig. 2.1. In addition, some gauges contain microprocessors or other means to convert the count rate to soil water content electronically by means of a theoretical or experimental calibration curve. All fast neutron sources used in neutron moisture gauges are mixtures of an alpha emitter and a fine powder of beryllium. Americium and radium are commonly used to provide the alpha particles that are absorbed by the nucleus of beryllium which subsequently emits a neutron according to the reaction

$$^{9}_{4}\mathrm{Be} + ^{4}_{2}\mathrm{He} \rightarrow ^{1}_{0}\mathrm{n} + ^{12}_{6}\mathrm{C} + 5.7\ \mathrm{MeV}. \qquad (2.1)$$

The average surplus energy of reaction, as well as some of the energy of the alpha particle, is imparted to the emitted neutron. Because Ra–Be sources have a much higher output of gamma radiation than do Am–Be sources, the latter are preferred because they require less shielding for operator protec-

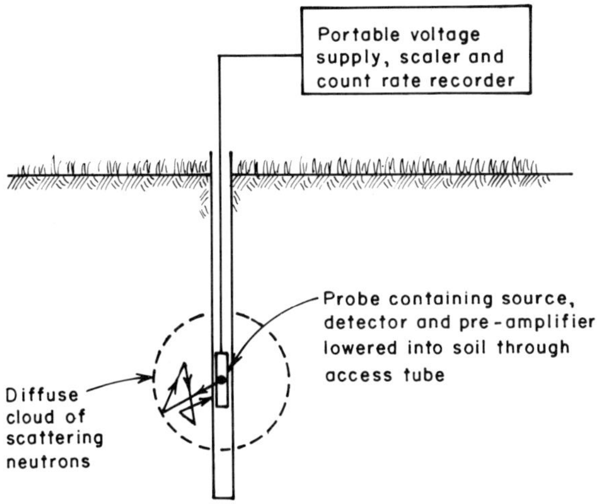

Fig. 2.1 Schematic diagram of the neutron moisture gauge.

tion. Gamma radiation comes from the alpha emitter, from radioactive impurities, and from ^{12}C initially in an excited state as a result of the nuclear reaction.

Slow neutron detectors contain a substance which has a high probability for slow neutron capture. With neutron capture, a charged particle is emitted and subsequently detected by some counting device. The boron trifluoride detector is based upon the reaction

$$^{10}_{5}B + ^{1}_{0}n \rightarrow ^{7}_{3}Li + ^{4}_{2}He + 2.8\ MeV, \tag{2.2}$$

while the ^{3}He detector is based upon

$$^{3}_{2}He + ^{1}_{0}n \rightarrow ^{1}_{1}p + ^{3}_{1}H + 760\ keV. \tag{2.3}$$

Because boron trifluoride detectors have a comparatively low slow neutron detection efficiency, the detection tube is usually relatively large (100 cm^3) compared with those filled with helium. These detectors, when provided with an appropriate, stable high voltage, produce small voltage pulses stemming from the above reactions. Instruments used to count these pulses are portable rate meters and scalers which amplify and discriminate between gamma-induced pulses and neutron-induced pulses. The count rate, usually expressed as pulses or counts per minute, may be read directly from a rate meter or ascertained with a scaler and a timer.

On each occasion that the neutron moisture gauge is used in the field, preferably both before and after use, a standard count rate is taken in a larger volume of water or in a suitable surrogate material rich in hydrogen. The standard count rate, N_{std}, serves to verify the performance of the neutron moisture gauge and to normalize the count rate, N, obtained in the soil. The gauge is calibrated in terms of the ratio n ($=N/N_{std}$) to compensate for any electronic drift which might occur as the gauge ages or for shifts in count rate caused by repairs or component replacements of the gauge. A calibration of the gauge is represented by a relationship between the volumetric soil water content θ (i.e. volume of water ÷ volume of soil) and the count rate ratio n shown in Fig. 2.2. Although calibration curves have been obtained from theoretical models describing the slowing down and the thermal diffusion of neutrons by the various constituents in soils (e.g. Couchat *et al.*, 1975), they are most frequently established by correlating values of n against θ, obtained by directly sampling the soil at the same depths that the gauge probe is inserted into the soil profile through an access tube. This procedure is not without difficulty owing to the fact that soils are not homogeneous and that volumetric soil samples are not unbiased. The composition of field soils are heterogeneous and spatially variable, giving rise to nonuniform distributions of water as well as other constituents. In addition to hydrogen, some constituents, such as boron and chlorine, are also efficient absorbers of slow neutrons. And, with the added complication that calibration is affected to some extent by the soil bulk density, there is an ever-present uncertainty in both n and θ. Hence, there are three sources of error of soil water content

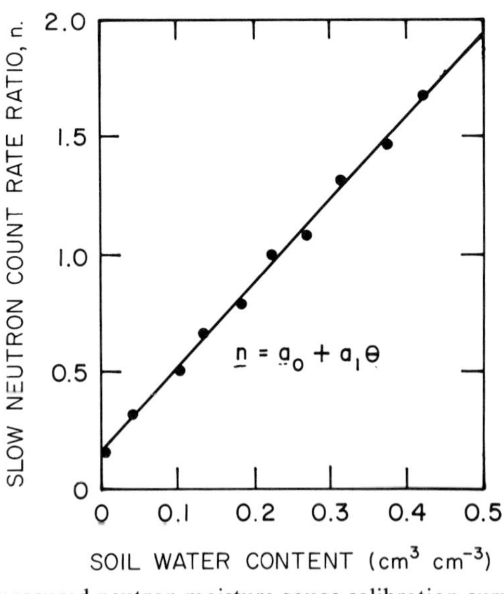

Fig. 2.2 Field-measured neutron moisture gauge calibration curve of slow neutron count rate ratio n or N/N_{std} against volumetric soil water content.

measurement: (i) variation in count rate due to variation in fast neutron emission from the source; (ii) errors associated with the detection and counting components of the gauge; (iii) errors associated with the establishment of the calibration curve. By taking count rates for sufficiently long times, the first error can be reduced to insignificance, although in practice this may not always be done. The remaining errors provide an uncertainty of θ in the order of 0.01, depending upon the particular gauge and nature of the soil profile and its particular water distribution. A comprehensive review of the theory and usage of neutron moisture gauges is given by the International Atomic Energy Agency (1970), while an article by Haverkamp *et al.* (1983) discusses error analysis.

B. Monitoring soil water content

The most common use of the neutron moisture gauge in agriculture is to monitor soil water accretion and depletion within soil profiles. Accretion as a result of rainfall or irrigation and depletion owing to evapotranspiration and percolation below the recall of crop roots are both monitored for crop management purposes. The neutron moisture gauge is particularly well suited for this usage because the full range of soil water contents manifested

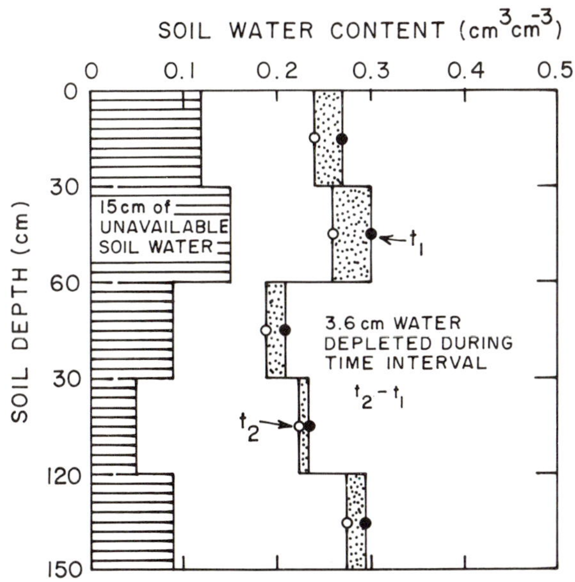

Fig. 2.3 Field-measured soil water content distribution measured at times t_1 (solid data points) and t_2 (open data points).

can be monitored repeatedly at exactly the same locations at which access tubes are placed in the field. The amount of water stored in a soil profile available to a crop can be measured as a function of time with the neutron moisture gauge. For example, Fig. 2.3 gives soil water content distributions measured at times t_1 and t_2 with a neutron moisture gauge to a soil depth of 150 cm. At times t_1 and t_2 there were 39.3 and 35.7 cm of water, respectively, within the profile. Recognizing that 15 cm of water remains in the profile unavailable to the crop when it permanently wilts, the amounts of crop available water in the profile at those times were 24.3 and 20.7 cm. And, during the period $(t_2 - t_1)$, 3.6 cm of water was lost as a result of evapotranspiration and deep percolation. If available soil water is monitored with the neutron moisture gauge and plotted against time, schedules for irrigation are readily obtained in relation to probable rainfall events.

C. Soil water properties

There are many sophisticated uses of the neutron moisture meter in agronomic research. A soil property of paramount importance to the water economy of a field is the soil hydraulic conductivity. The hydraulic conductivity K is strongly dependent upon the soil water content θ as well as the nature of the soil solution. Computer simulations of infiltration, drainage and evapotranspiration usually depend upon a quantitative relation between K and θ. The neutron moisture gauge is used in conjunction with tensiometers to ascertain $K(\theta)$, and, more recently, simplified procedures using only a neutron moisture gauge to allow adequate estimation of $K(\theta)$ have been developed. Such a procedure (Libardi *et al.*, 1980) is illustrated here for a small field plot containing an access tube for a neutron moisture gauge. The plot should be free of vegetation and without a shallow water table. After water has been continuously applied to the plot for a time sufficient for the soil profile to be wetted to the depth at which $K(\theta)$ is to be calculated, the soil water content throughout the profile is frequently monitored with the neutron moisture gauge during redistribution and internal drainage. The equation describing the redistribution, i.e.

$$\frac{\partial \theta}{\partial t} = \frac{\partial}{\partial z}\left[K(\theta)\frac{\partial H}{\partial z}\right], \tag{2.4}$$

is simplified by assuming that at soil depth $z = L$, the hydraulic gradient $\partial H/\partial z$ is unity and that

$$K(\theta) = K_0 \exp[\beta(\theta - \theta_0)], \tag{2.5}$$

where subscript zero corresponds to redistribution time t equal to zero. By

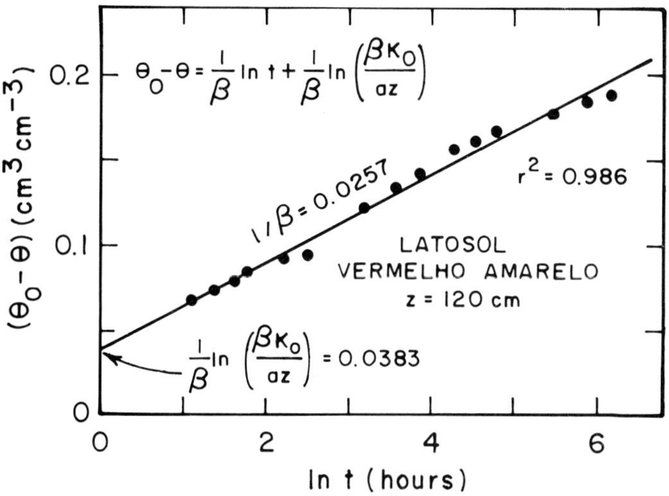

Fig. 2.4 A graph of neutron-measured values of $(\theta_0 - \theta)$ versus $\ln t$ used to calculate the parameters K_0 and β to ascertain the functional relation between K and θ by the method of Libardi *et al.* (1980).

assuming further that the average soil water content $\bar{\theta}$ from the soil surface to depth L is linearly related to θ at depth L $(\bar{\theta} = a + b)$, the solution of the redistribution equation is

$$\theta_0 - \theta = \frac{1}{\beta} \ln t + \frac{1}{\beta} \ln (\beta K_0/az). \tag{2.6}$$

Hence, plotting values of $\theta_0 - \theta$ measured with the neutron moisture gauge against $\ln t$, the slope and intercept at $t = 1$ provide values of β and K_0 that define the $K(\theta)$ function. An example of such an analysis is given in Fig. 2.4. Details of the analysis are given by Libardi *et al.* (1980).

III. Gamma radiation attenuation

The interaction of gamma radiation with soil provides a method for measuring the total density of soil. And, if the soil matrix remains invariant with the exception of its soil water content, gamma radiation can also be used to measure soil water content.

A. Principle of method

Gamma rays are absorbed in soil as a function of their energy, the atomic number of the constituent elements of the soil and the density of the soil. The most important types of interaction are photoelectric absorption, Compton absorption and scattering and pair production. The extent to which each of the interactions exists depends upon the energy of the gamma radiation. A beam of monoenergetic gamma rays is absorbed exponentially as a function of the thickness of the soil as absorbing material. For a beam of intensity I, the change in intensity per unit absorber thickness $\mathrm{d}I/\mathrm{d}x$ is proportional to the intensity of the beam. Hence,

$$\frac{\mathrm{d}I}{\mathrm{d}x} = -\mu_x I, \tag{2.7}$$

where μ_x is the total linear attenuation coefficient accounting for both absorption and scattering. If an oven-dried soil is placed between a collimated source of monoenergetic gamma radiation of intensity I_0 and a detector, the intensity I passing through the soil beam can be calculated by

$$I = I_0 \exp(-\mu\rho x), \tag{2.8}$$

where I is the count rate of unscattered gamma rays passing through the soil, I_0 the counting rate without soil (only air), μ the total mass attenuation coefficient (in square centimeters per gram), ρ the total soil bulk density (in grams per cubic centimeter) and x the thickness of soil (in centimeters) (see Fig. 2.5). The term $\mu\rho x$ is indeed the sum of many terms accounting for each and every constituent within the soil having its own value of μ, ρ and equivalent thickness x. Values of attenuation coefficients for many different materials are available (e.g. Davisson and Evans, 1952). In general, however, attenuation coefficients are determined experimentally for each soil whose

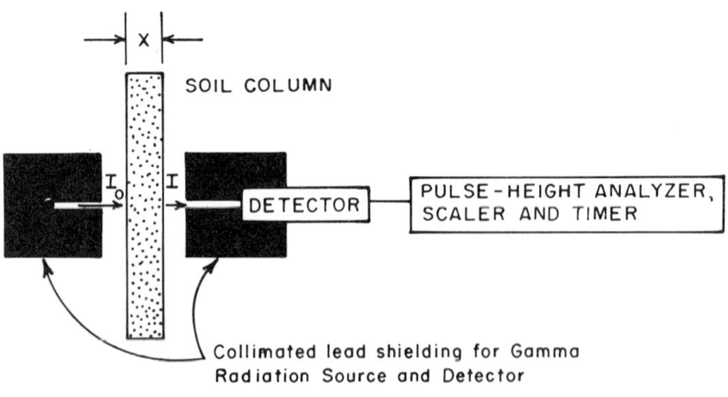

Fig. 2.5 Schematic diagram of the gamma-radiation attenuation method for measuring soil water content or soil density distributions within soil columns.

chemical composition is considered to be invariant with the exception of varying amounts of water. Hence, for an invariant water content, changes in the dry soil bulk density are reflected by changes in I, and vice versa, for invariant dry bulk density changes in soil water content are reflected by changes in I. The quantitative relation is given by

$$I = I_0 \exp\left[-(\mu_s \rho_s + \mu_w \rho_w \theta)x\right], \tag{2.9}$$

where the subscripts s and w refer to soil and water, respectively.

Low energy gamma radiation, such as that stemming from the decay of ^{137}Cs to ^{137}Ba, is easily absorbed by water and soil and easily counted. Hence, ^{137}Cs is virtually the only source of gamma radiation used in agronomic studies, even though other sources are available. Its utility is also enhanced owing to its 30-year half-life that makes day-to-day corrections for radioactive decay unnecessary and to its relatively inexpensive cost.

B. Monitoring soil density

Commercially available gamma attenuation apparatus consists of two probes at a fixed spacing that are lowered into the soil through parallel access tubes. One probe contains the gamma radiation source while the other contains a Geiger–Müller detector. Pulses from the detector are amplified, passed through a pulse-height analyzer to discriminate against radiation of energies less than that of the source and counted with a rate meter or scaler and timer. This double-probe technique allows the density of a soil horizon only a few centimeters in the vertical dimension to be measured. The density of the soil air is insignificant in comparison with those of the solid and water components, and can, therefore, be ignored. It is necessary, however, to determine the water content of the soil at sampling time in order to convert the total density of the soil to that on an oven-dry basis.

Similar to the use of the neutron moisture gauge, calibration of a two-probe gamma radiation attenuation apparatus relies upon a standard to ascertain the performance of the equipment, and has identical difficulties as regards uncertainties in counting rates and field-measured soil bulk density values.

Today there are a large variety of neutron and gamma radiation devices designed to measure both soil water content and soil bulk density concomitantly. The devices are based upon principles of scattering and attenuation for both kinds of radiation sources and are adapted to take measurements at the soil surface as well as throughout soil profiles to whatever depth desired. Used collectively, they allow changes in soil water retention and soil compaction to be monitored accurately and frequently in relation to soil and crop management alternatives.

IV. Beta radiation attenuation

Closely related to the radiation attenuation methods described above for soil profiles is the technique to use beta radiation attenuation for measuring changes in the water status of crop plants. Inasmuch as plant growth is controlled directly by the internal water stress of the plant tissue, it is sometimes advantageous to measure the water status of the crop in addition to that of the soil profile. Mederski (1961) introduced the beta radiation gauge as a nondestructive method for measuring relative water content (RWC) of plant leaves.

A. Principle of method

The attenuation of beta radiation from a source of suitable energy gives a measure of the total amount of material between the source and the detector. The equation describing this attenuation of beta radiation through a plant leaf is

$$I = I_0 \exp[-(\mu_o\rho_o + \mu_w\rho_w)x], \tag{2.10}$$

where I_0 is the intensity of the unattenuated beta radiation, x is the thickness of the plant leaf (in centimeters), μ the total mass attenuation coefficient (in square centimeters per gram) and ρ the density (in grams per cubic centimeter), with subscripts o and w referring to organic matter and water, respectively. Because the leaf thickness does not remain constant for different levels of turgidity, the above equation is modified to

$$I = I_0 \exp[-(\mu_o\,\delta_o + \mu_w\,\delta_w)], \tag{2.11}$$

where δ_o and δ_w are the mass per unit leaf area of the organic material and water, respectively. The extent of the radiation attenuated by the leaf varies with the fresh weight per unit area of leaf. The major portion of this weight consists of water with a much smaller part being organic matter. The equations describing the attenuation of the beta radiation by a completely dehydrated leaf and a leaf at full turgidity are

$$I_d = I_0 \exp[-\mu_o\,\delta_o], \tag{2.12}$$

$$I_t = I_0 \exp[-(\mu_o\,\delta_o + \mu_w\,\delta_{wt})], \tag{2.13}$$

where I_d is the count rate for a dehydrated leaf, and I_t and δ_{wt} are the count rate and water mass per unit leaf area, respectively, for a fully turgid leaf. Combining the above three equations yields

$$\ln(I/I_t) = \ln(I_d/I_t)[1 - \mathrm{RWC}], \tag{2.14}$$

where RWC is the relative water content, equal to δ_w/δ_{wt}. Hence, knowing the values of I_d and I_t for a dehydrated and turgid leaf, respectively, a calibration curve is immediately available for all RWC values. Obrigewitsch *et al.* (1975) provided the above analysis for obtaining the calibration of the beta attenuation method. Because of its simplicity, the pioneering work of Mederski (1961), as well as others (Nakayama and Erhler, 1964; Jarvis and Slatyer, 1966; Laing, 1966), has led to the practical usefulness of a beta radiation gauge to measure nondestructively the relative turgidity of plant leaves in the field.

B. Monitoring relative water content of plant leaves

Although beta attenuation gauges for monitoring relative turgidity or RWC of leaves are not commercially available, they may be easily constructed (Rolston and Horton, 1968). Components of the gauge are a source of beta radiation, a Geiger–Müller detector, a scaler and timer, and an apparatus to fix the position of the leaf between the source and detector. The choice of the radioisotope for the radiation source depends upon the half-life of the isotope and the fresh weight of the leaf. ^{14}C is suitable for fresh weights from 5 to 15 mg cm^{-2}, ^{147}Pm and ^{90}Tc are useful from 10 to 40 mg cm^{-2} and ^{204}Tl is appropriate for greater weights. Obrigewitsch *et al.* (1975) used a 6 μCi source of ^{14}C mounted in an area of about 2 cm diameter within a plastic holder attached to a movable rod, positioned immediately above a 2.8 cm diameter thin-window Geiger–Müller detector, as shown in Fig. 2.6. With the leaf inserted between the source and detector, separated at a fixed distance of about 3 mm, care is taken to ensure that the leaf completely covers the source. For narrow leaves, part of the source may have to be covered with metal plates to allow the detector to count only radiation that has passed through the leaf.

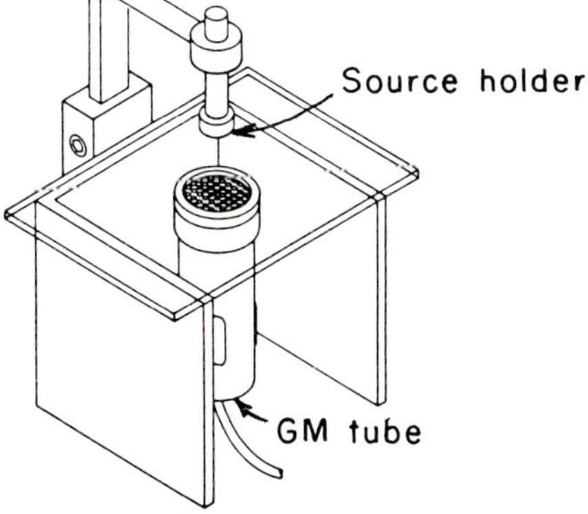

Fig. 2.6 Schematic diagram of beta-radiation gauge for alignment of plant leaf, detector and source (Obrigewitsch *et al.*, 1975).

Figure 2.7 shows a graph of ln (I/I_t) versus (1 − RWC) obtained by Obrigewitsch *et al.* (1975) for a sugar beet leaf (*Beta vulgaris* L.). The observation at (0, 0) was obtained for a turgid leaf that had been placed in distilled water for several hours at 21 °C and under 700 lux illumination. The observation (1, 6.51) was obtained for a dehydrated leaf that had been dried in a desiccator containing $CaCl_2$. Intermediate observations stemmed from radiation attenuation measurements at different levels of turgidity. The procedure for measuring the RWC should be as follows: (i) measure the radiation intensity that passes through leaves selected from desired leaves of plants growing in the field or growth chamber; (ii) place leaves in distilled water at constant temperature and light for several hours; (iii) measure radiation intensity I_t passing through the turgid leaves; (iv) measure radiation intensity I_d passing through leaves dried in a desiccator; (v) calculate RWC.

Because leaves differ in the amount of water which they contain at full turgidity according to their ages, relative positions on the plant and internal solute concentrations, a calibration curve must be established for each particular situation. The usefulness of beta radiation attenuation for

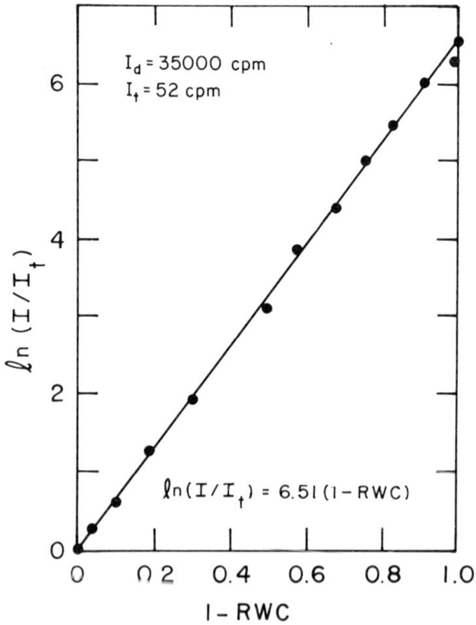

Fig. 2.7 Calibration curve of relative beta-radiation intensity passing through plant leaves versus their relative water content (Obrigewitsch *et al.*, 1975).

evaluating the water status of a crop rests on the fact that only three intensity measurements (I, I_d and I_t) are required in the absence of any other observation of the crop or its environment.

V. Tracers for soil water management

Field investigations of soil water and its constituents moving through the soil and ultimately into groundwater are enhanced by using tracer methods that offer alternative analyses not attainable by other methods. The tracer method is a technique used to investigate the actions of a substance by observing the behavior of the tracer. The substance traced is generally termed the tracer. The criteria for an ideal tracer are that it must be chemically and physically indistinguishable from the tracee and that the introduction of the tracer must not disturb the soil water system. Both of these criteria are almost perfectly met by radioactive or stable isotopic tracers. Isotopes have identical chemical properties and only slightly different physical properties (only slight mass differences), and they usually can be measured easily owing to their specific activity or tracer abundance. The introduction of an isotopic tracer generally adds negligible mass to the system and does not disturb its kinetics. In soil water investigations, there are two kinds of isotopic tracer – those that are purposely injected into a soil system to be later recovered after a reasonable time, and those that are not added artificially but occur naturally as a result of nuclear reactions and physically-based environmental processes. In soil systems, tracers are often used to identify alternative pathways and the ultimate fate of inorganic and organic soil and water constituents, fertilizers and pesticides as well as crop and animal residues. Because of its great importance, the movement of water in soils and geologic strata has been studied with a wide variety of substances used as a water tracer. Although isotopes of oxygen and hydrogen meet the criteria for ideal tracers of water, their usage to identify quantitatively the movement and retention of soil water is not without complications owing to the fact that water reacts chemically, physically and biologically within the soil system. Exhaustive studies have been conducted to identify suitable tracers for tracing water movement in soils and groundwater aquifers (see references in Freeze and Cherry, 1979).

A. Nonreactive tracers

Let us assume that we wish to determine, with the aid of a tracer, the rate at which a nonreactive fluid flows through a porous medium. If a pulse of radioactive tracer having a total activity C_0 (in microcuries) is injected at

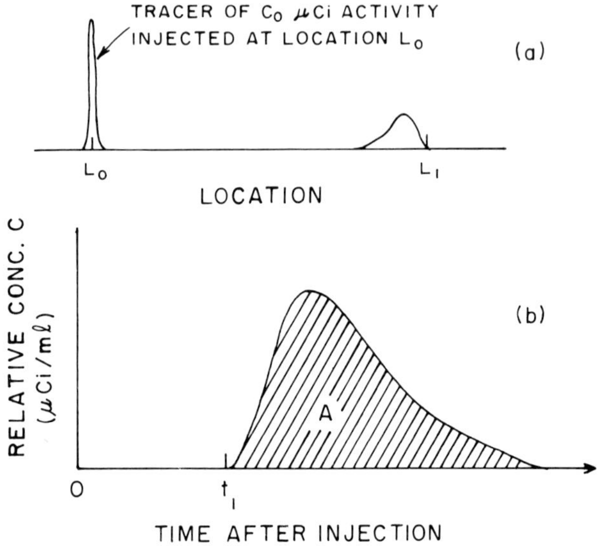

Fig. 2.8 Schematic diagram of the injection of a tracer at location L_0 being transported downstream to location L_1. At location L_1, the relative concentration of tracer measured as a function of time defines the shaded area A.

location L_0, its appearance observed at a downstream location L_1 will manifest a diffuse distribution owing to velocity distributions of the fluid within the pores as well as molecular diffusion. If the activity concentration C (in microcuries per milliliter) is measured at L_1 as a function of time, as shown in Fig. 2.8, the flow rate q (in milliliters per time period) can be calculated from the relation

$$q = C_0/A, \tag{2.15}$$

where A is the area under the curve to the right of t_1. The equation is simple because it is based upon the conservation of matter and that no tracer is lost or generated between the two locations as the fluid moves in only one direction. Experimentally, sufficient observations of C must be taken to define the curve $C(t)$ accurately. Corey and Horton (1968) displaced 170 cm^3 of tracer water containing 160 μCi of ^3H as well as labels of ^2H and ^{18}O through a 186 cm long water-saturated column of acidic kaolinitic soil under steady state conditions in the laboratory. Their observations of the isotopes measured in the effluent from the column as a function of time are given in Fig. 2.9. With the area A under the ^3H curve being 2.87 μCi h cm^{-3}, the flow of water q from the above equation is

$$q = \frac{160\ \mu\text{Ci}}{2.87\ \mu\text{Ci h cm}^{-3}} = 55.7\ \text{cm}^3\ \text{h}^{-1}, \tag{2.16}$$

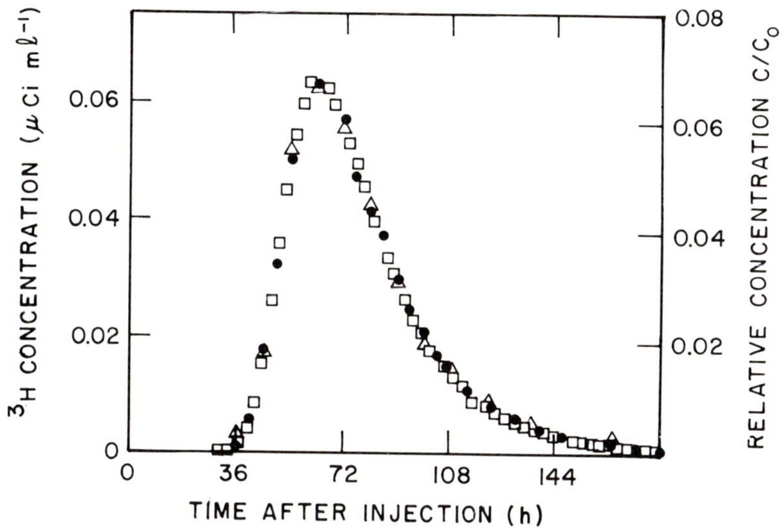

Fig. 2.9 Relative concentrations of ^{3}H (□), ^{2}H (●) and ^{18}O (△) measured 180 cm downstream in a soil column as a function of time following their application at the soil surface (Corey and Horton, 1968).

which compares favorably with 55.8 cm^3 h^{-1} measured volumetrically. Because the ^{2}H, ^{3}H and ^{18}O curves were virtually identical, and each represents nearly 170 cm^3, it may be concluded that, for the particular soil used, each of the isotopes serves as a suitable tracer of water.

B. Mean residence time

Another method of ascertaining the rate at which water flows through a soil is to calculate the time required for a tracer pulse to travel from its point of injection to a point of known distance, s, downstream. If a tracer pulse of concentration C_0 is injected into the soil water stream at time zero for a duration of t_0 (in hours), the water moves at a linear rate v (in centimeters per hour) through the soil according to the equation

$$v = s/t_s, \tag{2.17}$$

where $t_s = \bar{t} - t_0/2$, and the mean residence time $\bar{t}$ is given by

$$\bar{t} = \int_0^\infty tc(t)\,\mathrm{d}t \bigg/ \int_0^\infty c(t)\,\mathrm{d}t \tag{2.18}$$

where $c(t)$ is the concentration of the tracer measured at a distance s as a function of time. For instantaneous pulses ($t_0 \ll \bar{t}$), t_s is nearly equal to $\bar{t}$. For many field applications, the average residence time $\bar{t}$ is often approximated by the time t_p at which the maximum relative concentration $(c/c_0)_{max}$ is recorded at the distance s. Considering the data given in Fig. 2.9, s, t_0, t_p and $\bar{t}$ equal 180 cm, 3.05, 63.6 and 67.7 h, respectively. Hence, the value of v is 2.72 cm h^{-1}, which corresponds to a flux of 1.22 cm h^{-1} ($=v\theta$, where $\theta = 0.45$ cm^3 cm^{-3}).

C. Reactive tracers

Tracers of water do not, in general, behave ideally as shown in Fig. 2.9. Data presented in Fig. 2.10 represent the relative concentration of ^{36}Cl, tritiated water and the nematocide 1,2-dibromo-3-chloropropane (DBCP) in the soil solution, measured at the 150 cm depth within a field soil being leached by continuously ponding its surface with water following the application of 15 cm of water containing concentrations C_0 of each solute. Although the data resemble those in Fig. 2.9, several features are different. Tritiated water is leached more slowly and never attains as great a relative concentration as that of ^{36}Cl. Moreover, a comparison of the areas under the ^{3}H and ^{36}Cl curves reveals that some of the ^{3}H is not being leached readily and remains in the soil profile. By the time the soil solution passes the 150 cm depth, the relative DBCP concentration is no greater than 0.01, which indicates a much stronger interaction of the DBCP with the soil than that of ^{3}H or ^{36}Cl. When

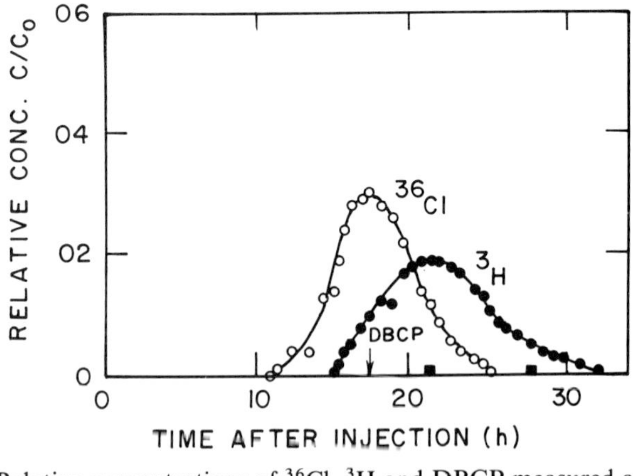

Fig. 2.10 Relative concentrations of ^{36}Cl, ^{3}H and DBCP measured at the 150 cm depth as a function of time following their application at the soil surface under water-ponded leaching conditions.

such interactions are manifested, the above equations are not valid. Virtually all tracers of water manifest some degree of interaction with the soil matrix as well as some tendency to mix with the tracee water. For example, movements of anionic and cationic tracers will be influenced by the electric fields surrounding the individual soil particles. These influences, together with the mixing of tracer and tracee by the process of diffusion, coupled with variations in pore water velocity, render the position and rates of movement of the tracer and tracee to be entirely different. Under such conditions, when tracers are used to identify pathways of either water or one of its solutes, the following equation has been used extensively for laboratory studies, and to a lesser extent for field conditions:

$$\frac{\partial \rho s}{\partial t} + \frac{\partial \theta c}{\partial t} = \frac{\partial}{\partial z}\left(\theta D \frac{\partial c}{\partial z}\right) - \frac{\partial J c}{\partial z} + \phi, \tag{2.19}$$

where ρ is the soil bulk density (in grams per cubic centimeter), s the solute associated with the soil particles (in grams per gram of soil), D the apparent diffusion coefficient (in square centimeters per time period), J the flux of water (in centimeters per time period), ϕ an irreversible source or sink of solute (in grams per cubic centimeter per time period), z the soil depth (in centimeters), and t time. Day (1956) used the above equation without its first and last terms to describe the leaching of a simple organic salt through sand under steady state water flux conditions. Van Genuchten (1981) has reviewed experimental and theoretical developments during the past 25 years associated with the above equation.

An example of the sophistication achieved utilizing equation (2.19) to trace the fate of a solute in a soil profile under controlled laboratory conditions is that of Wagenet *et al.* (1977). They extended the mathematical analysis of Cho (1971) to describe the fate of urea applied to a soil. The urea hydrolyzed to ammonium, the ammonium oxidized to nitrate, and the nitrate was reduced to dinitrogen gas. Using urea having a 95% enrichment of ^{15}N, they traced the above nitrogen species stemming from the applied urea fertilizer within soil columns being leached at controlled flow rates. Three equations similar to equation (2.19) were utilized:

$$\frac{\partial \rho s_1}{\partial t} + \frac{\partial \theta c_1}{\partial t} = \frac{\partial}{\partial z}\left(\theta D_1 \frac{\partial c_1}{\partial z}\right) - \frac{\partial J c_1}{\partial z} - k_1 c_1, \tag{2.20}$$

$$\frac{\partial \rho s_2}{\partial t} + \frac{\partial \theta c_2}{\partial t} = \frac{\partial}{\partial z}\left(\theta D_2 \frac{\partial c_2}{\partial z}\right) - \frac{\partial J c_2}{\partial z} + k_1 c_1 - k_2 c_2, \tag{2.21}$$

$$\frac{\partial \theta c_3}{\partial t} = \frac{\partial}{\partial z}\left(\theta D_3 \frac{\partial c_3}{\partial z}\right) - \frac{\partial J c_3}{\partial z} + k_2 c_2 - k_3 c_3. \tag{2.22}$$

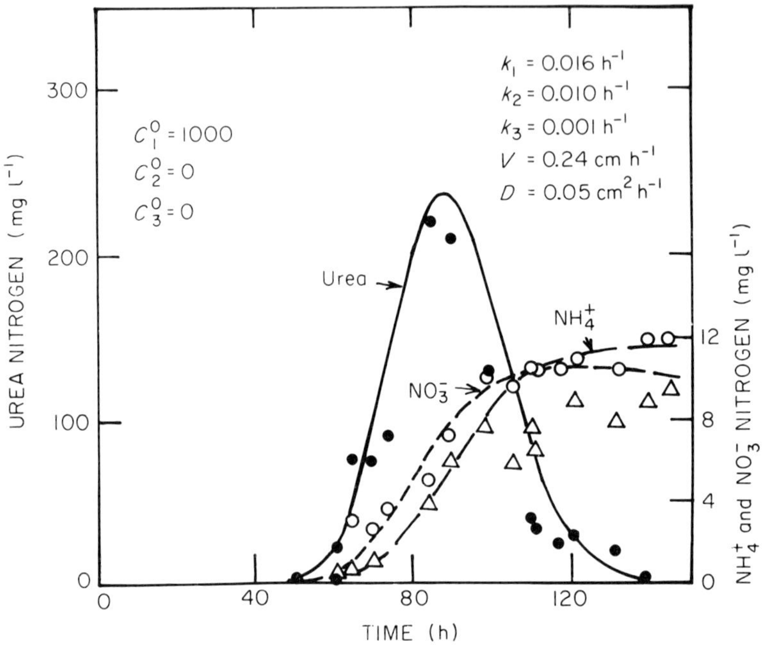

Fig. 2.11 Concentrations of N in urea, NH_4^+ and NO_3^- in the effluent of a 15 cm long soil column stemming from the application of ^{15}N-labeled urea (Wagenet *et al.*, 1977).

The subscripts 1, 2 and 3 refer to nitrogen in the form of urea, ammonium and nitrate, respectively, while the rate constants (k_i) describe the processes of urea hydrolysis, ammonium oxidation and nitrate reduction. The experimental technique described in detail by Wagenet and Starr (1977) yielded the data given in Fig. 2.11, resulting from the leaching of a pulse of urea through a 15 cm long soil column. The concentrations of N in the ammonium and nitrate shown in Fig. 2.11 stem solely from the ^{15}N-labeled urea. Such trace techniques allow an accurate assessment of the behavior of water and solutes in soil profiles under particular conditions established in the laboratory. Under such conditions, and especially using uniform soil columns made from sieved, homogeneous soil, the parameters in the above equations are easily ascertained. Similar analyses for field conditions are difficult owing to their heterogeneity and spatial variability. Such difficulty is manifested by the data given in Fig. 2.12 where the leaching of a pulse of ^3H_2O applied on the surface of a field soil was monitored by samples of the soil solution extracted through suction probes located at the depths of 33.5, 46.0 and 63.5 cm. The curves at the 33.5 cm and 63.5 cm depths appear at approximately the same times, and that at the 46.0 cm depth appears much

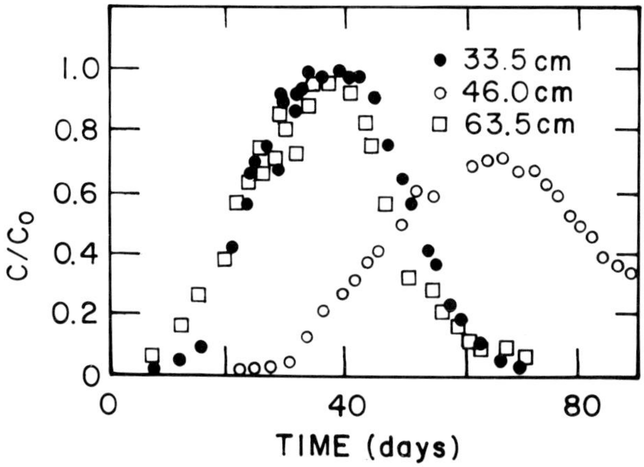

Fig. 2.12 Relative concentrations of ^3H measured at three depths of a field soil as a function of time following its application at the soil surface being leached under steady state conditions (Van De Pol *et al.*, 1977).

later. Using either equation (2.15) or equation (2.17) would reveal that the rate of movement to the greatest depth exceeds that to the shallower depths. These seemingly incongruous observations have been shown to be the result of spatial variability or to the difficulty of obtaining representative samples due to the concept that water moves preferentially through a limited area or limited set of soil pores. It has been known for at least a century that water and solutes, as well as tracers, are not displaced uniformly through field soils (Lawes *et al.*, 1881). Reasons for the relatively incomplete or irregular displacement are based on the presence of cracks and fissures, clay lenses, worm holes, root channels, textural variations, density and viscosity variations of the soil solution, and others. Van De Pol *et al.* (1977) showed that if many additional observations were made at each of the depths (33.5, 46.0 and 63.5 cm), the average water flow rate, based upon the displacement times of the ^3H_2O tracer, agreed with the infiltration rate measured volumetrically at the soil surface. Hence, a relatively larger number of observations are required to estimate adequately the magnitude of the term J in equation (2.19). Similarly, identifying the magnitudes of the other terms (e.g. D, θ, ϕ, etc.), as well as spatial and temporal averages of c and s, has seldom been attempted in the field except for highly instrumented, small field plots. Although there is a need and genuine interest in following the fate of surface-applied tracers through the soil and into surface and groundwaters, as well as their absorption by plant roots, the utility of equation (2.19) rests upon improving the technology of field sampling – a topic to be introduced at the end of this chapter.

D. Environmental tracers

Some environmental tracers are radioactive, and of these, tritium is particularly useful in estimating movement of soil water or the age of water in an aquifer. Tritium is produced naturally by cosmic radiation in the stratosphere and artificially by H-bomb detonation. With its concentrations in precipitation having been intensively studied during the past 30 years, and with it being infiltrated into the soil and leached with rainfall events, it serves as an indicator of the rate at which water moves through the unsaturated zone of soil towards the water table. In 1969, Smith *et al.* (1970) measured profiles of tritium in a clay soil to a depth of 4 m as shown in Fig. 2.13. The

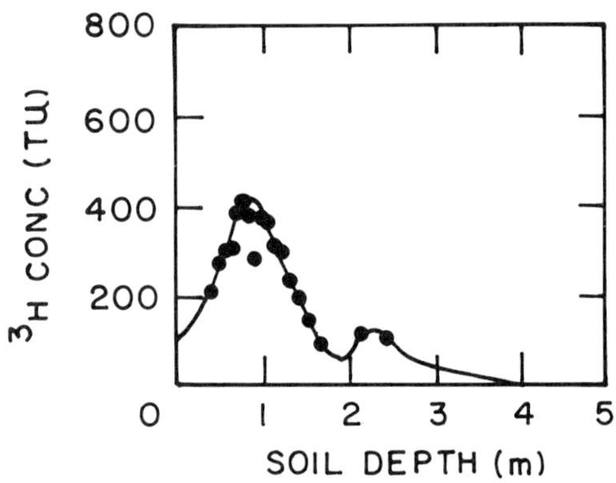

Fig. 2.13 Tritium concentration distribution in an Oxford clay profile. A tritium unit (TU) equals one atom of tritium per 10^{18} atoms of ^{1}H (Smith *et al.*, 1970).

two peaks of tritium concentration occurring at 0.9 and 2.25 m were caused by tritium concentrations in precipitation during the periods 1963–4 and 1958–9, respectively. Assuming an average soil water content of about 0.5 cm^3 cm^{-3}, and using equation (2.17), the soil water flux was estimated to be between 11.5 and 9.0 cm y^{-1}. The utility of using tritium in this manner to determine rates of water movement hinge upon observations of well defined peak concentrations in soil being correlated with appropriate peak concentrations of tritium occurring at known times earlier in the precipitation.

A well defined curve of concentration versus soil depth is oftentimes not attainable owing to spatial variability. For example, the data in Fig. 2.14 represent triplicate observations of tritium in a podsolic soil of Australia taken in 1971. Because of the variability of the data, the locations of any peak concentrations cannot be quantitatively ascertained. For such situations, Allison and Hughes (1974) showed that the mean annual recharge rate ($\bar{J}$)

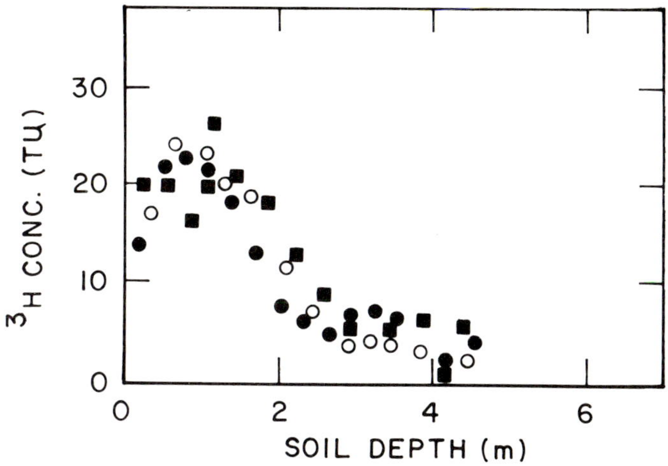

Fig. 2.14 Three tritium concentration distributions measured during July 1971, in an Australian soil to a depth of 6 m (Allison and Hughes, 1974).

could be estimated on the basis of the total amount of tritium remaining in the soil using the relation

$$\bar{J} = \int_0^{z_L} \theta C \, \mathrm{d}z \left[\sum_{n=0}^{N} C_n \exp(-\lambda n) \right]^{-1}, \tag{2.23}$$

where C is the concentration of tritium of the soil water, C_n the annual average concentration of tritium of the precipitation, λ the decay constant, n the year which is counted backwards from the sampling date, and z_L the depth below which no tritium is found. The magnitude of N is taken sufficiently large to allow the tritium to decay to nil. The above equation assumes that water passes through the soil slowly enough to allow the complete decay of the tritium before it reaches depth z_L. With this assumption, Allison and Hughes (1974) showed the annual recharge rate through the Australian soil to be about 5.0 cm y^{-1}.

VI. Sampling field soils

Unlike the application of radiation or tracer techniques in most other scientific disciplines, the interpretation of observations in soil water management studies must include a consideration of the spatial and temporal variations inherent in field soils. Field soils are heterogeneous. They vary in space and time owing to soil formation processes that have occurred daily through countless centuries. They also vary and continuously change owing to man's activities – cultivation, compaction, irrigation, land leveling and shaping, etc. And, because plant root distributions are dynamic, they extract soil water preferentially at different soil depths and at different horizontal distances depending upon the nature and density of the plant species within the ecological community. Hence, measurements of soil water or other soil properties often manifest geometric patterns of irregularities that may or may not be randomly distributed with respect to their observations on the landscape. Similarly, at a given location within a soil profile, observations of soil water content or other soil attributes ascertained repetitively by radiation techniques may or may not be related to the times at which they were made. In other words, we would like to have criteria to judge how far apart we should make measurements of soil water, how often we should make measurements, and how many measurements we should make to characterize a particular field or parcel of the landscape. The brief presentation that follows serves only as an introduction to the development of those criteria. It focuses on two concepts. The first concept considers the description of a field in terms of a set of observations reduced to their mean value and measures of their divergence without regard to their spatial distribution. The second

Table 2.1 Sequence of 160 values of soil water content measured with a neutron moisture gauge at the 50 cm depth along a horizontal transect[a]

0.101	0.102	0.105	0.107	0.114	0.114	0.116	0.120
0.120	0.133	0.142	0.143	0.135	0.131	0.129	0.120
0.124	0.139	0.135	0.138	0.140	0.141	0.140	0.126
0.126	0.125	0.132	0.137	0.139	0.138	0.136	0.135
0.128	0.139	0.144	0.142	0.135	0.132	0.129	0.132
0.137	0.139	0.142	0.142	0.148	0.145	0.144	0.138
0.135	0.135	0.129	0.116	0.113	0.116	0.123	0.127
0.125	0.124	0.127	0.129	0.121	0.116	0.113	0.117
0.119	0.122	0.127	0.133	0.126	0.124	0.123	0.123
0.130	0.137	0.137	0.139	0.137	0.130	0.127	0.125
0.123	0.121	0.124	0.128	0.141	0.145	0.147	0.151
0.150	0.155	0.156	0.158	0.156	0.153	0.127	0.127
0.131	0.133	0.134	0.134	0.137	0.142	0.144	0.140
0.138	0.139	0.147	0.173	0.173	0.172	0.169	0.166
0.156	0.147	0.146	0.144	0.143	0.143	0.140	0.141
0.141	0.142	0.147	0.152	0.153	0.152	0.152	0.153
0.156	0.155	0.152	0.148	0.147	0.152	0.156	0.145
0.145	0.153	0.156	0.157	0.157	0.157	0.157	0.159
0.160	0.155	0.151	0.146	0.146	0.140	0.135	0.126
0.115	0.112	0.104	0.101	0.097	0.093	0.092	0.093

[a] Observations were made 1 m apart and are ordered left to right.

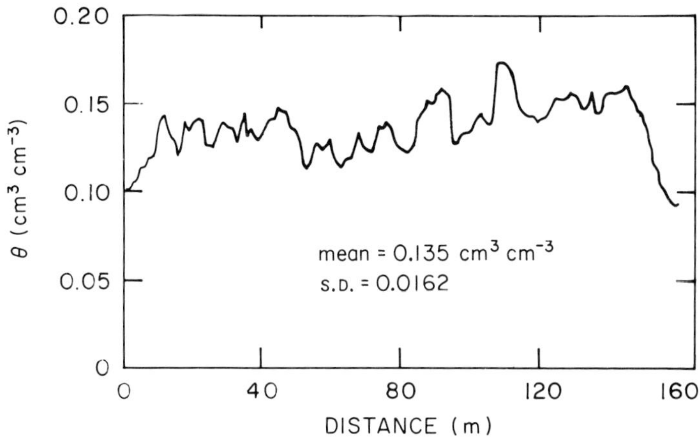

Fig. 2.15 Soil water content measured at 1 m intervals with neutron moisture gauge within a 160 m long transect.

considers the description of that portion of a field described by a single observation at a particular location. Each concept is equally useful and will be illustrated with a set of neutron moisture gauge observations of soil water content presented in Table 2.1. The data stem from 160 water content measurements made at the same soil depth at 1 m intervals along a horizontal transect within an agricultural field managed as a homogeneous unit (see Fig. 2.15). From a pedologic point of view, the data could have been taken from the same location at 160 different times.

A. Frequency distributions of soil observations

Relative to the first concept, the mean value μ of the soil water content can be estimated from the sum of the neutron moisture gauge observations of soil water content θ_i divided by the number of observations n:

$$m = \sum_{i=1}^{n} \theta_i/n, \tag{2.24}$$

where m is the sample mean. Notice that this calculation estimates the "true" mean value μ without considering the location of each of the observations within the transect. Nothing in the formula dictates where the observations should be made within the transect (a parcel of land 160 m long). The values given in Table 2.1 stem from equally spaced observations, but the formula would also apply to a set of observations of which the majority was taken near the center of the transect, or at one end of the transect, or at random positions throughout the transect. Because the locations of the observations are disregarded in this concept, the standard deviation s of the observations is calculated to provide a measure of the range or scatter of the soil water data. The value of s obtained by

$$s = \sum_{i=1}^{n} [(\theta_i - m)^2/(n-1)]^{1/2} \tag{2.25}$$

is an estimate of the "true" standard deviation σ of the soil water content of the transect. Small values of s would indicate that most of the observations were near the mean, while large values would indicate that at least a portion of the observations differed appreciably. The value of s depends collectively upon errors of measurement made by the observer which include calibration uncertainties illustrated in Fig. 2.2, as well as the spatial variations in soil water content that truly exist in field soils.

Although the above two statistics, m and s, are commonly calculated, they can be used much more effectively and rigorously if the frequency distribution of the population is estimated. Figure 2.16 is a histogram showing the distribution of the 160 observations of soil water content θ. The height of each vertical bar represents the number of observations found in a range $\Delta\theta$ of 0.01 $cm^3\ cm^{-3}$. The leftmost bar shows three observations to be within 0.085 and 0.095 $cm^3\ cm^{-3}$. Forty-three observations of θ were within 0.135 and 0.145 $cm^3\ cm^{-3}$. The estimated mean and standard deviation were 0.136

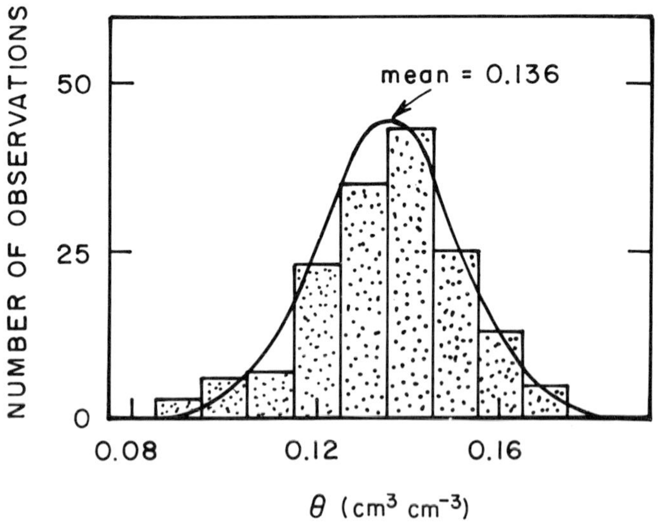

Fig. 2.16 The frequency distribution and histogram of the 160 soil water content values presented in Fig. 2.15.

and 0.0162 $cm^3\ cm^{-3}$, respectively. The smooth curve shown in the figure is for a normal probability density function

$$N = n\,\Delta\theta(2\pi s^2)^{-1/2} \exp\left[-(\theta - m)^2/2s^2\right], \tag{2.26}$$

where N is the number of observations of θ within $\theta \pm \Delta\theta$. Because the curve appears to fit the histogram, we conclude that the observations, as well as all values of soil water content capable of being measured within the transect, are normally distributed. Having made this conclusion, a determination of the number of observations to estimate the mean value of θ to be within any desired percentage of the true field value can be made for a given level of probability. The presentation of detailed statistical techniques for the vast variety of situations that can be examined using this first concept is neither appropriate nor possible in this chapter. We refer the reader to standard texts on statistical methods such as Fisher (1970), Ostle and Mensing (1979) and Steel and Torrie (1980). Such techniques are based upon the assumption that the frequency distribution is known or can be estimated from a set of observations. Examples of such distributions for soil water and other physical parameters of field soils have been published by Warrick and Nielsen (1980).

B. Spatial distributions of soil observations

The above analysis of the soil water content data given in Table 2.1 made no use of where the observations were made or how close together they might be. Intuitively, we would expect measurements made close together in the field to yield nearly equal values, while measurements made some distance apart would yield values more unrelated to each other. The framework for expressing such relationships has been developed in time and space series analyses (e.g. Haan, 1977). An analytical procedure for examining the above intuitive expression is the autocorrelation function r_h:

$$r_h = c_h/s^2, \tag{2.27}$$

where

$$c_h = \frac{1}{n-h-1}\sum_{i=1}^{n-h} (\theta_i - m)(\theta_{i+h} - m) \tag{2.28}$$

and h is the integer 0, 1, 2, 3, etc., that expresses the number of intervals (lag h) between measured values of θ. For example, when $h = 1$, r_1 is the value of the linear regression coefficient for lag 1 when values of θ_{i+1} are plotted against values of θ_i. In other words, nearest neighbors are plotted against each other. Similarly, when $h = 2$, the value of r_2 indicates the linear regression coefficient for values of θ_{i+2} plotted against other values θ_i located two intervals away (lag 2). For the data in Table 2.1, values of the autocorrelation coefficient r_h are plotted against lag h in Fig. 2.17. The figure shows that, as the distance between pairs of observations increases, the correlation between observations decreases. Had the observations of θ been spatially independent, all values of r_h ($h = 1, 2, 3, \ldots$) would have been near zero, and clearly this is not the case. The value of λ given in the equation, often called the autocorrelation length, defines the distance between observations beyond which the correlation is equal to or less than 1/e. In other words, anywhere along the transect, observations separated by a distance of 5.86 m or less would be expected to yield approximately equal values of θ. This second concept gives a measure of what length of the transect is represented by a single observation by taking into account all of the variability of the soil water content with respect to horizontal distance. Even though a neutron moisture gauge reading is the resultant of fast neutrons being slowed in the vicinity of a detector at a distance no greater than about 0.5 m, we can understand, after making the above analysis, that its reading represents the soil water content in a statistical sense to a distance 10 times greater than its deterministic value. And hence, autocorrelation lengths λ are often thought of as statistical scales of observation which may not necessarily relate to the physical dimensions of the sensor. Had the soil water data given in Table 2.1 not been observed in a transect, but been observed in a two-directional sampling scheme over a rectangular field, an

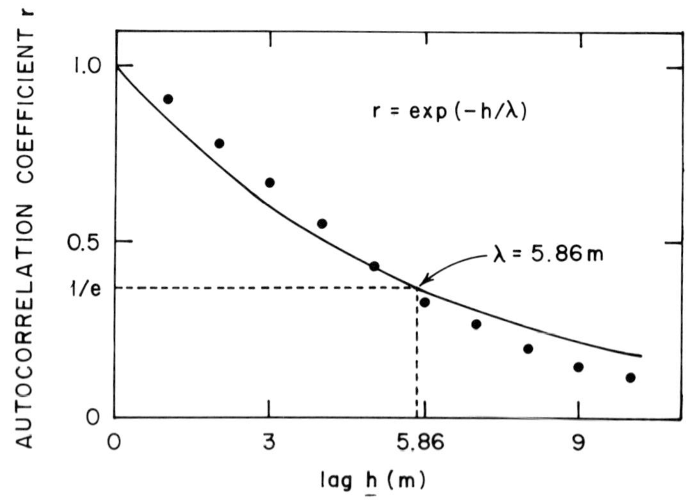

Fig. 2.17 The autocorrelogram of the 160 soil water content values given in Table 2.1. The value of r is calculated from equation (2.27).

extension of the above analysis would have provided scales of observation expressed in two directions or in a radial direction from a single observation. The technology that exists as a result of extending this second concept is far-reaching, and beyond the scope of this chapter. The technology includes interpolation procedures to identify contours of equal values across fields, procedures for correlating spatially two or more different kinds of observations, procedures for estimating the accuracy of an observation at a given location, and in general, procedures to deal more effectively with the development of sampling schemes relative to soil water management. The reader is referred to the textbooks of Davis (1973), Haan (1977), Journel and Huijbregts (1978) and Hillel (1980), as well as recent journal articles related to soil and water science (e.g. Webster and Cuanalo, 1975; Campbell, 1978; Webster, 1978; Vieira *et al.*, 1981; Nielsen *et al.*, 1982).

The utility of using both of the above concepts in soil water management studies is great. Neither is a substitute for the other. The frequency distribution of θ in Fig. 2.16 showed the observations to be approximately normal. Such a frequency distribution does not quantify the variability of the soil water content as regards their spatial arrangement but merely treats the values in terms of their magnitudes independent of their spatial position. The two concepts (frequency distribution and spatial variability) should not be confused. Using the first concept by assuming a frequency distribution is descriptive of the total population; it is possible to calculate the probability of an observation occurring within specified confidence intervals. These kinds of calculations, however, provide identical information for expected values for all locations. Expected values for specific locations cannot be obtained. On the other hand, using the second concept, the autocorrelation function provides information about the separation distances within which observed values are related to each other, and do not depend upon a knowledge of the frequency distribution. The spatial structure of a set of observations does not provide information about the frequency distribution of the observations. The fact that a set of observations manifests no spatial structure does not imply that the observations are normally distributed, and vice versa. The soil water content data in Table 2.1 constitute a clear example of a set of observations being normally distributed, yet these observations separated by less than 6 m are dependent on each other as expressed by λ in Fig. 2.17. The concept of randomness associated with the normal distribution should not lead to the conclusion that locations for field observations of soil water properties, using radiation techniques, could be selected randomly.

References

Allison, G. B. and Hughes, M. W. (1974). "Isotope Techniques in Groundwater Hydrology 1974", pp. 57–70, IAEA, Vienna.

Belcher, D. J., Cuykendall, T. R. and Sack, H. S. (1950). Technical Development Report No. 127, pp. 1–20, Civil Aeronautics Administration Technical Development and Evaluation Center, Indianapolis, Ind.

Campbell, J. B. (1978). *Soil Sci. Soc. Am. J.* **42**, 460–464.

Chadwick, J. (1932). *Proc. Roy. Soc. Lond. A* **136**, 692–708.

Cho, C. M. (1971). *Can. J. Soil Sci.* **51**, 339–350.

Corey, J. C. and Horton, J. H. (1968). *Soil Sci. Soc. Am. Proc.* **32**, 471–475.

Couchat, Ph., Carre, C., Marcesse, J. and LeHo, J. (1975). *In* "Nuclear Cross Sections and Technology" (R. A. Schrack and C. D. Bowman, eds), pp. 516–579, National Bureau of Standards, US Department of Commerce, Washington, D.C.

Davis, J. C. (1973). "Statistics and Data Analysis in Geology", p. 550, John Wiley and Sons, New York.

Davisson, C. M. and Evans, R. D. (1952). *Rev. Mod. Phys.* **24**, 79–85.

Day, P. R. (1956). *Trans. Am. Geophys. Union.* **37**, 595–601.

Fisher, R. A. (1970). "Statistical Methods for Research Workers", p. 362, Hafner Publishing Co., New York.

Freeze, R. A. and Cherry, J. A. (1979). "Groundwater", p. 604, Prentice-Hall, Englewood Cliffs, N.J.

Gardner, W. and Kirkham, D. (1952). *Soil Sci.* **73**, 391–401.

Haan, C. T. (1977). "Statistical Methods in Hydrology", p. 378, Iowa State University Press, Ames, Iowa.

Haverkamp, R., Vauclin, M. and Vachaud, G. (1983). *Soil Sci.* (in press).

Hillel, D. (1980). "Applications of Soil Physics", p. 385, Academic Press, New York.

International Atomic Energy Agency. (1970). "Neutron Moisture Gauges", p. 95, IAEA Technical Report Series no. 112, IAEA, Vienna.

Jarvis, P. G. and Slatyer, R. O. (1966). *Science* **153**, 78–79.

Journel, A. G. and Huijbregts, Ch. J. (1978). "Mining Geostatistics", p. 600, Academic Press, New York.

Laing, D. P. (1966). *Diss. Abst.* **66**, 6988; Ph.D. thesis, Iowa State University; University Microfilms, Ann Arbor, Mich.

Lawes, J. B., Gilbert, J. H. and Warington, R. (1881). *J. Roy. Agric. Soc. Engl. Ser 2* **17**, 249–279, 311–350.

Libardi, P. L., Reichardt, K., Nielsen, D. R. and Biggar, J. W. (1980). *Soil Sci. Soc. Am. J.* **44**, 3–7.

Mederski, H. D. (1961). *Soil Sci.* **92**, 143–146.

Nakayama, F. S. and Erhler, W. L. (1964). *Plant Physiol.* **39**, 95–98.

Nielsen, D. R., Biggar, J. W. and Wierenga, P. J. (1982). *In* "Nitrogen in Agricultural Soils" (F. J. Stevenson, ed.), pp. 423–448, American Society of Agronomy, Madison, Wisc.

Obrigewitsch, R. P., Rolston, D. E., Nielsen, D. R. and Nakayama, F. S. (1975). *Soil Sci. Soc. Am. J.* **67**, 729–732.

Ostle, B. and Mensing, R. W. (1979). "Statistics in Research", p. 596, Iowa State University Press, Ames, Iowa.

Rolston, D. E. and Horton, M. L. (1968). *Agron. J.* **60**, 333–336.

Smith, D. B., Wearn, P. L., Richards, H. J. and Rowe, P. C. (1970). "Isotope Hydrology 1970", pp. 73–87, IAEA, Vienna.

Spinks, J. W. T., Lane, D. A. and Torchinsky, B. B. (1951). *Can. J. Tech.* **29**, 371–374.
Steel, R. G. D. and Torrie, J. H. (1980). "Principles and Procedures in Statistics: A Biometrical Approach", p. 633, McGraw-Hill, New York.
Stone, J. F., Kirkham, D. and Read, A. A. (1955). *Soil Sci. Soc. Am. Proc.* **19**, 419–423.
Underwood, N., van Bavel, C. H. M. and Swanson, R. W. (1954). *Soil Sci.* **77**, 339–340.
Van De Pol, R. M., Wierenga, P. J. and Nielsen, D. R. (1977). *Soil Sci. Soc. Am. J.* **41**, 10–13.
van Genuchten, Th. M. (1981). "Non-equilibrium Transport Parameters from Miscible Displacement Experiments", p. 88, US Salinity Laboratory Research Report No. 119, US Salinity Laboratory, Riverside, Calif.
Vieira, S. R., Nielsen, D. R. and Biggar, J. W. (1981). *Soil Sci. Soc. Am. J.* **45**, 1040–1048.
Wagenet, R. J. and Starr, J. L. (1977). *Soil Sci. Soc. Am. J.* **41**, 658–659.
Wagenet, R. J., Biggar, J. W. and Nielsen, D. R. (1977). *Soil Sci. Soc. Am. J.* **41**, 896–902.
Warrick, A. W. and Nielsen, D. R. (1980). *In* "Applications of Soil Physics" (D. Hillel, ed.), pp. 319–344, Academic Press, New York.
Webster, R. (1978). *J. Soil Sci.* **29**, 388–402.
Webster, R. and Cuanalo, C. H. E. (1975). *J. Soil Sci.* **26**, 176–194.

3

Soil Organic Matter Studies

J. N. Ladd and J. K. Martin

I. Introduction

The beneficial effects of soil organic matter on plant productivity have long been established, and maintenance of adequate levels of organic matter in soils is a goal of major importance in productive agriculture (Allison, 1973; Russell, 1973). Its influence on soil fertility is due partly to its role as a reservoir from which plant nutrients, e.g. nitrogen, sulphur and phosphorus, can be released, following decomposition and metabolism of organic substrates by soil animals and microorganisms. Indirectly, soil organic matter influences soil fertility through its effects on the physicochemical environment. The cation exchange capacity of soils and the buffering capacity of soils against pH changes are due in part to reactive substituent groups of organic residues. Organic matter also has an important role in the formation of aggregates, and thus in stabilizing soil structure, facilitating root growth and improving both aeration and water infiltration and retention. The amounts of organic matter in soils also influence biological activity, and can determine the balance between saprophytic microorganisms and root pathogens, the survival of symbiotic microorganisms, and the activities of soil animals.

The accumulation of organic matter in soils is a resultant of the rates at which organic substrates enter the soil, and of the net rates of decomposition. Prime substrates are plant tissues, exudates and leachates, and animal tissues and excreta, which are decomposed and transformed by the concerted action of the soil microfauna and microflora (Fig. 3.1). Energy and growth substrates generated from catabolic reactions are utilized to increase the biomass of the decomposer populations. Their death subsequently provides substrates, together with other metabolic products, for further biological turnover. It should be emphasized that soil organic matter is a collective term and does not represent a well defined entity.

Cell growth accompanying decomposition of organic substrates creates demands also for nitrogen, phosphorus and sulphur. Thus concentrations in soils of these elements in forms suitable for plant growth are net values only; their release from organic residues (mineralization) is offset by their participation in other reactions, including synthetic reactions associated

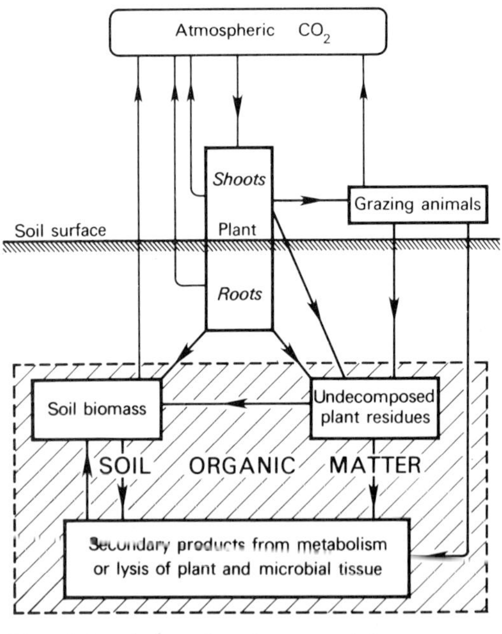

Fig. 3.1 Carbon flow in terrestrial ecosystems

mainly with microbial growth (immobilization). Transformations of nitrogenous substrates in mineralization–immobilization processes constitute the internal nitrogen cycle (Jansson, 1958). Reaction rates will be influenced by such factors as temperature, moisture, salinity, etc., which directly affect biological activity in soils, and by the amounts and amenability to attack of the diverse organic substrates which enter or are formed in soils. Whether mineralization or immobilization processes predominate will depend not only on the rates of substrate decomposition but also on the contents of the potential plant nutrient elements in substrates and organic products (see Chapter 4 in this volume).

In terms of productive agriculture, it is of major importance (i) to establish conditions which enable organic matter to accumulate in soils at suitable levels and (ii) to know the rates at which plant nutrients are incorporated into and released from organic matter due to biological activity. The diversity of origin, chemical nature and physical distribution of organic matter in soils are major factors in the dynamics of organic matter turnover. The use of isotopes has been invaluable not only for obtaining measures of organic carbon (and to some extent nitrogen) entering a soil annually, but also for enabling assessments to be made of the rates of decomposition of specific substrates, simple and complex, under natural or controlled conditions and for periods long after the initial substrate has been metabolized. Factors influencing decomposition rates, and the release and tie-up of nutrients can be assessed. Further, isotopic measurements, coupled with soil organic matter fractionation by chemical or physical means, have demonstrated differences in turnover rates of organic matter components. These aspects of the use of isotopes in organic matter studies are examined further, together with the use of labelled substrates to measure potential activities of the soil microflora and of soil enzymes. Valuable reviews of the use of isotopes in soil organic matter studies include those by Jansson (1966, 1971), Jenkinson (1971, 1981), Mayaudon (1971), Oberlander (1973), Oades and Ladd (1977), Paul and van Veen (1978) and Stout *et al.* (1981).

II. Input of organic carbon into soils

On a global basis, soil organic matter represents a major reservoir in the carbon cycle, both because of its magnitude and because it is a reservoir which is responsive to changes in climate and human activities, in contrast to the much larger reservoir of carbon contained at depths below the surface in sediments or deposits of fossil carbon (Table 3.1). Lower values for the carbon content of soil organic matter, ranging from 7×10^{14} to 10×10^{14} kg carbon have been given by other authors (reviewed by Woodwell *et al.*, 1978)

Table 3.1 Amounts of carbon in carbon reservoirs (Bohn, 1976).

Carbon reservoir	Amounts of carbon ($\times 10^{14}$ kg)
At Earth's surface	
Atmospheric CO_2	7
Biomass	4.8
Fresh water	2.5
Marine, above thermocline	5–8
Soil organic matter	30

but all estimates put the carbon in soil organic matter equal to, or greater than, the amount of carbon contained in atmospheric CO_2.

Plants are the primary source of soil organic matter, and almost all of the plant material entering soil is ultimately decomposed to CO_2 and water by the animal and microbial populations in the soil (referred to collectively as the soil biomass). For a soil under a stable vegetative cover and constant climatic conditions, the content of organic matter approaches a constant value, with the annual decomposition of soil organic matter to CO_2 balancing the annual input of organic carbon. The actual amount of organic matter present in a soil profile can vary over a very wide range depending on the ecosystem, its geographical location, the biological activity within the soil, and soil properties such as the amount and type of clay mineral. Table 3.2, adapted from a review by Schlesinger (1977), gives estimates for the amounts of carbon present in the soil profile and in the covering vegetation for a number of ecosystems. The values in Table 3.2 emphasize the very great difference in organic matter content which exists between soils, and they also show a sharp contrast between different ecosystems for the ratio (soil carbon/"plant carbon"). The smaller values for this ratio for the forest and tropical savanna ecosystems reflect a more complete decomposition of fresh plant material in these soils compared with the remainder. The true rate of decomposition or turnover time for the soil organic matter, given by the ratio (soil organic carbon/annual input of organic carbon), requires values for the total amount of organic carbon entering the soil annually from plant growth, both above and below ground. Valid estimates for primary production above ground can be obtained from litter collection within forests, or sampling by clipping and mowing grassland and cropping systems. Estimates of below-ground productivity have been obtained from the weights of crowns, rhizomes and roots washed out of soil cores but these methods will not account for epidermal tissue sloughed from the roots during growth, for soluble compounds released from roots by exudation or

Table 3.2 Carbon contained in the standing crop and in soil organic matter for major ecosystem types[a]

Ecosystem type	Mean total organic carbon in soil profile ($kg\ m^{-2}$)	Mean vegetation biomass of carbon ($kg\ m^{-2}$)	Soil carbon / "Plant carbon"
Tropical forest	10.2	18.8	0.54
Temperate forest	10.6	14.6	0.73
Boreal forest	12.9	9.0	1.43
Woodland shrubland	6.7	2.6	2.57
Tropical savanna	3.6	1.8	2.00
Temperate grassland	19.0	0.7	27.1
Tundra and alpine	21.1	0.3	70.3
Desert scrub	5.6	0.3	18.7
Cultivated soils	12.6	0.5	25.2
Swamp and marsh	67.0	6.8	9.85

[a] Recalculated from Schlesinger (1977).

following cell autolysis or for root tissue consumed by soil herbivores. The use of isotopes, particularly ^{14}C, offers the most accurate estimates for total carbon transfer below ground by plants.

By supplying $^{14}CO_2$ to the plant tops, Dahlman (1968) obtained values for the annual root production and turnover of the root system of native tall-grass prairie (*Andropogon–Sorghastrum–Sporobolus* association). The measured annual increment of ash-free root organic matter was at least $450\ g\ m^{-2}$, compared with $460\ g\ m^{-2}$ of shoot production. The average annual turnover of the root system was estimated to be 25% of the total underground biomass. Similarly, Warembourg and Paul (1977) pulse-labelled prairie grasses (*Agropyron–Koeleria* association) with $^{14}CO_2$ *in situ* and measured transfers of carbon from the foliage to the roots and estimated turnover rates for the ^{14}C-labelled root material. The seasonal shoot production was estimated to be 130 g of carbon per square metre while root growth represents 100 g of carbon per square metre. The half-life of the labelled roots was estimated to be 107 days.

A more direct measurement of the supply of organic matter to soil from plant roots was made by Shamoot *et al.* (1968). They supplied $^{14}CO_2$ to the shoots of several annual and perennial grasses and legumes from the seedling stage to maturity (up to 212 days). ^{14}C-labelled residues, left in the soil after removal of the intact roots, represented from 5 to 25% of the ^{14}C recovered in the tops and roots. This was equivalent to an addition of 160–1650 kg of organic matter per hectare, sampled to a depth of 15 cm. However, these values excluded root carbon converted to CO_2, either by root respiration or by decomposition of root-derived organic matter by the soil biomass. This latter component, termed rhizosphere CO_2, was shown by Sauerbeck and Johnen (1977) to be of a much greater magnitude than previously accepted, implying that earlier estimates of below-ground productivity were less than the true values. Sauerbeck and Johnen grew wheat (*Triticum aestivum*) to maturity with the shoots in an atmosphere containing $^{14}CO_2$ but with the roots growing in an air-tight container so that $^{14}CO_2$ released from the roots and associated soil biomass could be collected. ^{14}C recovered as rhizosphere CO_2 was more than three times greater than the ^{14}C activity contained in the roots washed out of the soil at harvest. Similar losses of carbon from the roots of actively growing wheat plants were reported by Martin (1977a). Wheat plants, grown at temperatures of 10 or 18 °C for periods from 3 to 8 weeks in an atmosphere containing $^{14}CO_2$, lost to the soil from 29 to 44 % of the ^{14}C translocated to the roots.

The results from the growth cabinet experiments are supported by field data. Jenkinson and Rayner (1977) used the enrichment of soil organic matter carbon with ^{14}C, produced in the atmosphere during thermonuclear bomb tests, to estimate the amount of fresh organic matter entering the soil under wheat rotations at Rothamsted. They estimated that 1200 kg of organic carbon per hectare annually entered the top 23 cm of soil for the non-fertilized plots. This compares with earlier estimates of 300 kg of organic carbon per hectare from the dry weight of roots present in plots of unfertilized winter wheat 6 weeks before harvest (Welbank *et al.*, 1974).

There is increasing evidence that the transfer of carbon through the rhizosphere is a major component in the total carbon budget for most, if not all, terrestrial ecosystems. The proportion of photosynthate transferred to roots probably ranges from 30 % for cereals to 50 % for pastures and up to 60 % for some trees (Ågren *et al.*, 1981).

The amounts of nitrogen which enter soils due to the process of symbiotic N_2 fixation also can be determined by isotopic methods knowing (i) the total nitrogen of the plant parts of the N_2 fixer which are returned to soils, and (ii) the proportions of the plant nitrogen due directly to the fixation process. The use of a ^{15}N-dilution technique is potentially useful for measuring N_2 fixation, especially by agricultural legumes as indicated in Chapter 4 in this volume and in the review of Knowles (1981). The technique requires a comparison of the ^{15}N atom % enrichments of legume nitrogen, and nitrogen of a reference non-fixing plant, both of which are grown in a soil whose available nitrogen pool is also labelled. This latter may be achieved by regular additions of small quantities of enriched inorganic nitrogen, or by labelling the soil organic nitrogen pool. The ^{15}N-dilution technique provides an integrated measure of cumulative N_2 fixation on a daily or seasonal basis.

III. Decomposition of isotope-labelled materials in soils

Compounds labelled with ^{14}C have been most commonly used, and decomposition has been followed either by measuring rates of evolution of $^{14}CO_2$ or by rates of decrease in the amounts of organic ^{14}C residues. Substrates have ranged from simple, well defined compounds such as glucose, acetate, amino acids, etc. (and these may be uniformly labelled or labelled in specific carbon atoms) to complex materials such as plant tissues, animal excreta or synthetic high molecular weight derivatives of amino acids and peptides. Experiments may be very brief (< 1 h) or they may extend for years.

Basically, labelled substrates can be added to soils and their metabolism followed against a background of decomposition of other substrates present in relatively greater quantities. However, because of turnover of organic ^{14}C in newly synthesized microbial residues, measurements of $^{14}CO_2$ evolution (or of total residual organic ^{14}C) do not reflect the decomposition of the added substrate exclusively, except perhaps in the briefest of assays. Measurements specifically of substrate disappearance are free of this complication but these are confined to a limited number of simple compounds. With complex substrates such as plant material, it is difficult to distinguish organic compounds of relatively resistant substrate components from those of accumulating microbial residues.

A. Radiorespirometry

Wang (1971) defined radiorespirometry as a technique by which respiratory activities of biological systems are determined from the rate and extent of conversion of a ^{14}C-labelled carbon atom of a substrate to $^{14}CO_2$. Applied to soils, activities can be measured of the heterotrophic microflora towards specific substrates under conditions approximating those of their natural habitat (although soil disturbance is involved). Activities towards a range of substrates may reflect the general biological activities of soils, and measurements of the extent to which activities are influenced by various chemical and physical factors may be useful in understanding the biological fertility of soils and their capacity to support plant growth (Mayaudon, 1971).

Although radiorespirometry is concerned primarily with measuring continuously the evolution of $^{14}CO_2$ during brief (1–2 h) periods of incubation, its use does not preclude analyses of the soil for early products of microbial synthesis. Nevertheless, emphasis is placed on establishing the initial velocity of respiration of a given substrate, this parameter being considered the most valid measure of the activity of the microflora in their natural habitats in the soil. Initial velocities are calculated from reaction rates at time zero, and are established under first-order kinetic conditions, i.e. when the decomposition rate of a ^{14}C-labelled substrate at a given time is directly proportional to the residual substrate concentration (Mayaudon, 1971).

As an example radiorespirometry has been used to compare glycolytic activities, i.e. conversion of [3,4-^{14}C]glucose to $^{14}CO_2$, of soils which are fresh, air-dried or heated (Fig. 3.2). Drying, and especially heating, decreased respiratory rates expressed on a soil dry weight or soil carbon basis but increased rates when expressed as $^{14}CO_2$ released per minute per number of viable microbial cells remaining in the soils. It would appear that heating and drying of soils were more effective in reducing the populations of those cells with little or no glycolytic activity (Mayaudon, 1971). An examination of the reaction velocities, changing with time over periods extending to the conclusion of the rapid phase of substrate decomposition, permitted speculation on the types of microorganisms responsible for glycolytic activities of the fresh and treated soils.

Radiorespirometry has been used to compare the effects of a variety of treatments on the respiratory activities of soils, e.g. the effects of imposed variations in temperature, pH, aerobiosis, moisture content, substrate and potential inhibitor concentrations, etc. Domsch *et al.* (1973) measured the effects of fungicides on the rates of [^{14}C]glucose oxidation to $^{14}CO_2$, in an attempt to determine the contributions of fungi to soil oxidative activity. Measurements of the rates of release of $^{14}CO_2$ from glucose substrates, specifically labelled at carbon atoms 1, 2, 3:4 or 6, have led to an analysis of the importance of various pathways of glucose metabolism by the soil microflora.

Radiorespirometry is not confined to the use of glucose as substrate nor

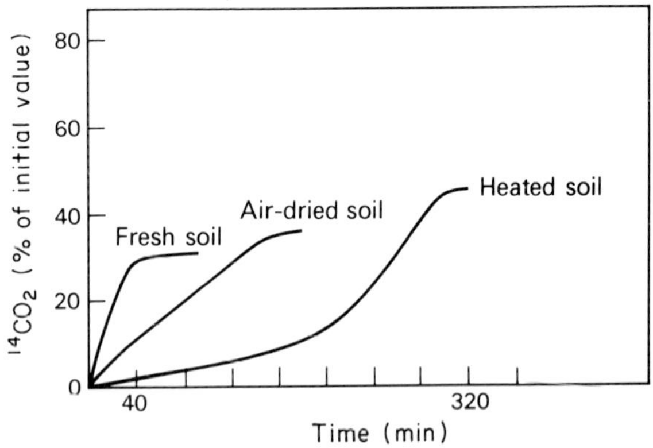

Fig 3.2 Radiorespirometric measurements of [3,4-^{14}C]glucose oxidation in soils (Mayaudon, 1971).

indeed to measuring activities of whole soils. The conversion of urea to CO_2 and NH_3 by soil urease does not involve an oxidation of substrate carbon. Nevertheless, $^{14}CO_2$ evolution accompanying hydrolysis of ^{14}C-labelled urea has formed a basis of assay of this most important soil enzyme (Skujins and McLaren, 1969). In this instance assay conditions ensure zero-order kinetics over periods of several hours.

B. Decomposition of defined organic compounds

Whereas the use of radiorespirometry permits a comparison of the initial rates of substrate oxidation to CO_2, the greatest use of isotope-labelled substrates has been in soil incubation studies of relatively long duration. The conversion of substrate ^{14}C to $^{14}CO_2$ (or of substrate ^{15}N to inorganic ^{15}N) is often measured, but most frequently decomposition has been assessed from the proportions of isotope-labelled material which remain as residues in the soil (especially in studies of decomposition under natural conditions in the field).

Simple low molecular weight compounds like glucose (e.g. Persson, 1968), inositol (L'Annunziata and Gonzalez, 1977), acetate (Sorensen and Paul, 1971), amino acids (Mayaudon and Simonart, 1965) and aromatic compounds related to lignin building blocks (Haider and Martin, 1975) are rapidly metabolized when incubated with soils. The initial substrate often has completely disappeared within a few days whereas products of metabolism are decomposed far more slowly.

Simonart and Mayaudon (1958) fractionated ^{14}C-labelled ryegrass (*Lolium* spp.) plants into several groups of compounds which were then incubated separately in soil for several months. Soil amended with ^{14}C-labelled glucose was an additional treatment. Glucose was most rapidly metabolized, followed by hemicellulose, and then the cellulosic fraction (which contained some lignin); ^{14}C-labelled carotenoid and chlorophyl pigments decomposed very slowly. Plant proteins decomposed readily, although less extensively than amino acids of a protein hydrolysate, supporting the notion that some plant proteins may be directly stabilized in soil, perhaps by adsorption to clays or humic compounds without prior metabolism by microorganisms (Simonart and Mayaudon, 1961). Lignin was decomposed relatively slowly when compared with cellulose (Mayaudon and Simonart, 1959).

For compounds such as glucose, amino acids, some proteins and some polysaccharides, about 15–20% of substrate carbon remains in the soil as relatively stable residues after incubation for periods ranging from a few weeks to several months. For aromatic compounds of low molecular weight, representatives of lignum constituents, the extent to which carbon accumulates in organic residues depends not only upon the compound but also upon the location of the carbon atom within the substrate's chemical structure. Carboxyl and methoxyl groups attached to the ring are extensively converted to CO_2, whereas ring carbon is retained in organic residues to a far greater extent (Martin *et al.*, 1974; Haider and Martin, 1975, 1981). Factors influencing stabilization of organic ^{14}C residues can be assessed and mechanisms of stabilization can be inferred. In some cases stabilization of substrate ^{14}C appears to be due to reactions involving the substrate directly, rather than those involving metabolic products exclusively (Batistic and Mayaudon, 1970; Haider and Martin, 1975; L'Annunziata and Gonzalez, 1977).

Studies that have used substrate derivatives, specific ^{14}C-labelled compounds (phenolic acids, amino acids, proteins) linked into complexes with phenolic polymers to represent models of soil humic compounds have demonstrated the potential importance of such processes in stabilizing soil organic residues (Haider and Martin, 1975; Verma *et al.*, 1975) (Fig. 3.3). In

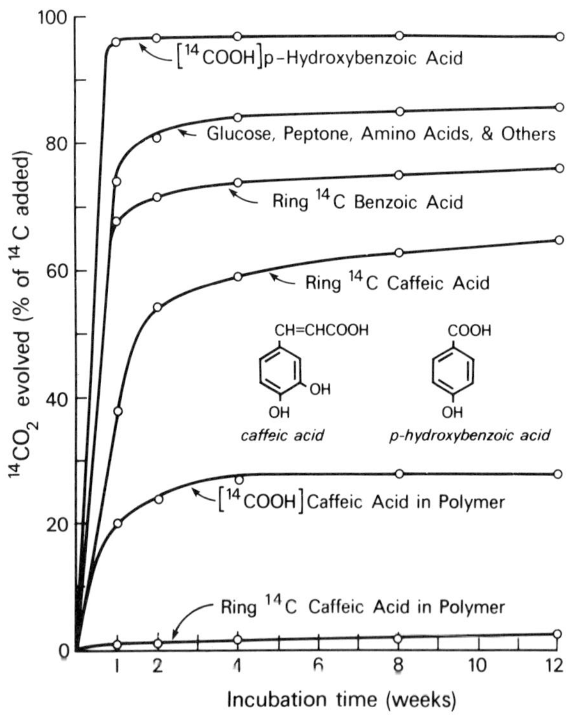

Fig. 3.3 Decomposition of specifically-labelled ^{14}C substrates in soil (Haider and Martin, 1975).

other studies clay has been clearly implicated in the stabilization process (Sorensen, 1972b, 1975).

C. Decomposition of microbial material

The involvement of microorganisms in decomposition processes and in the turnover of substrate carbon has led to measurements of the rates at which ^{14}C-labelled cells or cell fractions are decomposed when added to soil (Mayaudon and Simonart, 1963; Zeller *et al.*, 1966; Hurst and Wagner, 1969; Nelson *et al.*, 1979). Differences in rates have been noted when comparing different organisms, or the same fractions from different organisms, or different fractions of the same organism. For example, cell walls of a melanic fungus were more resistant to decomposition than was its cytoplasm, whereas the cytoplasm and especially the cell walls of a hyaline fungus (*Aspergillus niger*) were decomposed relatively rapidly (Hurst and Wagner, 1969). Melanins from some fungi were almost as resistant to decomposition as soil humic compounds (Linhares and Martin, 1978). Martin *et al.* (1974) showed that polysaccharides from several microorganisms also differed in their susceptibility to decomposition. A preparation from *Chromobacterium violaceum* was most resistant (32% of added ^{14}C was mineralized in 16 weeks), whereas a polysaccharide from *Leuconostoc dextranicus* was least resistant (86% of added ^{14}C was mineralized).

Whilst the addition to soils of labelled compounds – pure, or as complexes, or as fractions of plant and microbial tissues – have been of great value; it is important to recognize that the fates and rates of decomposition of these substrates may differ from those of the same compounds added as intact plant remains or present in soils as microbial cells and products generated *in situ*. Thus, Cheshire *et al.* (1975) showed that a ^{14}C-labelled polysaccharide extracted from soil decomposed more rapidly when added to soil than did the same polysaccharide formed *in situ*; its persistence was not due intrinsically to a stable chemical structure.

As a further complication, methods of fractionation of plant and microbial material may themselves induce changes in the chemical and physical properties of the substrates, and hence influence their susceptibility to biological attack (Cheshire *et al.*, 1975).

D. Decomposition of plant material

Field studies of the decomposition of uniformly labelled whole plant material, especially over several years, have direct relevance to the accumulation of organic matter in soils under natural conditions.

Grossbard (1969) used autoradiography (see Volumes 1 and 2, Chapter 5) to obtain a visual record of the decomposition of ^{14}C-labelled fragments of grasses (e.g. *Lolium perenne*) and rye (*Secale cereale*), either placed on the soil surface or incorporated. She observed a disappearance of ^{14}C from some small discrete sites, dispersion of ^{14}C from the initial locations of the plant residues, and the presence of microorganisms, particularly fungi, which became labelled as the plant residues decomposed. The participation of soil mesofauna was also inferred from the presence of ^{14}C-labelled faecal pellets. Quantification of decomposition based on image densities of recognizable structures showed that fungi were numerous in the first 3 months, when there was a relatively rapid breakdown of the added plant material. Faecal pellets and resting fungal structures concentrated the ^{14}C of the residues and were relatively stable.

Most studies involving decomposition of isotope-labelled plant material have sought to quantify the process by measuring the amounts of isotope either in the total organic residues or in selected organic fractions. Early studies by Jenkinson (1965), Fuhr and Sauerbeck (1968) and Oberlander and Roth (1968), using ^{14}C-labelled materials, have been continued by these workers and extended by others, e.g. Smith and Douglas (1971), Shields and Paul (1973), Nyhan (1975), Ladd *et al.* (1981a) to include a variety of plant materials and environments. Animal manures prepared from labelled plant materials have also been used (Sauerbeck, 1968b; Oberlander, 1973).

Characteristically, within the first few months of decomposition, there is a relatively rapid decline in the concentrations of residual organic ^{14}C; thereafter the rates of decline slow considerably (Fig. 3.4). For many plant materials studied, e.g. rye grass, maize (*Zea mays*), wheat straw, legume (*Medicago littoralis*), etc., about two-thirds of substrate ^{14}C is lost within 1 year of their addition to soils. Not surprisingly, however, the net rates of decline of organic ^{14}C in residues do differ, according to plant type and part (e.g. Nyhan, 1975), soil properties (e.g. Jenkinson, 1977b, 1981; Zunino *et al.*, 1982) and climate (e.g. Jenkinson and Ayanaba, 1977; Paul and van Veen, 1978). Differences in the amounts of residual organic ^{14}C are often small and are most apparent early in the decomposition phase; the differences may persist throughout extended periods, e.g. the effect of clay (Jenkinson, 1977b), or they may disappear within a year or so, e.g. the effect of plant part (Jenkinson, 1965), or of soil pH (Jenkinson, 1977b). Percentage decomposition is approximately independent of the amount of plant material added (Jenkinson, 1977a; Ladd *et al.*, 1983b).

Where different materials are compared, differences in the amounts of organic ^{14}C of residues may result almost entirely from differences in the proportions of the initial substrate ^{14}C present in readily decomposable forms. Oberlander (1973), for example, demonstrated that the percentage

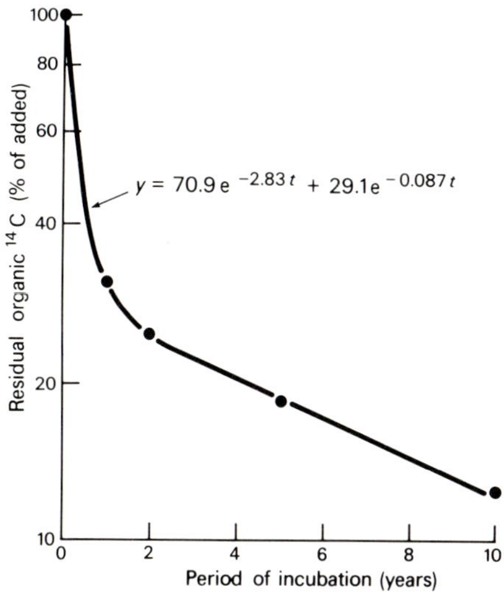

Fig. 3.4 Decomposition of ^{14}C-labelled plant material in soil (modified from Jenkinson, 1977b).

retention of substrate ^{14}C in organic residues decreased in the order farmyard manure, mature wheat straw and green maize, when considered over the first year of decomposition. Most of the readily decomposable components of materials which formed manure had been metabolized before its addition to the soil (Sauerbeck, 1968b). After 1 year, ^{14}C residues from all three substrates decomposed at similar rates.

The decomposition of labelled residues under field conditions has usually been followed in soils kept bare of vegetation. The effects of cropping on decomposition rates have been variable and usually not statistically significant. However, Jenkinson (1977b) found that decomposition of rye grass residues in soil under grass was slower than in the bare soil, due possibly to less favourable temperature–moisture regimes. Some recent evidence by Sparling *et al.* (1982) suggests that growing plants may compete with soil organisms for organic products of small molecular weight, produced as intermediates in the decomposition of added plant residues in soil.

It is important to re-emphasize that the amounts of residual organic ^{14}C accumulated in soils do not truly reflect the decay of the added substrate ^{14}C exclusively, but are the net resultant of catabolic and synthetic processes; ^{14}C of plant components directly contribute ever decreasing proportions of the total ^{14}C remaining. A model proposed by Paul and van Veen (1978) has sought to describe the true rates of plant material decomposition. Decay constants are required for individual plant components, for the microbial biomass and for microbial metabolites. Values for the efficiency of microbial synthesis and maintenance were assigned. Nevertheless, the amounts of microbial residues formed from the metabolism of a given carbon source will be determined by the pathways of decomposition and the nature of the products; they are set to a large extent in the early stages of decomposition of available substrate (Jenkinson, 1981). Also, the rates of decay of the microbial biomass will be influenced by several factors, including cell type, soil properties and climate. Clay content of soils may influence survival of microorganisms present (Ladd *et al.*, 1981a).

The addition to soils of ^{15}N-labelled plant materials and manures has also permitted measurements of the rates of change in the concentrations of nitrogen of organic residues. The proportions of added ^{15}N retained in organic forms will be influenced by the same factors important in organic matter decomposition. Since the nitrogen of plant materials appears to enter the immobilization–mineralization turnover cycle rapidly and extensively (Bartholomew, 1965), the percentage of substrate ^{15}N accumulating mainly in microbial residues will be strongly influenced by the relative amounts of available carbon and nitrogen, both of the substrate and of soil organic matter, and by the extent to which nitrogen released from each source equilibrates before immobilization occurs, induced by the metabolism of carbon from both sources.

Ladd *et al.* (1981a) showed that following the addition of ^{14}C, ^{15}N-labelled legume material of C:N ratio 15.3:1 to four soils, residual organic ^{15}N accounted for 60–65 % of input after 0.6 years, and for 45–50 % after 4 years. Thus net ^{15}N mineralization was relatively rapid within 6 months of substrate addition to soils but, as for residual ^{14}C, the net rates of decrease of ^{15}N-labelled residues slowed considerably thereafter. Broadbent and Nakashima (1974) showed that after 5 years decomposition, 38 % of the ^{15}N of barley (*Hordeum vulgare*) tops (C:N ratio, 17.3:1) was retained as organic residues, compared with 57 % of the ^{15}N of barley roots (C:N ratio, 35.6:1). Because nitrogen released from plant and soil sources may partly or completely equilibrate before entering immobilization reactions, it is possible for ^{15}N of plant residues to be released as inorganic ^{15}N to some extent at a time when net immobilization of soil-derived nitrogen is occurring. This was observed by Kanazawa and Yoneyama (1980) for ^{15}N-labelled rice tops (C:N ratio, 59:1) decomposing under aerobic conditions for 2 years.

The decomposition of organic compounds in soils is generally assumed to follow first-order kinetics whereby concentrations of organic carbon decrease exponentially. Jenkinson (1977b) demonstrated that following application of ^{14}C-labelled plant material, changes in residual organic ^{14}C over a 10 year period were well described by a two-compartment model in which about 70 % of the added carbon was decomposed with a half-life of 0.25 years, and 30 % with a half-life of 8 years (Fig. 3.4). Others have calculated half-lives and decay constants for organic residues formed from a variety of labelled plant materials or animal manures and decomposing in a range of environments. Half-lives of "stabilized" residues generally have ranged from 3 to 8 years.

Overall rates of decomposition of labelled residues, followed for up to 10 years, are faster than those of the unlabelled soil organic matter present at the time of substrate addition. Thus decay constants, based on first-order kinetics and calculated for the net decomposition of the "stabilized", labelled organic residues in soils in experiments ranging from 2 to 10 years duration, cannot be used to predict equilibrium levels of organic matter in soils receiving known annual additions of organic materials (Jenkinson and Ayanaba, 1977; Sauerbeck and Gonzalez, 1977). Predicted equilibrium levels were too low. In the same way, annual inputs of plant carbon calculated to be required to maintain known soil organic levels were higher than those considered likely in practice, and in some cases were impossibly high even allowing for increased assessments of carbon input due to root decomposition.

It may reasonably be expected that the decay rates of total labelled residues should decrease with time as the proportions of relatively resistant components of the residues increase, but this is not generally apparent within the first decade of decomposition (Sauerbeck and Gonzalez, 1977). Although the two-compartment exponential model describes the total organic ^{14}C decline accurately within the first 10 years of decomposition, it nevertheless oversimplifies the decay processes (Jenkinson, 1977b). In particular, it does not allow for organic carbon turnover nor for the fact that different parts of the soil organic matter decompose at different rates, some very slowly, as revealed by ^{14}C dating techniques. More sophisticated models have been devised to accommodate these facets of organic substrate decomposition (e.g. Jenkinson and Rayner, 1977).

IV. Characterization of soil organic matter

Soil organic matter is chemically complex. Less than 50 % of soil organic carbon, 50 % of organic nitrogen and 40–70 % of organic phosphorus can be chemically defined, and then only as products of various degradative procedures. Analysis of products thus enables assumptions to be made regarding the classes of parent compounds from which they are derived, but does not distinguish their sources. Organic compounds in soils are present in a range of components, including partially decayed plant and animal tissues, microbial cell debris, the biomass of soil organisms, ephemeral products of decomposition, and relatively stable residues bearing no trace of the anatomical structures of the material from which they were derived.

The previous section described the use of isotopes, particularly ^{14}C, to study the decomposition processes occurring in soil, following the addition of fresh plant material or pure compounds corresponding to major components of plant tissue. A number of studies have been made to characterize the chemical nature of the labelled residues and also their physical distribution within the soil fabric, since there is considerable evidence that physical constraints are as important in influencing the decomposition of soil organic matter as chemical structure. The major findings from the different approaches are discussed below.

A. Soil biomass

Although there is some transformation of substrates by extracellular enzymes, and by metal-catalysed oxidation, etc., almost all decomposition of organic material added to soil results from the metabolic activity of the soil flora and fauna (soil biomass). Some of the substrate is decomposed by catabolic reactions, supplying energy to the biomass, and is accompanied by the evolution of CO_2; some of the substrate carbon is retained within the biomass; some of the original substrate carbon is found in the soil outside cells in molecules representing the products of decay of the organism which first ingested the substrate; and some is found in products resulting from the growth of succeeding generations of soil organisms.

This sequence of reactions was shown very clearly in a laboratory experiment by Ladd and Paul (1973). After incubating soil with added [^{14}C]glucose for 1.5 days it was shown that 96.4% of the glucose had been metabolized but only 31.2% of the ^{14}C had evolved as CO_2, demonstrating that 65% of the added glucose-carbon was present in the biomass or metabolites at the cessation of the active decomposition phase. Similar results were obtained when [^{14}C]glucose and $^{15}NH_4{}^{15}NO_3$ were added to a field soil (Shields *et al.*, 1973). Soil extracts taken 3 days after addition of the [^{14}C]glucose contained barely detectable amounts of glucose although 63% of the ^{14}C remained in the soil. Also, only 18% of the ^{15}N remained in the mineral form after 3 days, indicating that approximately 80% had been synthesized into microbial material during the period of rapid glucose decomposition.

Jenkinson (1966) showed that partial sterilization of field soil, in which ^{14}C-labelled ryegrass had been allowed to decompose for up to 4 years, accelerated the release of $^{14}CO_2$ from the soil and increased the specific activity of the $^{14}CO_2$. After 1 year of decomposition of labelled plant material, 16% of the CO_2 released after chloroform treatment was labelled compared with a value of 1.8% labelling of the total soil organic carbon. Jenkinson postulated that the heavily labelled fraction in the soil, rendered decomposable by the chloroform treatment, was the fraction of the soil biomass involved in decomposition of the ^{14}C-labelled ryegrass. About 23% of the organic ^{14}C which remained in soil after decomposition for 1 year, was present in the microbial biomass (recalculated using a *Kc* factor of 0.45); this proportion decreased to about 13% of the residual organic ^{14}C after 4 years of decomposition. Unlabelled biomass carbon accounted for about 2% of total soil organic carbon. Ladd *et al.* (1977c, 1981a) showed that the ^{14}C and ^{15}N mineralized from labelled biomass (which formed from decomposing ^{14}C,^{15}N-labelled legume material and from immobilized $^{15}NO_3^-$), also declined disproportionately more rapidly than did total residual organic ^{14}C or ^{15}N during periods of net mineralization.

The proportions of residues accounted for in microbial biomass differed between soils both in field and laboratory experiments and sometimes when the total amounts of residual ^{14}C in soils were similar (Malik and Azam, 1980; Cerri and Jenkinson, 1981; Kassim *et al.*, 1981; Ladd *et al.*, 1981a). Soil pH, salinity and clay contents may be important factors in determining the percentage of residues as biomass; possibly they affect the turnover of biomass by influencing efficiency of biomass synthesis and/or biomass viability. In a very acid soil (pH 3.6) the low proportion of residues found as biomass may be due to its underestimation by the fumigation technique (Cerri and Jenkinson, 1981).

Kassim *et al.* (1981) also used the fumigation technique to demonstrate the incorporation of carbon from a wide variety of ^{14}C-labelled substrates into soil biomass. The type of substrate and the position of ^{14}C label affected both the extent to which $^{14}CO_2$ was evolved and the proportion of residual organic ^{14}C accounted for in biomass. Readily available substrates such as glucose, acetate, pyruvate, uracil, uridine, amino acids and some polysaccharides were rapidly metabolized by the soil organisms with 80% of the ^{14}C evolved as CO_2 after 12 weeks incubation. The fumigation technique showed that 4–8% of the initial substrate carbon was present in the biomass. Glucosamine, cysteine, cellulose, protein and some plant and microbial polysaccharides were degraded more slowly, with 3–9% of the initial substrate carbon present in the biomass at 12 weeks. Microbial cells and plant residues such as wheat straw were decomposed to a still lesser extent with an incorporation of 2–5% of the supplied carbon present in biomass after 12 weeks incubation. Various mathematical models describing the dynamics of carbon and nitrogen in soils include a consideration of the amounts and flow of these elements in and through the soil microbial biomass (Jenkinson and Rayner, 1977; Paul and van Veen, 1978; Paul and Voroney, 1980; Juma and Paul, 1981; Paul and Juma, 1981; van Veen and Frissel, 1981).

B. Chemical fractions

1. *Classical fractionation*

Work in the late 18th and early 19th centuries showed that some of the soil organic matter was dissolved by alkaline solutions and the extracted material could be fractionated by acidification of the extract. The organic material which was soluble in alkali but precipitated in acid solution (pH 1.0), termed *humic acid*, could represent up to 70 % of the soil organic matter in a fertile, well aerated soil. The fraction which was acid and alkali soluble was termed *fulvic acid*, while the organic matter insoluble in alkali was termed *humin*. The humus materials are amorphous polyelectrolytes with a wide range of molecular weights (MWs). Typically, fulvic acids have MWs below 10^3 daltons while humic acids have apparent MWs from 5×10^3 to $>1 \times 10^6$ daltons. Schnitzer (1977) demonstrated both the aliphatic and aromatic characters of these fractions; the major building blocks of humic substances are complex phenolic and benzene carboxylic structures.

Most of the early studies on the decomposition of ^{14}C-labelled substrates added to soil included a measurement of ^{14}C activity present in the fulvic acid, humic acid and humin fractions of the soil organic matter. Simonart and Mayaudon (1958) added ^{14}C-labelled glucose, cellulose, hemicellulose and ryegrass to soil and found that after 6–8 weeks incubation 6–16 % of the initial activity was recovered from the soil in the α and β humus and humin fractions. Similarly, ^{14}C-labelled barley straw and cellulose, hemicellulose and lignin fractions prepared from the straw were added to soil and allowed to decompose for 100 days (Sorensen, 1963). The ^{14}C activity recovered in the fulvic acid, humic acid and humin fractions from the straw-amended soil represented, respectively, approximately 12, 24 and 51 % of the residual ^{14}C activity in the soil. Similar results were obtained by Sorensen for the other substrates, except lignin, where the corresponding values were 2, 37 and 33 %.

Other studies involving the metabolism in soil of added defined chemical compounds (e.g. Ivarson and Stevenson, 1964; Wagner, 1968; Igel, 1969; Freytag and Igel, 1970; McGill *et al.*, 1975), of plant materials (e.g. Sauerbeck, 1968a; Sauerbeck and Führ, 1968; Oberlander and Roth, 1974) and of microbial cells (e.g. Simonart and Mayaudon, 1963; Zeller *et al.*, 1966; Martin *et al.*, 1974) demonstrated that labelled residues are distributed amongst all of the classical fractions, the pattern of distribution being influenced by the nature of the added substrate and the extent of decomposition.

However, such results are now considered to be of limited value since it has been shown that when complex substrates such as labelled plant material or microbial cells were mixed with soil and extracted immediately, the isotope was found in all of the classical fractions and that the distribution of label was affected by the soil. Sauerbeck and Führ (1968) extracted ^{14}C-labelled barley straw and green rape (*Brassica* spp.) plants by standard procedures with and without soil. Undecomposed plant carbon not only appeared in the fulvic acid fraction but also in the humic acid fraction. In the presence of soil, the ^{14}C activity of the humic acid fraction was higher, indicating that chemical reactions were probably occurring during the alkaline extraction between soil humus and soluble plant constituents.

Simonart and Mayaudon (1963), using ^{14}C-labelled *Aspergillus* and *Azotobacter* cells, demonstrated that the classical chemical scheme fractionates microbial tissue also, and that the presence of soil alters this fractionation. During subsequent incubation, ^{14}C residues from the added microbial cells decreased in all soil fractions, but particularly in the soluble (80 % ethanol) and fulvic acid fractions. McGill *et al.* (1975) also examined changes in the ^{14}C of fulvic acid and humic acid fractions, but extracted from a soil amended with [^{14}C]acetate and $^{15}NH_4^+$. These substrates were metabolized by successive microbial populations, dominated initially by fungi and then by bacteria and actinomycetes. Extracellular microbial metabolites in the fulvic acid fraction were rapidly attacked, but the percentage decreases in the fulvic acid ^{14}C residues were, overall, less than those observed for ^{14}C residues of the humic acid fraction, and especially of those labelled components separated from the humic acid by phenol extraction.

Thus, changes in the distribution of isotope-labelled residues in the classical humic fractions may appear to be inconsistent, and difficult to interpret in terms of metabolic sequences and biological turnover. Nevertheless, the results of McGill *et al.* (1975) clearly demonstrated a similarity in the distribution and net changes of the ^{14}C and ^{15}N of microbial residues in the fractions, which suggested a high dependence of nitrogen transformation on carbon metabolism. Comparable results were obtained by Freytag and Igel (1970).

Chu and Knowles (1966), working with ^{15}N-labelled cells of a *Pseudomonas* sp., also indicated the limitations of chemical fractionation techniques. They showed that when the cells were added to a sandy loam, at time zero before metabolic changes occurred, only 55 % of the ^{15}N content of the cells was recovered in the amino sugar and amino acid fractions, while 39 % of the ^{15}N was recovered in a humin fraction, resistant to hydrolysis by hot acid. The distribution of label differed from that observed when the cells were added to a raw humus soil and also extracted at time zero.

2. *Proximate analysis*

This method, developed by Waksman, isolates fractions of the soil organic matter with similar chemical composition, e.g. fats and waxes, sugars, amino acids. The advent of ^{14}C and ^{15}N allowed studies of the transformation of isotope-labelled organic materials added to soil into other classes of compounds, and also provided measurements of the rate of formation or decomposition of specific compounds.

Sorensen (1967, 1972a) showed that ^{14}C added to soil as [^{14}C]cellulose was incorporated into amino acids, and Wagner and Mutatkar (1968) also found that after a 6 month incubation of [^{14}C]glucose in soil, the amino compounds released from the soil by acid hydrolysis contained 17–22% of the ^{14}C activity remaining in the soil. Although all 13 compounds isolated were labelled, the specific activities were not the same. The highest specific activities were found in amino acids known to predominate in fungal and bacterial cell walls. Oades and Wagner (1971) showed that [^{14}C]glucose was also transformed into other sugars. There was rapid incorporation of ^{14}C, added as glucose or dextran, into galactose and mannose but slow incorporation, if at all, into xylose, arabinose, rhamnose and fucose. Of the residual ^{14}C remaining in the soil after 28 days incubation, about 6% was present in saccharides in the glucose treatment, with a corresponding value of 15% for the dextran treatment.

A more detailed study of the formation of amino acids and carbohydrates was made by Sorensen and Paul (1971) using [^{14}C]acetate, which was easily separated from its metabolic products. Maximal incorporation of acetate-carbon into the soil fractions was observed after 4 days when 19% of the residual ^{14}C was present in carbohydrates and 29% in amino acids. The soil incubation was continued for 220 days, by which time about 18% of the initial ^{14}C activity remained in the soil. Of this, 12% was found in carbohydrates and 40% in amino acids, indicating different rates of decomposition for the two fractions. The half-lives of the ^{14}C-labelled metabolites left in the soil after 90 days were calculated to be 800 days for the carbohydrates and 1600 days for the amino acids.

The transformation of acetate into sugars and amino acids or of glucose into amino acids and other sugars is almost certainly due to microbial synthesis. It is also probable that the greatest part of the ^{14}C-labelled sugars and amino acids released by acid hydrolysis of soil, shortly (1–10 days) after addition of ^{14}C-labelled acetate or glucose, have been released from viable microbial cells. It is uncertain whether the ^{14}C-labelled sugars and amino acids recovered from soil after prolonged incubation result from the continued cycling of the initial substrate carbon through successive generations of microorganisms, or whether they represent secondary metabolites or decay products from the first generation of microorganisms which have become stabilized against decomposition by virtue of their chemical structure or by their physical environment.

Sorensen (1972b) showed that montmorillonite clay in soil had no effect on the rates of decay of carbohydrate substrates but tended to stabilize metabolic products hydrolysable to amino acids. The amounts of such products increased with increasing concentrations of silt plus clay in soil; this order was established early in the incubation when metabolic activity was high, and persisted for at least 2 years (Sorensen, 1975).

There is very limited information on the *in situ* formation of specific fractions of soil organic matter, derived from live plant roots, rather than from pure chemicals or plant residues mixed uniformly through the soil. Martin (1977a) grew wheat plants continuously in an atmosphere of $^{14}CO_2$ and measured the proportion of the fixed ^{14}C which was released from the root chamber as CO_2 or remained in the soil after the removal of roots. ^{14}C activity in the "root-free" soil represented 17.5% of net carbon fixed by plants grown in non-sterile soil and 28.5% of fixed carbon for plants grown axenically. The chemical nature of the ^{14}C-labelled organic matter in the sterile and non-sterile soils was determined by procedures based on solubility, molecular size and ion-exchange chromatography (Martin, 1977b). Three fractions were obtained from each soil; they behaved on ion-exchange resins as neutral sugars, amino acids and carboxylic acids + phosphate esters. The recovery of ^{14}C in these three fractions represented 34.4, 14.2 and 4.9% of total soil ^{14}C for the sterile treatment and 41.7, 11.0 and 7.6% of total soil ^{14}C for the non-sterile treatment, respectively. Distribution of ^{14}C activity in the fulvic acid and humic acid fractions was 8.9 and 22.6% of soil ^{14}C for the sterile treatment and 7.2 and 17.9% of soil ^{14}C for the non-sterile treatment, respectively. The similarity in the values for the sterile and non-sterile treatments was thought to be due to root autolysis stimulated by water stress in both treatments, with the ability of microorganisms in the non-sterile soil to degrade the soluble substrates produced by root lysis being curtailed by water limitations.

Fractionation of soil organic nitrogen by chemical methods usually has involved hot acid (6 M HCl) hydrolysis of soils and separation of the acid-soluble nitrogen into NH_4^+, amino acids, amino sugars and an unidentified nitrogenous remainder. Such studies with ^{15}N-labelled substrates most frequently have been concerned with immobilization of inorganic ^{15}N into microbial residues, and have demonstrated that ^{15}N becomes distributed amongst all fractions of the soil hydrolysate, as well as in nitrogenous compounds resistant to degradation by the hot acid treatment (e.g. Cheng and Kurtz, 1963).

The amounts of nitrogen in all chemical fractions decline during periods of net nitrogen mineralization. Amino acid ^{15}N is quantitatively the most important and, in short-term experiments at least, generally shows the greatest percentage decline during periods of mineral nitrogen release (e.g. Stewart *et al.*, 1963; Kai *et al.*, 1973). Similar results have also been reported for changes in the nitrogenous constituents of immobilized fertilizer ^{15}N in longer term (4 year) field experiments (Allen *et al.*, 1973). By contrast, other experiments (e.g. McGill, 1971) have demonstrated a greater susceptibility of the amino sugar ^{15}N to biological degradation; and the results of several long-term field trials (Keeney and Bremner, 1964, 1966) and experiments (Broadbent, 1968a,b) have demonstrated that no one nitrogenous chemical fraction consistently exhibits a greater biological availability than any other.

3. *Biological fractionation*

From ecological considerations, Jansson (1967) distinguished three main fractions of soil organic matter: (i) fresh debris, primarily autotrophic; (ii) heterotrophic biomass; and (iii) waste materials and stabilized residues. Persson (1968) compared different chemical fractionation procedures in an effort to separate these three classes of material more successfully than was possible with either the classical separation into fulvic acid, humic acid and humin or by Waksman's proximate analysis, which did not differentiate sugars or amino acids released from living microbial tissue, plant residues or humic acids. Persson studied two contrasting systems – soil with added [^{14}C]glucose and soil with added ^{14}C-labelled straw. In the first case, after a short period of incubation, the ^{14}C remaining in the soil would be only in microbial cells or products of their metabolism and decay. In the straw-amended soil, in addition to ^{14}C-labelled biomass and its products, there would be undecomposed straw and fractions such as lignin which were resistant to decomposition. Persson's recommended procedure involved extraction of soil with increasing concentrations of H_2SO_4 followed by NaOH. Cold 0.5 N H_2SO_4 extracted substances readily decomposed by the soil biomass, e.g. sugars and amino acids released by air-drying; hot 0.5 N H_2SO_4 was thought to extract most of the carbon contained within the soil biomass; while hot 3.5 N H_2SO_4 extracted mainly secondary products of decomposition of the labelled substrate. Concentrated H_2SO_4 (25 N) extracted cellulose from undecomposed plant material. Cold 0.5 N NaOH extraction following the sequence of acid extractions recovered that part of the original substrate which had been incorporated into the stable humic acid fraction, while hot 0.5 N NaOH extracted lignin present in the undecomposed plant material.

Other more specific extractions have been used to isolate fractions of biological interest. Gupta and Sowden (1964) employed Schweitzer's reagent to extract cellulose from soil, while Simonart *et al.* (1967) and Biederbeck and Paul (1973) recovered a protein-rich fraction by treating humic acid fractions with phenolic solvents. Extraction of soils with buffers or sodium bicarbonate solutions is considered to be sufficiently mild to remove free or loosely adsorbed compounds with minimal damage to biomass. Such extracts may be enzymically active (Ladd and Paul, 1973; Nannipieri *et al.*, 1974). Bicarbonate-extractable nitrogen is correlated with plant-available nitrogen (Jenkinson, 1968). Acid-hydrolysable, amino acid nitrogen of bicarbonate extracts was shown in ^{15}N turnover studies to be biologically labile, decreasing relatively rapidly as microbial activity diminished and before onset of substantial net mineralization from the total labelled organic ^{15}N pool (Ladd and Paul, 1973).

C. Physical fractions

Ford and Greenland (1968) described a procedure for the separation of plant fragments from soil, based on flotation in a high density liquid following disruption of soil aggregates with ultrasonic energy. The procedure was refined and extended by Turchenek and Oades (1979) to include a particle size fractionation, giving fractions containing plant fragments, intact microorganisms, microbial fragments and extracellular products associated with inorganic and organic colloids.

Physical fractionation has been used in association with isotopic tracers to follow the distribution of ^{14}C and ^{15}N, added to soil either singly or in combination, within microbial biomass and in soil particles of different size and density (e.g. Ladd and Paul, 1973; McGill *et al.*, 1975; McGill and Paul, 1976; Ladd *et al.*, 1977a,b,c; Ladd and Amato, 1980; Cheshire and Mundie, 1981).

Ladd and Paul (1973) added glucose and $Na^{15}NO_3$ to soil and determined the distribution of ^{15}N at regular time intervals up to 63 days. ^{15}N-labelled acid-soluble amino acid nitrogen was measured in soluble fractions and in particulate fractions $>2.0\ \mu m$, $0.2–2.0\ \mu m$ and $0.04–0.2\ \mu m$. During the incubation period from 2.5 to 35.5 days the sum of the ^{15}N-labelled amino nitrogen in all fractions remained relatively constant. Increases in a fraction thought to contain intact organisms balanced a decrease in other fractions. A decrease of amino ^{15}N in fractions containing protein, peptides and amino acids, free or in association with soil colloids ($<0.2\ \mu m$), occurred during the active phase of oxidation of microbial metabolites, as measured by the release of $^{14}CO_2$.

The results of further experiments, using ^{15}N added to a calcareous soil as NO_3^- together with glucose or wheat straw (Ladd *et al.*, 1977a,b,c; Ladd and Amato, 1980), or as leaf material labelled with ^{14}C and ^{15}N, and where the distribution of isotope within physically separated fractions was determined, have been summarized by Amato and Ladd (1980). During the decomposition of ^{14}C- and ^{15}N-labelled plant material and the immobilization of $^{15}NO_3^-$ in soil, newly formed microorganisms and their products become distributed amongst soil components of widely varying particle size. The organic ^{14}C and ^{15}N associated with material of fine clay size is particularly readily transferred or decomposed, the decrease being due to net decreases in both the biomass and non-biomass pools of this fraction. The percentage decline in the amounts of ^{14}C and ^{15}N in the biomass pool is slightly faster than in the non-biomass pool. In the long term, isotope-labelled material will continue to decline disproportionately rapidly from the fine clay size components and from the biomass pool compared with the non-biomass residues; whereas silt size components will tend to accumulate higher proportions of the more resistant residues. ^{15}N immobilized by wheat straw metabolism is also partly associated with plant fragments, separated from soil as a light fraction using a solvent of specific gravity 1.6. This fraction also contains ^{15}N-labelled microbial biomass plus metabolites. Decreases in the ^{15}N of this fraction, either due to transfer or metabolism, occur relatively rapidly during periods of net nitrogen mineralization. By contrast, ^{15}N immobilized by glucose metabolism appeared in substantial amounts in a light fraction only when the separation was carried out using a solvent of higher specific gravity (2.1). This light fraction was derived mainly from silt-size components of the soil. Substantial changes occurred in the organic ^{15}N of the light fraction from silt size particles during nitrogen turnover in soils, but these were matched by similar changes in the organic ^{15}N of the complementary heavy fractions.

The decomposition of ^{14}C-labelled glucose or plant material added to soil, followed by physical fractionation, has been used to study the contribution made by plant and microbial residues to soil organic matter fractions. Murayama *et al.* (1979) used a separation based on density, with a further chemical fractionation of the ^{14}C associated with the dense soil particles, to study the decomposition of [^{14}C]glucose and [^{14}C]ryegrass. Most of the ^{14}C-labelled sugars were recovered in the light fraction which represented either undecomposed plant material or labelled microbial tissue from the decomposers. Even after a year of incubation the light fraction still appeared to contain a mixture of unmetabolized ryegrass sugars and microbial products. Attempts to separate these types of material were only partially successful. It was concluded that, for the soil studied, fractionation by density and filtration did not satisfactorily distinguish soil organic matter derived from different sources. A later experiment (Cheshire and Mundie, 1981) utilized a fractionation method based on particle size. It was concluded that size fractionation provides a good method of isolating carbohydrates present in residual plant fragments in soil. The results also confirmed earlier conclusions that the rate of decomposition of plant fragments may be more dependent on their disposition in soil than on their chemical composition.

D. Age of soil organic matter

^{14}C, produced by the action of cosmic ray-induced neutrons on atmospheric N_2, has been incorporated into soil organic matter over very long periods of time by the decomposition of plant tissue containing photosynthetically fixed ^{14}C. A radiocarbon dating technique developed by Libby (1952), has been refined to allow for (i) variations in the $^{14}CO_2$ activity of the atmosphere within the past 7000 years; (ii) a decrease in the specific activity of the atmospheric $^{14}CO_2$ due to the release of CO_2 from combustion of fossil fuels, which are almost completely devoid of ^{14}C activity; and (iii) a sharp increase in atmospheric $^{14}CO_2$ following the detonation of thermonuclear bombs from 1959 onwards. Because of the long half-life of ^{14}C (5730 $\pm$ 40 years), ^{14}C-dating techniques can be extended to periods about 40 000 years before the present (BP).

^{14}C occurs naturally in very low concentrations so that several grams of organic carbon are required for dating soil organic matter. This has resulted in a strong bias to studies of soils or soil fractions rich in organic matter, e.g. peats and the organic horizon of podsols. Interpretation of the ^{14}C data for a whole soil presents considerable difficulty since the soil organic matter is a heterogenous mixture ranging from fresh plant residues and microbial tissue to fractions which are very resistant to decomposition. Even a single fraction, e.g. humic acid, will represent an accumulation over many years. Thus values are frequently presented as mean residence times. A comparison of the ^{14}C age of soils in different climatic zones or with very different physical and mineralogical properties is of very limited value since the inputs of carbon and the rates of the decomposition processes can be so different (cf. Table 3.2).

Radiocarbon dates for peats and soils from Alaska range from 1800–11 000 years BP for samples taken at 1 m depth to 25 000 years BP for a sample collected at a depth of 2 m (Brown, 1965). These values contrast with samples of organic matter from a Swedish forest soil where the radiocarbon ages ranged from modern for material in the litter and topsoil to 370 $\pm$ 100 years for the B horizon (Tamm and Ostlund, 1960). There is also a very considerable variation in the radiocarbon age of fractions from the same soil. Jenkinson and Rayner (1977) found that while the total organic matter from a Rothamsted soil sampled in 1881 had a mean age of 1450 years, material in a 6 M HCl hydrolysate of the soil had an age of 515 years and the residue had an age of 2560 years. Stout *et al.* (1981) presented a description of the techniques for radiocarbon dating, the use of the stable isotopes ^{12}C and ^{13}C

to determine soil organic matter transformations (see also Chapter 3 in Volume 2), and the range of ages obtained for soils with differing climates and vegetation.

V. Availability of nutrients released from organic residues

The recovery by crops of nutrients released in soils from decomposing plant residues or animal manures can be followed using isotopes as tracers. The effects of soil and plant properties, climate and management practices on the transfer of nutrient from residues to crop can be measured in experiments analogous to the many devised for determining the availability of isotope-labelled fertilizers.

Most studies have employed ^{15}N-labelled sources. In one of the earliest ^{15}N studies Norman and Werkman (1943) showed that soybeans (*Glycine max*) grown in pots recovered 26.5% of the nitrogen of labelled soybean material incorporated in the soil. Ladd *et al.* (1981b) found that wheat crops at three field sites ranged in grain yields from 1.1 to 4.9 t ha^{-1} and recovered only 11–17% of the ^{15}N of legume material (C:N ratio, 14.9:1) incorporated in topsoils 8 months before sowing. At another site in contrasting growth seasons, first wheat crops (4.7 and 1.5 t ha^{-1} of grain, respectively) took up 28 and 20% of the ^{15}N of added legume material (C:N ratio, 11:1); a second crop took up the equivalent of <5% of the ^{15}N input (Ladd *et al.*, 1983a). Thus the organic ^{15}N residues from legume material plus first crop wheat roots had attained within a year a stability similar to that demonstrated for immobilized fertilizer ^{15}N (e.g. Legg and Allison, 1967; Jansson, 1968). The recoveries of ^{15}N by both first and second wheat crops were approximately proportional to legume ^{15}N inputs. The value of the isotopic measurements lay in demonstrating that only a minor proportion of the nitrogen of incorporated legume material was available to even first wheat crops, and further that for a normal range of legume inputs for the regions considered, the contributions of legume residues to the available nitrogen pool were relatively small (<10%) compared with those of the nitrogen released from decomposing soil organic matter (Ladd *et al.*, 1983a). Of course the relative contributions from plant residues, manures and accumulated soil organic matter may vary enormously in different situations (e.g. Yaacob and Blair, 1980).

Although isotopic analysis of crops clearly demonstrates the availability of nutrients (^{15}N) released from decomposing organic material, it is important to recognize that inorganic nitrogen formed in decomposition processes may also participate in other reactions, e.g. immobilization, denitrification, both before and during crop growth. Such reactions are a resultant of the metabolism of carbon compounds derived from the added organic materials, from soil organic matter, and from the crop itself due to turnover of some root tissue. To the extent that ^{15}N from plant residues or manures substitutes for unlabelled nitrogen in such reactions, promoted by the metabolism of carbon from soil and crop sources, so an additional amount of soil-derived nitrogen will be available for uptake by the crop.

The potential for such reactions to occur has been offered as an explanation for the greater uptake of unlabelled, soil-derived inorganic nitrogen by crops, frequently observed when ^{15}N-labelled nitrogen fertilizers are added; that is, for the so-called "priming" effect as revealed by isotope analysis (see Chapter 4 in this volume). The dilemma lies in deciding the extent to which these reactions may account for the priming effect, i.e. in distinguishing the effects of these reactions from perhaps a genuine increase in the rates of organic nitrogen decomposition or in the crop's capacity to utilize soil inorganic nitrogen, as stimulated by fertilizer nitrogen addition.

In contrast to ^{15}N-labelled fertilizers, ^{15}N-labelled plant materials and manures are of course sources of both carbon and nitrogen. Just as ^{15}N released from organic residues may enter processes initiated by the metabolism of carbon from soil or crop sources, so too can unlabelled nitrogen from the latter sources be involved in reactions initiated by the metabolism of carbon from the added labelled residues. Thus, the possibility of obtaining a "priming" effect explicable entirely in terms of biological interchange reactions involving ^{15}N and ^{14}N is less likely with labelled organic residues than with inorganic fertilizer ^{15}N, especially if the C:N ratios of the added organic materials are similar to those of the decomposable sources from soil and crop.

Indeed there is the possibility that with added ^{15}N-labelled plant materials of high C:N ratios, their value as sources of nitrogen for crops may be overstated when based on analysis of crop ^{15}N uptake. Because of interchange of labelled and unlabelled nitrogen from plant residues and from soil sources, respectively, some net mineralization (Kanazawa and Yoneyama, 1980) and uptake by crops of ^{15}N may occur under conditions when plant material addition may actually cause some net immobilization of soil-derived nitrogen.

Moore (1974) demonstrated that less nitrogen was transferred to successive grass (*Chloris gayana*) crops from surface-placed plant materials than when these were incorporated in soils. Additions of inorganic nitrogen increased the proportions of plant residue ^{15}N taken up by the grass crops, from about 22% of input to 29%. Krishnappa and Shinde (1980) also found that fertilizer nitrogen increased the recovery by field-grown rice (*Oryza sativa*) crops of ^{15}N, released from decomposing rice straw. Wallace and

Smith (1954) earlier had demonstrated that barley crops recovered higher proportions of ^{15}N released from orange leaves when NO_3^- was added to soils.

The results of Wallace and Smith (1954) are of particular interest. In the absence of NO_3^- addition, the amounts of crop nitrogen attributable to nitrogen transfer from labelled plant materials were very similar, whether calculated from crop ^{15}N contents or by the difference method, viz. by comparing crop nitrogen contents when grown in soil with and without plant residue addition. The results thus indicated that no priming effect had occurred under these conditions, although of course turnover and biological exchange of ^{14}N and ^{15}N may still have taken place. The addition of unlabelled NO_3^-, however, increased the uptake of ^{15}N by the crop to an extent that it exceeded the amounts of nitrogen attributable to the added plant materials, as calculated by the difference method. It seems likely that in this experiment, and in those cited above, the effect of added inorganic nitrogen in stimulating ^{15}N uptake from decomposing organic residues may be due at least partly to the unlabelled inorganic nitrogen substituting for labelled ^{15}N in immobilization reactions, etc., thus enabling higher amounts of labelled inorganic nitrogen to be available for uptake by the crop.

Studies of the availability of nitrogen from organic residues are not restricted to the use of ^{15}N-labelled plant materials or animal manures – the supply to successive crops of nitrogen from immobilized fertilizer nitrogen is of major importance and has been studied extensively using isotopic techniques (see Chapter 1 in this volume). ^{32}P- and ^{35}S-labelled organic materials have been used in experiments analogous to those employing ^{15}N (Hashimoto and Tsuji, 1972; Dalal, 1979; Tsuji and Goh, 1979; Till *et al.*, 1982).

References

Ågren, G. I., Axelsson, B., Flower-Ellis, J. G. K., Linder, S., Persson, H., Staff, H. and Troeng, E. (1981). *Ecol. Bull.* (*Stockholm*) **32**, 307–313.

Allen, A. L., Stevenson, F. J. and Durtz, L. T. (1973). *J. Environ. Qual.* **2**, 120–124.

Allison, F. E. (1973). "Soil Organic Matter and Its Role in Crop Production", Elsevier, New York.

Amato, M. and Ladd, J. N. (1980). *Soil Biol. Biochem.* **12**, 405–411.

Bartholomew, W. V. (1965). *In* "Soil Nitrogen" (W. V. Bartholomew and F. E. Clark, eds), pp. 285–306, American Society of Agronomy, Madison, Wisc.

Batistic, L. and Mayaudon, J. (1970). *Ann. Inst. Pasteur* (*Paris*) **118**, 199–206.

Biederbeck, V. O. and Paul, E. A. (1973). *Soil Sci.* **115**, 357–366.

Bohn, H. L. (1976). *Soil Sci. Soc. Am. J.* **40**, 468–470.

Broadbent, F. E. (1968a). *Pontif. Acad. Sci. Scr. Varia* **32**, 61–82.

Broadbent, F. E. (1968b). *In* "Isotopes and Radiation in Soil Organic Matter Studies", pp. 131–142, IAEA, Vienna.

Broadbent, F. E. and Nakashima, T. (1974). *Soil Sci. Soc. Am. Proc.* **38**, 313–315.
Brown, J. (1965). *Arctic* **18**, 37–48.
Cerri, C. C. and Jenkinson, D. S. (1981). *J. Soil Sci.* **32**, 619–626.
Cheng, H. H. and Kurtz, L. T. (1963). *Soil Sci. Soc. Am. Proc.* **27**, 312–316.
Cheshire, M. V. and Mundie, C. M. (1981). *J. Soil Sci.* **32**, 605–618.
Cheshire, M. V., Greaves, M. P. and Mundie, C. M. (1975). *Proc. Int. Coll. Biodegradation and Humification, Nancy, 1974*, pp. 148–156, Sarreguemines, Pierron.
Chu, J. P.-H. and Knowles, R. (1966). *Soil Sci. Soc. Am. Proc.* **30**, 210–213.
Dahlman, R. C. (1968). *In* "Methods of Productivity Studies in Root Systems and Rhizosphere Organisms", pp. 11–21, USSR Academy of Sciences, Moscow.
Dalal, R. C. (1979). *Soil Sci. Soc. Am. J.* **43**, 913–916.
Domsch, K. H., Anderson, J. P. E. and Ahlers, R. (1973). *Appl. Microbiol.* **25**, 819–824.
Ford, G. W. and Greenland, D. J. (1968). *Trans. 9th Int. Congr. Soil Sci.* **2**, 403–410.
Freytag, H. E. and Igel, H. (1970). *Albrecht-Thaer-Archiv* **14**, 495–507.
Fuhr, F. and Sauerbeck, D. (1968). *In* "Isotopes and Radiation in Soil Organic Matter Studies", pp. 241–250, IAEA, Vienna.
Grossbard, E. (1969). *J. Soil Sci.* **20**, 38–51.
Gupta, U. C. and Sowden, F. J. (1964). *Soil Sci.* **97**, 328–333.
Haider, K. and Martin, J. P. (1975). *Soil Sci. Soc. Am. Proc.* **39**, 657–662.
Haider, K. and Martin, J. P. (1981). *Soil Biol. Biochem.* **13**, 447–450.
Hashimoto, H. and Tsuji, T. (1972). *Soil Sci. Plant Nutr.* **20**, 104.
Hurst, H. M. and Wagner, G. H. (1969). *Soil Sci. Soc. Am. Proc.* **33**, 707–711.
Igel, H. (1969). *Albrecht-Thaer-Archiv* **13**, 267–282.
Ivarson, K. C. and Stevenson, I. L. (1964). *Can. J. Microbiol.* **10**, 677–682.
Jansson, S. L. (1958). *Lantbrukshogsk. Ann.* **24**, 101–361.
Jansson, S. L. (1966). *In* "The Use of Isotopes in Soil Organic Matter Studies", pp. 283–296, FAO/IAEA Symposium, Brunswick-Volkenrode, Pergamon Press, Oxford.
Jansson, S. L. (1967). *In* "Soil Chemistry and Fertility" (G. V. Jacks, ed.), pp. 1–10, Aberdeen.
Jansson, S. L. (1968). *In* "Festkrift til H. L. Jensen", pp. 49–60, Gadgaard Nielson, Lemvig.
Jansson, S. L. (1971). *In* "Soil Biochemistry", Vol. II (A. D. McLaren and J. J. Skujins, eds), pp. 129–166, Marcel Dekker, New York.
Jenkinson, D. S. (1965). *J. Soil Sci.* **16**, 104–115.
Jenkinson, D. S. (1966). *J. Soil Sci.* **17**, 280–302.
Jenkinson, D. S. (1968). *J. Sci. Food Agric.* **19**, 160–168.
Jenkinson, D. S. (1971). *Soil Sci.* **111**, 64–70.
Jenkinson, D. S. (1977a). *J. Soil Sci.* **28**, 417–423.
Jenkinson, D. S. (1977b). *J. Soil Sci.* **28**, 424–434.
Jenkinson, D. S. (1981). *In* "The Chemistry of Soil Processes" (D. J. Greenland and M. H. B. Hayes, eds), pp. 505–561, John Wiley and Sons, New York and Chichester.
Jenkinson, D. S. and Ayanaba, A. (1977). *Soil Sci. Soc. Am. J.* **41**, 912–915.
Jenkinson, D. S. and Rayner, J. H. (1977). *Soil Sci.* **123**, 298–305.
Juma, N. G. and Paul, E. A. (1981). *In* "Simulation of Nitrogen Behaviour of Soil–Plant Systems" (M. J. Frissel and J. A. van Veen, eds), pp. 145–154, Centre for Agricultural Publishing and Documentation (PUDOC), Wageningen.

Kai, H., Ahmad, Z. and Harada, T. (1973). *Soil Sci. Plant Nutr.* **19**, 275–286.
Kanazawa, S. and Yoneyama, T. (1980). *Soil Sci. Plant Nutr.* **26**, 241–254.
Kassim, G., Martin, J. P. and Haider, K. (1981). *Soil Sci. Soc. Am. J.* **45**, 1106–1112.
Keeney, D. R. and Bremner, J. M. (1964). *Soil Sci. Soc. Am. Proc.* **28**, 653–656.
Keeney, D. R. and Bremner, J. M. (1966). *Soil Sci. Soc. Am. Proc.* **30**, 714–719.
Knowles, R. (1981). *In* "Current Perspectives in Nitrogen Fixation" (A. H. Gibson and W. E. Newton, eds), pp. 327–333, Australian Academy of Sciences, Canberra.
Krishnappa, A. M. and Shinde, J. E. (1980). *J. Indian Soc. Soil Sci.* **28**, 474–479.
Ladd, J. N. and Amato, M. (1980). *Soil Biol. Biochem.* **12**, 185–189.
Ladd, J. N. and Paul, E. A. (1973). *Soil Biol. Biochem.* **5**, 825–840.
Ladd, J. N., Parsons, J. W. and Amato, M. (1977a). *Soil Biol. Biochem.* **9**, 309–318.
Ladd, J. N., Parsons, J. W. and Amato, M. (1977b). *Soil Biol. Biochem.* **9**, 319–325.
Ladd, J. N., Amato, M. and Parsons, J. W. (1977c). *In* "Soil Organic Matter Studies", Vol. I, pp. 301–310, IAEA, Vienna.
Ladd, J. N., Oades, J. M. and Amato, M. (1981a). *Soil Biol. Biochem.* **13**, 119–126.
Ladd, J. N., Oades, J. M. and Amato, M. (1981b). *Soil Biol. Biochem.* **13**, 251–256.
Ladd, J. N., Amato, M., Jackson, R. B. and Butler, J. H. A. (1983a). *Soil Biol. Biochem.* **15**, 231–238.
Ladd, J. N., Jackson, R. B., Amato, M. and Butler, J. H. A. (1983b). *Aust. J. Soil Res.* **21** (in press).
L'Annunziata, M. F. and Gonzalez, J. (1977). *In* "Soil Organic Matter Studies", Vol. I, pp. 239–253, IAEA, Vienna.
Legg, J. O. and Allison, F. E. (1967). *Soil Sci. Soc. Am. Proc.* **31**, 403–406.
Libby, W. F. (1952). "Radiocarbon Dating", University of Chicago Press, Chicago.
Linhares, L. F. and Martin, J. P. (1978). *Soil Sci. Soc. Am. J.* **42**, 738–743.
McGill, W. B. (1971). PhD thesis, University of Saskatchewan, Saskatoon.
McGill, W. B. and Paul, E. A. (1976). *Can. J. Soil Sci.* **56**, 203–212.
McGill, W. B., Shields, J. A. and Paul, E. A. (1975). *Soil Biol. Biochem.* **7**, 57–63.
Malik, K. A. and Azam, F. (1980). *Pak. J. Bot.* **12**, 117–127.
Martin, J. K. (1977a). *Soil Biol. Biochem.* **9**, 1–7.
Martin, J. K. (1977b). *In* "Soil Organic Matter Studies", Vol. I, pp. 197–203, IAEA, Vienna.
Martin, J. P., Haider, K., Farmer, W. J. and Fustec-Mathon, E. (1974). *Soil Biol. Biochem.* **6**, 221–230.
Mayaudon, J. (1971). *In* "Soil Biochemistry", Vol. II (A. D. McLaren, and J. J. Skujins, eds), pp. 202–256, Marcel Dekker, New York.
Mayaudon, J. and Simonart, P. (1959). *Plant Soil* **11**, 181–192.
Mayaudon, J. and Simonart, P. (1963). *Ann. Inst. Pasteur* (*Paris*) **105,** 257–266.
Mayaudon, J. and Simonart, P. (1965). *Suppl. Ann. Inst. Pasteur* (*Paris*) **109**, 224–234.
Moore, A. W. (1974). *Soil Biol. Biochem.* **6**, 249–255.
Murayama, S., Cheshire, M. V., Mundie, C. M., Sparling, G. P. and Shepherd, H. (1979). *J. Sci. Food Agric.* **30**, 1025–1034.
Nannipieri, P., Ceccanti, B., Cervelli, S. and Sequi, P. (1974). *Soil Biol. Biochem.* **6**, 359–362.
Nelson, D. W., Martin, J. P. and Ervin, J. O. (1979). *Soil Sci. Soc. Am. J.* **43**, 84–88.
Norman, A. G. and Werkman, C. H. (1943). *J. Am. Soc. Agron.* **35**, 1023–1025.
Nyhan, J. W. (1975). *Soil Sci. Soc. Am. Proc.* **39**, 643–648.
Oades, J. M. and Ladd, J. N. (1977). *In* "Soil Factors in Crop Production in a Semi-arid Environment" (J. S. Russell and E. L. Greacen, eds), pp. 127–160, University of Queensland Press, Adelaide.

Oades, J. M. and Wagner, G. H. (1971). *Soil Sci. Soc. Am. Proc.* **35**, 914–917.
Oberlander, H. E. (1973). *Pontif. Acad. Sci. Scr. Varia.* **38**, 1001–1071.
Oberlander, H. E. and Roth, K. (1968). *In* "Isotopes and Radiation in Soil Organic Matter Studies", pp. 251–261, IAEA, Vienna.
Oberlander, H. E. and Roth, K. (1974). *Bodenkultur* **25**, 111–129.
Paul, E. A. and Juma, N. G. (1981). *Ecol. Bull.* (*Stockholm*) **33**, 179–195.
Paul, E. A. and van Veen, J. A. (1978). *Proc. 11th Int. Congr. Soil Sci., Edmonton* **3**, 1–43.
Paul, E. A. and Voroney, R. P. (1980). *In* "Contemporary Microbial Ecology" (D. C. Ellwood *et al.*, eds), pp. 215–237, Academic Press, New York and London.
Persson, J. (1968). *Lantbrukshogsk. Ann.* **34**, 81–217.
Russell, E. W. (1973). "Soil Conditions and Plant Growth", 10th edn, Longman, London.
Sauerbeck, D. R. (1968a). *In* "Isotopes and Radiation in Soil Organic Matter Studies", pp. 57–66, IAEA, Vienna.
Sauerbeck, D. R. (1968b). *In* "Isotopes and Radiation in Soil Organic Matter Studies", pp. 219–225, IAEA, Vienna.
Sauerbeck, D. R. and Führ, F. (1968). *In* "Isotopes and Radiation in Soil Organic Matter Studies", pp. 3–11, IAEA, Vienna.
Sauerbeck, D. R. and Gonzalez, M. A. (1977). *In* "Soil Organic Matter Studies", Vol. I, pp. 159–170, IAEA, Vienna.
Sauerbeck, D. R. and Johnen, B. G. (1977). *In* "Soil Organic Matter Studies", Vol. I, pp. 144–148, IAEA, Vienna.
Schlesinger, W. H. (1977). *Annu. Rev. Ecol. Syst.* **8**, 51–81.
Schnitzer, M. (1977). *In* "Soil Organic Matter Studies", Vol. II, pp. 117–132, IAEA, Vienna.
Shamoot, S., McDonald, I. and Bartholomew, W. V. (1968). *Soil Sci. Soc. Am. Proc.* **32**, 817–820.
Shields, J. A. and Paul, E. A. (1973). *Can. J. Soil Sci.* **53**, 297–306.
Shields, J. A., Paul, E. A., Lowe, W. E. and Parkinson, D. (1973). *Soil Biol. Biochem.* **5**, 753–764.
Simonart, P. and Mayaudon, J. (1958). *Plant Soil* **9**, 367–375.
Simonart, P. and Mayaudon, J. (1961). *Pedologie Symp. Intern. 2 Appl. Sci. mid Ped.* 91–103.
Simonart, P. and Mayaudon, J. (1963). *Ann. Inst. Pasteur* (*Paris*) **105**, 257–266.
Simonart, P., Batistic, L. and Mayaudon, J. (1967). *Plant Soil* **27**, 153–161.
Skujins, J. J. and McLaren, A. D. (1969). *Soil Biol. Biochem.* **1**, 89–99.
Smith, J. H. and Douglas, C. L. (1971). *Soil Sci. Soc. Am. Proc.* **35**, 269–273.
Sorensen, L. H. (1963). *Soil Sci.* **95**, 45–51.
Sorensen, L. H. (1967). *Soil Sci.* **104**, 234–241.
Sorensen, L. H. (1972a). *Soil Biol. Biochem.* **4**, 245–255.
Sorensen, L. H. (1972b). *Soil Sci.* **114**, 5–11.
Sorensen, L. H. (1975). *Soil Biol. Biochem.* **7**, 171–177.
Sorensen, L. H. and Paul, E. A. (1971). *Soil Biol. Biochem.* **3**, 173–180.
Sparling, G. P., Cheshire, M. V. and Mundie, C. M. (1982). *J. Soil Sci.* **33**, 89–100.
Stewart, B. A., Porter, L. K. and Johnson, D. D. (1963). *Soil Sci. Soc. Am. Proc.* **27**, 302–304.
Stout, J. D., Goh, K. M. and Rafter, T. A. (1981). *In* "Soil Biochemistry", Vol. V (E. A. Paul and J. N. Ladd, eds), pp. 1–73, Marcel Dekker, New York.
Tamm, C. O. and Ostlund, H. G. (1960). *Nature* (*Lond.*) **185**, 706–707.

Till, A. R., Blair, G. J. and Dalal, R. C. (1982). *In* "The Cycling of Carbon, Nitrogen, Sulfur and Phosphorus in Terrestrial and Aquatic Ecosystems" (I. E. Galbally and J. R. Freney, eds), pp. 51–59, Australian Academy of Sciences, Canberra.

Tsuji, T. and Goh, K. M. (1979). *N.Z. J. Agric. Res.* **22**, 595–602.

Turchenek, L. W. and Oades, J. M. (1979). *Geoderma* **21**, 311–343.

van Veen, J. A. and Frissel, M. J. (1981). *In* "Simulation of Nitrogen Behaviour of Soil–Plant Systems" (M. J. Frissel and J. A. van Veen, eds), pp. 126–144, Centre for Agricultural Publishing and Documentation (PUDOC), Wageningen.

Verma, L., Martin, J. P. and Haider, K. (1975). *Soil Sci. Soc. Am. Proc.* **39**, 279–284.

Wagner, G. H. (1968). *In* "Isotopes and Radiation in Soil Organic Matter Studies", pp. 197–205, IAEA, Vienna.

Wagner, G. H. and Mutatkar, V. V. (1968). *Soil Sci. Soc. Am. Proc.* **32**, 683–686.

Wallace, A. and Smith, R. L. (1954). *Soil Sci.* **78**, 231–242.

Wang, C. H. (1971). *In* "Methods in Microbiology", Vol. VIB (T. Bergan and J. R. Norris, eds), pp. 185–230, Academic Press, New York and London.

Warembourg, F. R. and Paul, E. A. (1977). *Soil Biol. Biochem.* **9**, 295–301.

Welbank, P. J., Gibb, M. J., Taylor, P. J. and Williams, E. D. (1974). *Rothamsted Exp. Stn Rep. Part 2*, **1974**, 22–66.

Woodwell, G. M., Whittaker, R. H., Reiners, W. A., Likens, G. E., Delwiche, C. C. and Botkin, D. B. (1978). *Science* **199**, 141–146.

Yaacob, O. and Blair, G. J. (1980). *Plant Soil* **57**, 237–247.

Zeller, A., Oberlander, H. E., Roth, K. and Stadler, I. (1966). *In* "The Use of Isotopes in Soil Organic Matter Studies", pp. 275–279, FAO/IAEA Symposium Braunschweig-Volkenrode, Pergamon Press, Oxford.

Zunino, H., Borie, F., Aguilera, S., Martin, J. P. and Haider, K. (1982). *Soil Biol. Biochem.* **14**, 37–43.

4

Soil Microbiology

D. C. Wolf and J. O. Legg

I. Introduction

Both stable and radioactive isotopes have been used to study microbiological and biochemical reactions in soil. The use of radioactive isotopes such as ^{3}H, ^{13}N, ^{14}C and ^{32}P have enabled scientists to understand better many reactions involved in organic matter decomposition, pesticide degradation and biosynthesis of numerous microbial products important in humus formation. Extensive studies have been conducted using the stable isotope ^{15}N to elucidate nitrogen transformations such as symbiotic, nonsymbiotic and associative nitrogen fixation, denitrification, mineralization and immobilization.

The purpose of this chapter is to summarize the major areas of soil microbiological and biochemical research which have involved both stable and radioactive isotopes. The primary emphasis will be to develop an overview of the various types of research that have been conducted. For details in the techniques used in the studies described here, we refer you to the literature cited.

II. Soil sampling and processing

By its very nature, much of the research involving isotopes requires laboratory incubation studies. Many problems can be encountered in soil sampling and processing for incubation studies since the soil contains a large and diverse microbial population (Table 4.1). Soil samples must be collected in the field and returned to the laboratory for these studies. Anything that disrupts or destroys the soil microbial population will adversely influence or invalidate subsequent experiments.

Table 4.1 The microbial population of a medium textured soil with a near neutral pH value

Microorganism	Population (number per gram of dry soil)
Viruses	Unknown
Bacteria	10^{6}–10^{9}
Actinomycetes	10^{5}–10^{8}
Fungi	10^{3}–10^{5}
Protozoa	10^{3}–10^{5}
Algae	10^{3}–10^{5}
Nematodes	10^{0}–10^{2}

To collect a representative soil sample from an area, several individual soil cores should be collected and combined. The soil series should be reported as well as the depth of sampling and the horizon or horizons sampled. The sample should not be air-dried, but rather it should be sieved in the field-moist condition since certain biochemical reactions may be altered by air-drying the soil. It is also important to establish a water release curve for each soil studied so that the moisture level for the soil can be expressed as the moisture potential in addition to the percentage of water on a weight basis. Sommers *et al.* (1981) have presented a recent discussion on the influence of soil moisture in the decomposition processes in soils.

III. Decomposition of naturally occurring materials

Microbial decomposition is initiated when organic components of micro-organisms, plants and animals are added to the soil in various quantities and at various times. The easily degraded or labile substances are utilized first by the microbial population, and a large portion of the carbon is respired as CO_2 under aerobic conditions (Fig. 4.1). A smaller portion of the carbon is incorporated into the microbial biomass and metabolic products, and a portion undergoes partial degradation and reacts to form an amorphous dark material known as soil humus. Soil humus is a complex mixture of polymers of phenolic units, polysaccharides and proteins and can be chemically separated into three general components known as fulvic acid, humic acid and humin.

In most cases, the organic material added to the soil consists of both simple and complex substrates subject to decomposition by microbial activity.

A. Simple substrates

Many researchers have used ^{14}C-labeled sugars, amino acids and nucleosides to quantitate the decomposition rate of such materials in soil. To conduct aerobic decomposition studies, it is essential that molecular oxygen levels not be a limiting factor. To insure aerobic conditions a CO_2-free air flow-through system such as the one described by Stotzky (1965) is commonly used. In the system, evolved $^{14}CO_2$ from the labeled substrate is trapped in a base such as KOH or NaOH and assayed using liquid scintillation techniques. This system has also been used in several studies to evaluate pesticide decomposition. Other incubation systems which are not flow-through have been used and are illustrated in Fig. 4.2.

Mayaudon (1971) has presented a review of the use of radiorespirometry

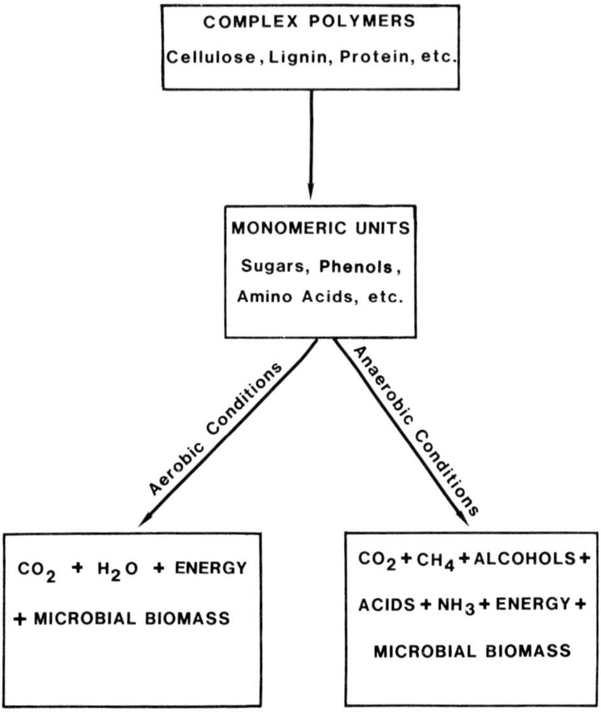

Fig. 4.1 General reactions involved in microbial decomposition processes.

in soil microbiology and biochemistry. In addition to a discussion of the methodology, several applications are presented and the author notes that radiorespirometric techniques can aid in determining the dynamic state and activity of soil microorganisms.

Martin and coworkers used a Greenfield sandy loam, pH 7.0, to study the decomposition rate of a number of simple organic substrates such as glucose, alanine, and uracil (Bondietti *et al.*, 1972; Wolf and Martin, 1974; Verma *et al.*, 1975; Martin and Haider, 1976). The results indicated that these materials undergo very rapid decomposition with a large portion of the carbon being evolved as $^{14}CO_2$ (Table 4.2). A similar rate of $^{14}CO_2$ evolution from soil amended with labeled glucose was reported by Chahal and Wagner (1965). They noted that the rate of glucose decomposition during the first 14 days of incubation varied according to the inorganic nitrogen content of the soils, but after 3 months of incubation 75% of the [^{14}C]glucose was evolved as $^{14}CO_2$, and the remaining 25% had become incorporated into microbial biomass or soil organic matter.

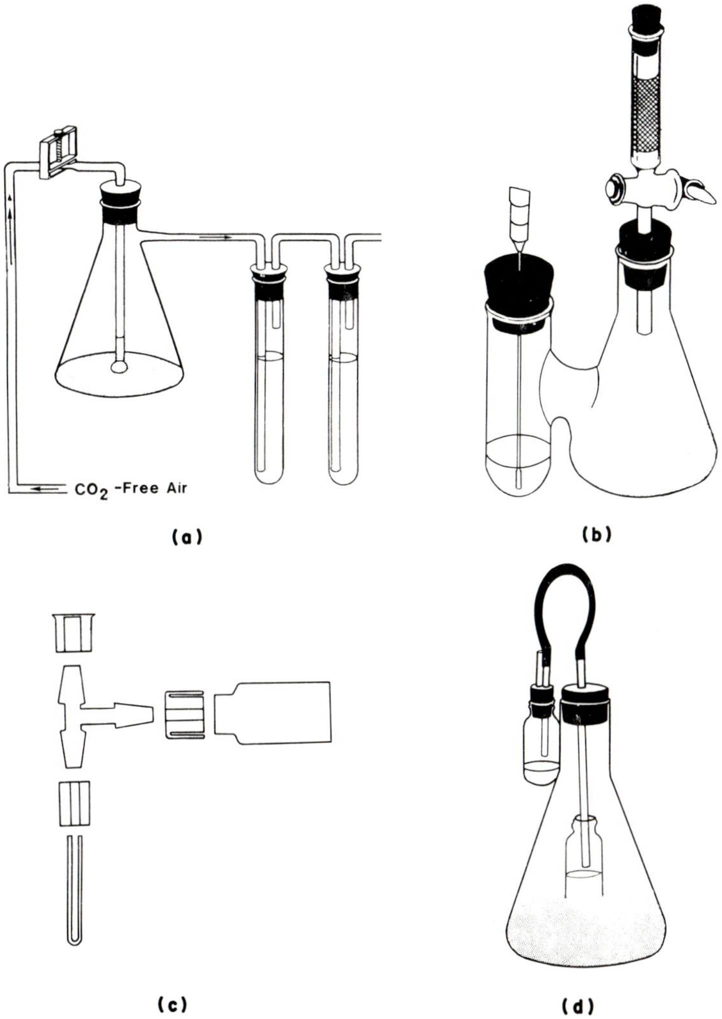

Fig. 4.2 Various units used for the collection of $^{14}CO_2$ in soil respiration studies: (a) flow-through system (Atlas and Bartha, 1972); (b) biometer flask (Bartha and Pramer, 1965); (c) tube with attached scintillation vial (Sissons, 1976); (d) static flask unit (Loos *et al.*, 1980).

In a microbial ecology study of ^{14}C-labeled glucose decomposition, Behera and Wagner (1974) showed that bacteria were the primary microorganisms involved in decomposition of glucose added to a Mexico silt loam. Fungal growth was noted at the time the bacterial count declined.

Table 4.2 Decomposition rate of various ^{14}C substrates in a Greenfield sandy loam[a]

Substrate	Position of ^{14}C	Decomposition (%)[b]				
		Weeks				
		1	2	4	8	12
Glucose	Uniform	75	82	84	86	87
Acetic acid	$^{14}COOH$	74	80	84	86	86
	$^{14}CH_3$	52	58	63	67	71
Pyruvic acid	$^{14}COOH$	83	87	90	92	93
	^{14}CO	57	65	74	78	80
	$^{14}CH_3$	47	55	61	68	70
Glycine	Uniform	75	83	86	88	89
Alanine	Uniform	75	76	79	82	83
	$^{14}COOH$	90	92	94	95	95
Aspartic acid	Uniform	76	81	85	87	88
Uracil	2-^{14}C	11	64	85	90	93
Glucosamine	Uniform	55	60	65	69	73

[a] Data adapted from Bondietti *et al.* (1972), Martin and Haider (1976), Verma *et al.* (1975) and Wolf and Martin (1974).
[b] Decomposition over time calculated as the percentage of applied ^{14}C evolved as $^{14}CO_2$.

Radioactive isotopes have been used also to show the rapid decomposition of nucleosides in soil. D. C. Wolf and J. P. Martin (unpublished data) added 2-^{14}C-labeled uracil, uridine and cytosine to a Greenfield sandy loam and found that greater than 90% of the added material was evolved as $^{14}CO_2$ in 4 weeks.

Various aromatic compounds such as *p*-hydroxybenzoic acid, vanillic acid, syringic acid and benzoic acid have been labeled with ^{14}C either in the benzene ring or the side chains and the decomposition rate studied in soil. Haider and Martin (1975) showed that the side chain carbons such as carboxyl, methyl and methoxy groups were rapidly evolved as $^{14}CO_2$ whereas the carbon in the benzene rings was much more resistant to decomposition (Table 4.3). The data showed that approximately 30% of the ^{14}C remaining in the soil following the 12 week incubation was isolated in the humic acid fraction and 20–30% was found in the fulvic acid fraction. Scott *et al.* (1982) also demonstrated the rapid decomposition of [^{14}C]phenol under aerobic conditions and showed that, as the phenol concentration increased, the time required for 50% decomposition, $t_{1/2}$, increased due to the antimicrobial properties of phenol.

In general, the studies on the decomposition of simple organic compounds have shown a rapid evolution of $^{14}CO_2$ from the labeled substrate. Often,

Table 4.3 Decomposition of ^{14}C-labeled aromatic compounds added to a Greenfield sandy loam at a rate of 100 μg per gram of loam (Haider and Martin, 1975; Martin and Haider, 1976)

Phenolic acid	Position of ^{14}C	Decomposition (%)[a]				
		Weeks				
		1	2	4	8	12
Vanillic 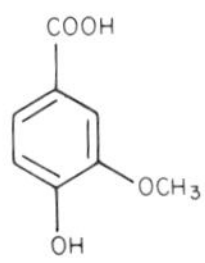	COOH	83	85	87	89	91
	Ring	52	61	65	66	67
p-Hydroxybenzoic 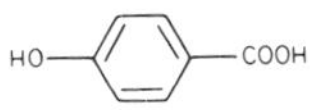	COOH	94	95	96	97	97
Syringic 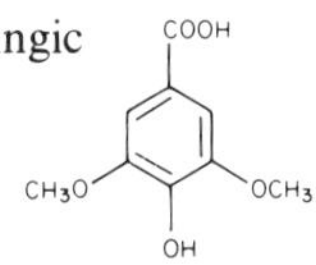	OCH_3	58	62	65	67	68
	COOH	87	94	95	95	96
Benzoic 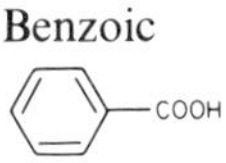	Ring	68	71	74	75	76
Ferulic 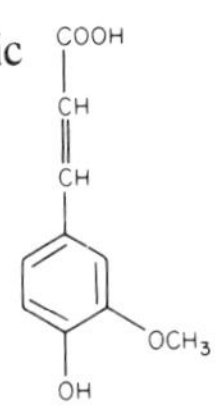	$—H^{14}C{=}CHCOOH$	50	55	60	64	66
	OCH_3	50	69	72	73	75
	Ring	36	57	61	65	67

[a] Decomposition over time calculated as a percentage of applied ^{14}C evolved as $^{14}CO_2$.

half of the substrate is lost in a matter of hours; however, a portion of the substrate carbon is not rapidly lost and has been shown to be incorporated into the microbial biomass and fractions of soil organic matter.

The decomposition of more complex substrates has also been studied in soil. Two ^{14}C-labeled lipopolysaccharides were added to soil and their decomposition determined (Cortez, 1977). The results indicated that complexation of metal ions and clays by one of the lipopolysaccharides protected it from decomposition. Martin and Haider's coworkers used ^{14}C-labeled glucosamine and chitosan (Bondietti *et al.*, 1972) and proteins, peptides and amino acids (Verma *et al.*, 1975) to demonstrate the increased resistance to decomposition afforded these materials by humic-type polymers similar to those found in soil humic acid. Bondietti *et al.* (1972) found that during an 8 week incubation, 69 % of the ^{14}C in glucosamine was evolved as $^{14}CO_2$ in a neutral sandy loam, but only 8–19 % of the activity was collected as $^{14}CO_2$ when the glucosamine was incorporated into phenolic polymers similar to humic acid. Comparable values for chitosan were 53 % when added to the soil alone and 31 % when incorporated in the humic-type polymer. From 71 to 95 % of the ^{14}C-labeled proteins, peptides and amino acids applied to a Greenfield sandy loam was evolved as $^{14}CO_2$ during laboratory incubations of 4–12 weeks. Linkage of the proteins, peptides or single amino acids into humic-type polymers reduced degradation by 80–90 %. When peptides containing two or three amino acid units were enzymatically linked into phenolic polymers, the N-terminal unit which was attached to the phenolic polymer was more resistant to decomposition than the COOH-terminal amino acid unit. These results are important in establishing the role of soil microorganisms in synthesizing the various components of soil organic matter and in explaining the decomposition rate of soil organic matter, especially the stabilization of microbially produced compounds.

B. Complex substrates

1. *Crop residue*

The production of ^{14}C-labeled plant material can be accomplished by growing the plant in an atmosphere of $^{14}CO_2$. This technique has been used to study the decomposition of labeled plant material in various soils. Extensive research of this nature has been conducted by Jenkinson and coworkers. In a 10 year study using small outdoor lysimeters, ^{14}C-labeled ryegrass (*Lolium multiflorum*) and five different Rothamsted soils, Jenkinson (1977) found that approximately two-thirds of the added carbon in the ryegrass was evolved as CO_2 within the first year (Fig. 4.3). The data were also used to develop a model for the turnover of organic matter in soil (Jenkinson and Rayner, 1977). In a similar study conducted in the forest zone of Nigeria, Jenkinson and Ayanaba (1977) measured the decomposition rate of ^{14}C-labeled ryegrass and maize (*Zea mays*) under field conditions in eight contrasting soils. After 1 year, they found 20 % of the ^{14}C-labeled ryegrass originally added still remained in the soils. This value decreased to 14 % after 2 years. Both plant materials exhibited similar patterns of decomposition in the Nigerian soils studied. However, the ryegrass decomposition rate was four times faster in the Nigerian soils when compared to the earlier study conducted in soils at Rothamsted, England.

Jenkinson's research demonstrates that a substantial portion of added plant residue is rapidly decomposed, but as much as 13 % is relatively stable

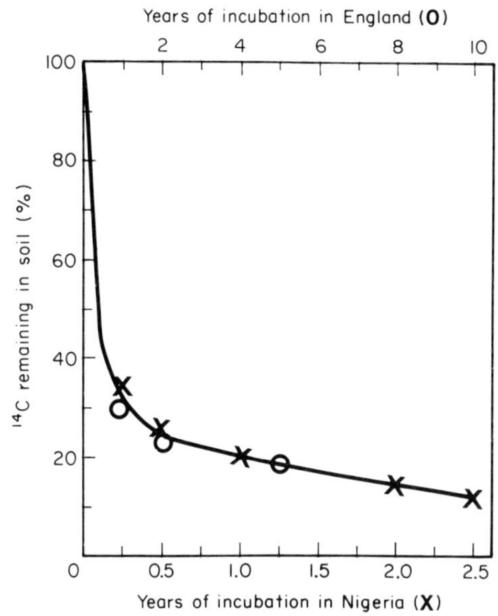

Fig. 4.3 Decomposition of uniformly ^{14}C-labeled ryegrass incubated with soils in outdoor lysimeters (adapted from Jenkinson and Ayanaba, 1977).

and resistant to decomposition. The data also demonstrate the impact of increasing temperature on increasing the decomposition rate of plant residue in soil.

Side-chain and ring ^{14}C-labeled *p*-coumaric acids were synthesized and injected into growing corn plants as precursors to prepare a labeled lignin (Haider *et al.*, 1977). The plant material was extracted with boiling ethanol to remove labeled low molecular weight compounds, and the residue was added to a neutral sandy loam and adjusted to 0.33 bar moisture. After 28 weeks of incubation, the percentage of ^{14}C in the CH_2OH, the number 2 position of the side chain, and ring of the coumaryl alcohol linked into corn stalk lignins that was evolved as $^{14}CO_2$ averaged 38, 29 and 28%, respectively. Under similar conditions, uniformly ^{14}C-labeled wheat straw evolved 62% of the ^{14}C as $^{14}CO_2$ in 12 weeks (Table 4.4). Of the ^{14}C activity from the corn stalk lignin that remained in the soil after 28 weeks, about 33 and 15% were isolated in the humic and fulvic acid fractions of the soil organic matter, respectively. The data presented by Haider *et al.* (1977) indicate that lignin is not as inert a substance as once thought. Martin and Focht (1977) have presented possible pathways which indicate that the degradation products of lignin and microbially synthesized compounds react to form the humic polymers found in soil humic acid (Fig. 4.4).

Table 4.4 Decomposition of ^{14}C-labeled plant material added to a Greenfield sandy loam[a]

Plant material	Decomposition (%)[b]				
	Weeks				
	1	2	4	8	12
Corn stalks	21	48	54	58	63
Oat straw	21	32	46	68	
Wheat straw	30	38	48	59	62

[a] Data adapted from Haider *et al.* (1977) and Verma and Martin (1976).
[b] Decomposition over time calculated as a percentage of applied ^{14}C evolved as $^{14}CO_2$.

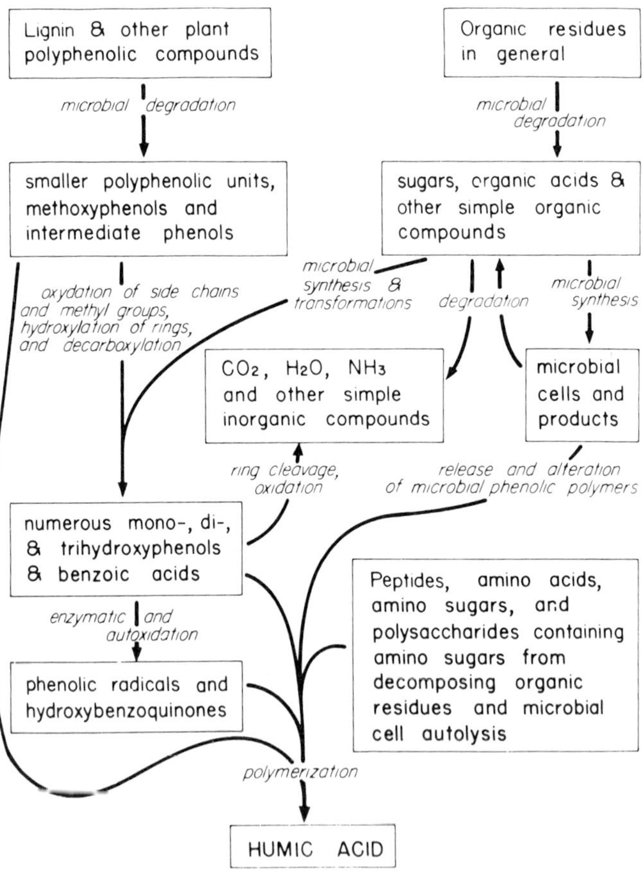

Fig. 4.4 Possible pathways of microbial synthesis of soil humic acid (Martin and Focht, 1977).

2. *Microbial cells*

A portion of the CO_2 evolved from soil is due to the decomposition of dead microbial cells and microbially produced residues. The transformations and resistance to decomposition of residues of bacteria, actinomycetes, fungi, algae and protozoa in soil have been reviewed by Webley and Jones (1971).

It is not unexpected that considerable variation has been reported in the decomposition rate of microbial residues because the residues consist of a wide range of biochemical compounds, some of which are very labile or easily degraded. Others are very resistant to decomposition. Zeller *et al.* (1966) labeled the fungus *Aspergillus niger* with ^{14}C and studied the decomposition of the mycelium in soil. They reported that approximately 55% of the ^{14}C was evolved as $^{14}CO_2$ during the 3 month incubation. Hurst and Wagner (1969) studied the decomposition of the same fungus and a sterile melanic or pigmented fungus. The fungal mycelia were separated into the cell wall and cytoplasmic fractions which were added to the soil, and the $^{14}CO_2$ evolved was determined. During the 6 month incubation, 71% of the carbon from the non-pigmented or hyaline fungal cell walls of *A. niger* was evolved as $^{14}CO_2$, compared to a value of 35% for the melanic fungus. The percentage decomposition of the cytoplasmic fractions was 62 and 48% for *A. niger* and the melanic fungus, respectively.

Wagner and Medynska-Rabinska (1971) studied the decomposition of the cell wall and cytoplasmic fractions of sclerotia from the melanic fungus *Macrophomina phaseoli.* They reported that following 84 days of incubation in a silt loam soil, the amount of ^{14}C recovered as $^{14}CO_2$ from the cell wall and cytoplasmic fractions of the sclerotia was 12 and 48%, respectively.

The authors concluded that cell walls of dark colored or melanic fungi are relatively resistant to decomposition; consequently, this fraction would be preferentially accumulated in soil organic matter. They suggested that partial degradation of the fungal cell wall was followed by recombination of the resistant components into polymers resembling humic material in soil organic matter.

The decomposition of two species of ^{14}C-labeled green algae, a unicellular *Chlorella* sp. and a filamentous *Oedogonium* sp., was reported by Wagner and Krzywicka (1975). After 46 days of incubation in a silt loam soil, the percentage of the added ^{14}C recovered as $^{14}CO_2$ for the *Oedogonium* sp. and the *Chlorella* sp. was 51 and 41%, respectively. By using pedoscope slides, the authors showed an increase in fungal development adjacent to the algae tissues which the authors suggested was responsible, at least in part, for decomposition of the algae.

The decomposition of six blue-green algae and one green alga in soil was reported by Verma and Martin (1976). They used ^{14}C-labeled whole cells, cell walls and cytoplasmic constituents and reported that after 22 weeks of incubation from 61 to 81% of the ^{14}C was collected as $^{14}CO_2$. When the cell-wall fractions of *Nostoc muscorum* and *Anabaena flos-aquae* were chemically combined into phenolic polymers, cell-wall decomposition was reduced from 70 to 40% and from 61 to 35%, respectively. *N. muscorum* and *A. flos-aquae* cytoplasm decomposition was reduced from 75 to 27% and from 76 to 18%, respectively, when combined with phenolic polymers. Also included in the study were the mycelia from a melanic fungus (*Stachybotrys chartarum*) and an unnamed hyaline fungus, cells of the bacterium *Chromobacterium violaceum*, and cells of the yeast *Hansenula holstii*. The ^{14}C collected as $^{14}CO_2$ following the 22 week incubation of the above listed materials was 49, 68, 71 and 72%, respectively. Although not directly comparable, the data from the two fungi confirm the observations of Hurst and Wagner (1969) about resistance of cell walls to decomposition.

In addition to cellular components of microorganisms, many fungi and actinomycetes synthesize dark-colored polymers known as melanins which contribute to soil humic acid formation (Martin and Haider, 1971). *Hendersonula toruloidea* was shown to produce some 40 different phenolic substances, most of which become constituents of the melanins (Martin *et al.*, 1972). Proteins, amino acids and various cellular components may be linked into the melanins during the polymerization reactions. The resulting polymer has been shown to be highly resistant to decomposition with only 5% of the carbon being evolved as CO_2 during a 30 week incubation study (Martin *et al.*, 1972). Other fungi and actinomycetes have been shown to produce melanins which are resistant to decomposition, with commonly reported decomposition values of 5–25% following 12 weeks of incubation.

3. *Pesticides*

The use of ^{14}C-labeled compounds in the study of pesticide degradation is one of the most important areas of research involving the use of radio-isotopes. To elucidate the degradation pathway and rate and to facilitate isolation of metabolites, labeled pesticide material is commonly used. Several publications are available which discuss the subject of pesticide degradation (e.g. Kaufman, 1974; Kearney and Kaufman, 1975). The example presented here will illustrate the use of radioisotopes in the study of pesticide degradation.

One of the first organic herbicides to be studied was 2,4-dichloro-phenoxyacetic acid, commonly known as 2,4-D. In addition to using isotopes to define the degradation pathway of 2,4-D (Fig. 4.5) numerous researchers have used ^{14}C-labeled 2,4-D to elucidate the influence of temperature, moisture, pH and concentration on the rate of 2,4-D degradation. Wilson and Cheng (1978) incubated a Naff silt loam at 25 °C and 22% moisture for 73 and 58 days for carboxyl- and ring-labeled 2,4-D, respectively. At the end

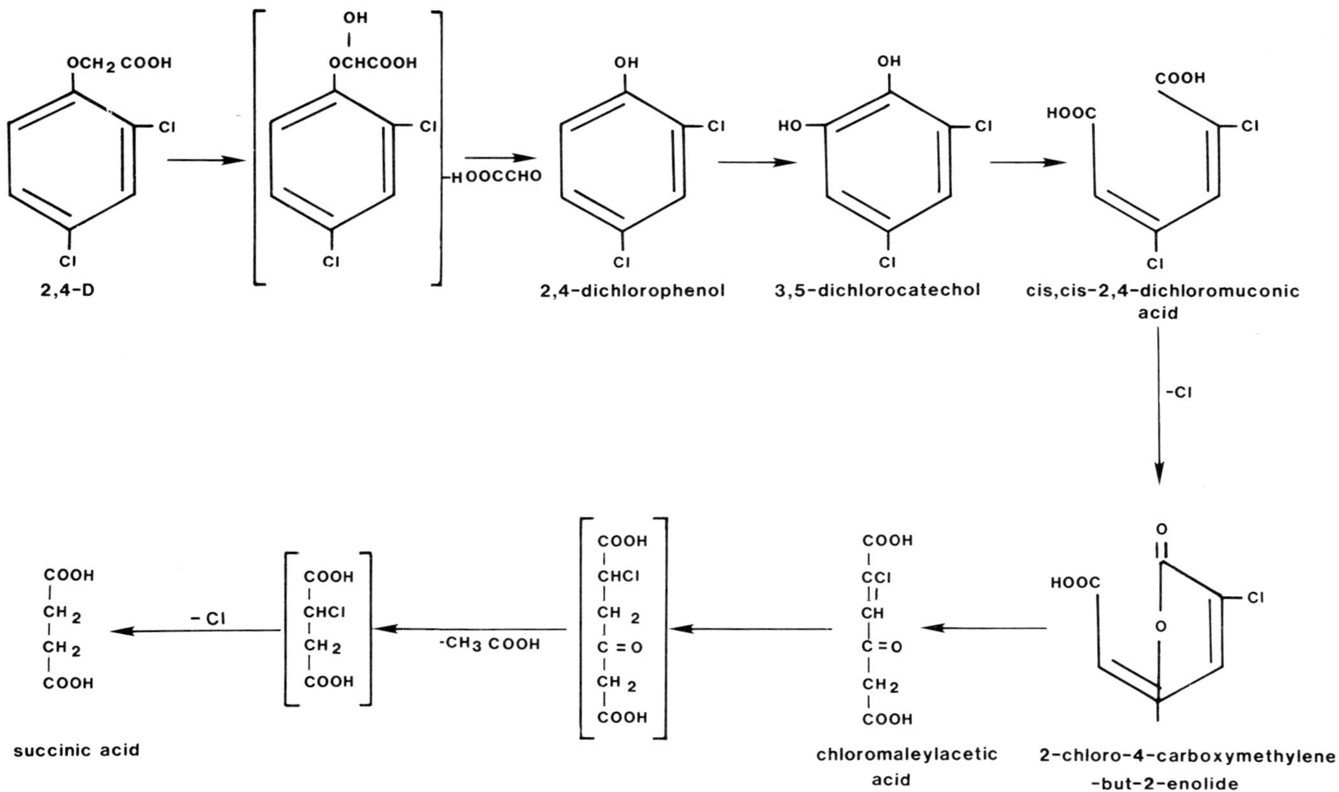

Fig. 4.5 Degradation pathway of the herbicide 2,4-D (Alexander, 1977).

of the incubation, 60–80% of the ^{14}C from labeled 2,4-D was recovered as $^{14}CO_2$, 19–31% was found as ^{14}C in the soil and described as degradation products or part of the soil organic matter or biomass, and approximately 1% was unaltered 2,4-D. The influence of soil pH on the degradation of carboxyl-labeled 2,4-D was reported by Moreale and Van Bladel (1980). After 1 month of incubation, 80–95% of the ^{14}C from labeled 2,4-D was collected as $^{14}CO_2$ in soils with a pH greater than 6, but less than 10% of the ^{14}C was collected as $^{14}CO_2$ in soils with a pH less than 5.6.

Limited research has indicated that the side chain carbon in 2,4-D was incorporated into the general microbial biomass, and ring carbon was incorporated into cellular products such as fungal melanins which were relatively resistant to decomposition (Wolf and Martin, 1976).

IV. Microbial biomass

In the decomposition of substrates by soil microorganisms, a large portion of the carbon is evolved as CO_2; however, a portion of the carbon is used by the microbes for the synthesis of new cells. Traditionally, it was difficult to obtain an estimate of the amount of carbon contained in the soil microbial biomass or the living portion of the organic matter. Jenkinson and Ladd (1981) recently discussed the measurement and turnover of microbial biomass in soil. They presented data and calculations which showed that as much as 20% of the carbon in ^{14}C-labeled glucose could be found in the soil microbial biomass after 3 months of incubation. With more complex substrates such as ryegrass and straw, the general observation was that, at the time of maximum microbial biomass, about 10% of the original substrate carbon was in the biomass.

It appears that the residual ^{14}C from complex substrates such as lignin and fungal melanins does not appear as biomass carbon to any great degree but exists in soil primarily as the humic acid portion of soil humus.

If an incubation study is conducted with a ^{14}C-labeled substrate, and the amount of $^{14}CO_2$ evolved or the amount of ^{14}C remaining in the soil is measured, it is not possible to determine the actual decomposition rate of the labeled substrate because of the ^{14}C incorporated into the biomass. Paul and Van Veen (1978) have discussed the problem of microbial biomass and efficiency as it relates to the dynamic nature of soil organic matter. They presented data from a number of sources which showed that if the carbon incorporated into biomass and metabolites is not considered, the decomposition rate could be underestimated by as much as a factor of 7.5 for an efficiency of 60% during short-term incubation studies.

Recently, data were presented which showed that as the concentration of

^{14}C-labeled glucose increased from 10 to 10 000 ppm, the percentage of residual ^{14}C recovered in microbial biomass after 6 months of incubation decreased (Kassim *et al.*, 1982). When expressed as a percentage of the applied ^{14}C, the amount of $^{14}CO_2$ evolved increased as the amount of glucose added increased. The same authors evaluated a number of ^{14}C-labeled substrates including amino acids, cellulose and straw to determine incorporation of substrate carbon into soil biomass. Following a 12 week incubation in which approximately 80% of the added ^{14}C was evolved as $^{14}CO_2$, some 20–40% of the carbon remaining in the soil was present as biomass when substrates such as acetate, pyruvate and several amino acids were tested (Kassim *et al.*, 1981).

Microbial biomass represents an active portion of the organic carbon in soil which heretofore has been difficult to evaluate, but recent data suggest that a substantial amount of carbon from labile compounds may be incorporated into the cells of the soil microbial population.

V. Interactions of plants and microbes

A. Rhizosphere

The presence of a plant root growing in the soil has a marked influence on the microbial population in the general area of the root. The soil area that is influenced by the root is known as the rhizosphere (Rovira and McDougall, 1967). Microbial numbers in the rhizosphere are generally much larger than numbers in soil devoid of plant roots. For soil bacteria the microbial numbers in the rhizosphere may be 100 times larger. One major reason for the increased microbial numbers in the rhizosphere is that the plant root provides an available supply of carbon in the form of sloughed-off cells and exuded soluble organic compounds.

Martens (1982) used an artificial root which consisted of a membrane filter to introduce three concentrations of synthetic root exudates containing ^{14}C-labeled glucose into soil. He then estimated the microbial biomass at distances of 0–52 mm from the artificial root, using the procedure of Jenkinson and Powlson (1976). The results indicated that as the concentration of glucose was increased, the increase in microbial biomass extended further away from the plant root. In general, the microbial biomass was highest in the soil from 0 to 2 mm away from the root and rapidly decreased with distance from the root.

Warembourg (1975) used ^{14}C techniques to separate root-respired $^{14}CO_2$ from $^{14}CO_2$ produced by the microbial utilization of root exudates. He grew wheat plants (*Triticum aestivum*) under axenic and nonaxenic conditions and exposed them briefly to a $^{14}CO_2$ atmosphere. Under axenic conditions, the evolution of $^{14}CO_2$ followed a bell-shaped curve. In the presence of rhizosphere microorganisms, there was a second peak of $^{14}CO_2$ evolution which was attributed to microbial decomposition of the root exudates. By exposing plants to $^{14}CO_2$ for short periods of time and using gel filtration techniques for isolated metabolites, Billes (1978) was able to distinguish root exudates from newly synthesized microbial metabolites. The differences were based on changes in molecular weight and specific activity of the recovered compounds.

The influence of the rhizosphere may be to reduce significantly the persistence of some pesticides in soil. Hsu and Bartha (1979) reported that mineralization of the ^{14}C-labeled insecticides diazinon and parathion was 5.0 and 7.8%, respectively, by soil without roots but increased to 12.9 and 17.9%, respectively, in rhizosphere soil.

From a practical viewpoint, laboratory studies involving plant root metabolism or uptake of labeled compounds must not neglect the microorganisms generally associated with the root. In a study using ^{14}C-labeled glucose, Zwarun (1972) showed that to obtain meaningful data from excised root experiments, the microbial population must be eliminated or reduced to an insignificant level. The presence of an active microbial population surrounding the plant root could also be important in studies of plant root uptake or metabolism of various pesticides and may be a point often neglected.

B. Mycorrhizae

The use of radioisotopes such as ^{32}P, ^{35}S and ^{65}Zn in soil microbiology research is much more limited than that of ^{14}C. Problems involving such characteristics as the much shorter half-lives and higher energies of radiation have resulted in less common utilization of these isotopes.

Much of the research using ^{32}P has dealt with vesicular-arbuscular mycorrhizae. This group of fungi infects plant roots and has been shown to be important in plant nutrition, especially of phosphorus under deficiency conditions. Research conducted in Gerdemann's laboratory provided evidence that the endomycorrhizal fungus, *Endogone mosseae*, enables plants to remove phosphate from a larger volume of soil and thus the plants exhibit greater phosphorus uptake (Hattingh *et al.*, 1973). The authors reported that the ability of onion (*Allium cepa*) roots to retrieve phosphate was increased due to the increased absorbing surface provided by the hyphae of the endomycorrhizal fungus. In subsequent work, Rhodes and Gerdemann (1978) used the radioactive isotope ^{35}S to demonstrate that onion roots infected with the endomycorrhizal fungus *G. fasciculatus* absorbed more ^{35}S per unit dry weight than non-mycorrhizal roots when the $^{35}SO_4^{2-}$ was injected into soil 3–6 cm from the roots.

Recent work has indicated that vesicular-arbuscular mycorrhizae are important in mediating nutrient transfer between plants. Leaves of an annual grass, *Plantago erecta*, were treated with ^{32}P and the results showed that ^{32}P could be detected in about 20% of the plants in close proximity to the treated plant (Chiariello *et al.*, 1982). The mycorrhizae were shown to connect the root systems of neighbors of different species and it was suggested that they may mediate nutrient transfers among the plants under conditions of low phosphorus fertility.

VI. Denitrification

Denitrification is generally defined as the biological reduction of NO_3^- or NO_2^- to gaseous end products, usually N_2O or N_2. This process can result in significant losses of combined nitrogen in agricultural systems, and the N_2O released to the atmosphere is also of concern due to its effect on the ozone layer.

True denitrification is limited to certain bacteria, chiefly the genera *Pseudomonas*, *Achromobacter*, *Bacillus* and *Micrococcus*, although other genera are involved to a lesser extent. *Pseudomonas* and *Achromobacter* are the dominant genera in soils. Growth of the microorganisms involved in denitrification is not dependent upon the reduction of NO_3^- since the bacteria are also active in proteolysis, ammonification and other nitrogen transformations. The presence of a large denitrifying population, therefore, does not necessarily mean that soil conditions are suitable for denitrification, but it does indicate that the potential for denitrification exists.

Quantitative measures of denitrification losses may be made directly by determining the amounts of N_2 and N_2O evolved, or indirectly by determining the nitrogen inputs and outputs during an experimental period, with any nitrogen unaccounted for being attributed to denitrification. Most of the data on denitrification losses have been obtained by indirect methods. In either case, isotopic tracers have made many types of denitrification experiments possible and have increased the precision of measurements.

A. Indirect measurement of nitrogen losses

It is possible to determine nitrogen gains and losses in the field in non-tracer experiments, but the results are generally imprecise (Allison, 1955, 1966). The early use of ^{15}N as a tracer in soil–plant research enabled researchers to determine nitrogen balances for the applied ^{15}N-labeled fertilizers and to observe any losses that might occur under different conditions. Since most of the early research was conducted in greenhouse pots, leaching and NH_3 volatilization were controlled, and losses were generally attributable to denitrification. For example, MacVicar *et al.* (1950) found that it was not possible to account for all of the ^{15}N-labeled fertilizer added, either as residual nitrogen in the soil or in the plant material. The deficits varied from about 1% at a low nitrogen level to about 10–15% at a higher nitrogen level on the same cropped soil. There were lower losses from a virgin soil, presumably because more of the added ^{15}N was immobilized in the higher organic matter soil. All of the observed losses were attributed to denitrification.

In a study on the effect of partial pressure of oxygen on soil nitrogen transformations, Broadbent and Stojanovic (1952) found that during a 10 day soil incubation period as much as 80% of the ^{15}N added as KNO_3 was lost when there was 0.1% oxygen, and 16–26% was lost at 19% oxygen content. At the end of the 10 day incubation period, as much as 6% of the ^{15}N was recovered as NH_4^+ and 30% as organic nitrogen. This was an early demonstration not only of the effect of oxygen on denitrification, but also that failure to recover the added ^{15}N as NO_3^- was not necessarily an indication of nitrogen loss.

MacRae *et al.* (1968) studied nitrogen transformations in six Philippine soils during 6 weeks of submergence after applying $K^{15}NO_3$. In two soils having the highest native organic matter content, the added ^{15}N was completely lost after 2 weeks. From 4.5 to 39% of the ^{15}N was recovered from the organic fraction of the four remaining soils after submergence for 6 weeks, and this result was attributed to assimilatory nitrate reduction.

In laboratory studies, Stanford *et al.* (1975) found that rates of NO_3^- disappearance under anaerobic conditions from six widely differing soils gradually increased and then became relatively constant when glucose was added in increments supplying 0.04 to 1.5 mg of carbon per gram of soil. Maximum rates of NO_3^- loss ranged from approximately 9.3 to 13.7% h^{-1}. Results based on ^{15}N-labeled $Ca(NO_3)_2$ showed that substantial amounts of NO_3^- were reduced to NH_4^+ after 4 h of anaerobic incubation, and $^{15}NH_4^+$ continued to increase until all or most of the $^{15}NO_3^-$ had been converted to NH_4^+, organic nitrogen and gaseous forms (N_2 and N_2O). These results indicated that dissimilatory NO_3^- reduction was taking place, as contrasted to assimilatory reduction by the soil microorganisms, which embodies not only the conversion of NO_3^- to N_2 and N_2O but also the reduction of NO_3^- to NH_4^+ under anaerobic conditions. The sequence of reductions may take the following pathways (Campbell and Lees, 1967):

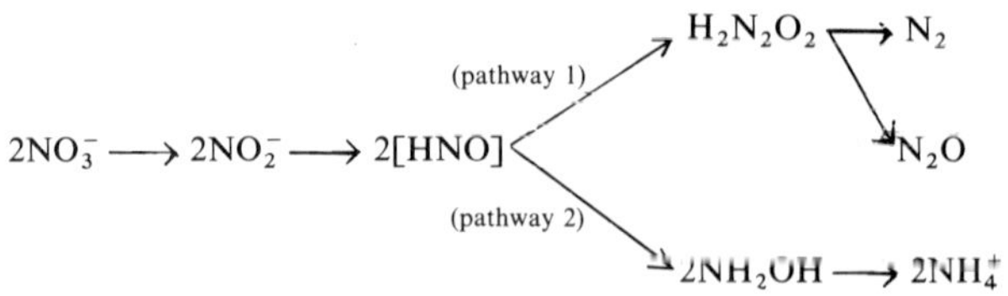

Although it has been known for a number of years that *Bacillus lichenformis* is capable of reducing NO_3^- through both pathways, the list of other organisms with this capability is still incomplete.

In order to detect dissimilatory reduction of NO_3^- to NH_4^+, one must consider (i) the presence of a sufficient population of the proper denitrifying microorganisms in the soil; (ii) a readily available carbon source, either from the indigenous soil organic matter or from an added source; (iii) measurements over relatively short time periods (hours to a few days); and (iv) use of ^{15}N-labeled NO_3^- to distinguish between NH_4^+ arising from indigenous organic nitrogen and that from the reduced NO_3^-. The NH_4^+ quite often is subsequently immobilized into organic forms and may go undetected in long-term experiments.

A large number of nitrogen balance experiments have been conducted in the field, using ^{15}N as a tracer, and some of the results have been reviewed recently by Legg and Meisinger (1982) and Hauck and Tanji (1982). To construct a nitrogen balance, it is necessary to make careful measurements of plant uptake, residual nitrogen in the soil and possible leaching losses. Efforts should also be made to control NH_3 volatilization from the soil. There is a possibility that nitrogen losses can occur from plant foliage (Stutte *et al.*, 1979), but generally this has not been taken into account in nitrogen balance studies. Possible foliar losses and imprecise measurements of residual nitrogen are the main influences on the precision of measurements of denitrification by difference. Carter *et al.* (1967) showed that sampling the entire volume of soils in cylinders containing ^{15}N was much superior to taking core samples. In larger microplots where it is not possible to sample the entire soil mass, extreme care must be used in soil sampling methods.

Although indirect measures of denitrification may suffer from lack of precision, they offer the best estimates of nitrogen losses under a variety of field conditions that are currently available. Data from ^{15}N experiments indicate a wide range in nitrogen losses, varying from 0 to 50% or greater. Hauck (1971) estimated that on the average about 15% of ^{15}N-labeled fertilizers is lost from the soil–plant system.

A good example of a field study of nitrogen balance is seen in a lysimeter experiment conducted by Chichester and Smith (1978). They applied 336 kg of nitrogen per hectare as $Ca(^{15}NO_3)_2$ to four monolith lysimeters in which corn was grown under mulch tillage and conventional tillage. In the second year of the experiment no nitrogen fertilizer was applied, but in the third year 178 kg ha^{-1} of unlabeled nitrogen was added. The ^{15}N contents of the runoff, percolate and crops were determined, and after 3 years the residual ^{15}N in the soil was estimated from core samples taken to 2.4 m. Unaccounted for losses of ^{15}N from the lysimeters ranged from 5 to 26%, depending on treatment, and averaged 15.5% for the 3 year period. One would expect that most of the ^{15}N loss by denitrification occurred during the year of ^{15}N-fertilizer application, although slow losses over long periods may also be quite important (Broadbent and Clark, 1965).

B. Direct measurement of nitrogen losses

Direct measurements of denitrification can be carried out in the laboratory with relatively simple systems. Nommik (1956), for example, used a small vacuum flask, containing soil and a tube of ascarite, that was connected directly to a mass spectrometer. A somewhat more complex system was constructed by Cady and Bartholomew (1960) with an internal pumping system to circulate the gaseous phase. Generally, the apparatus is designed for a specific purpose, and the initial atmospheric nitrogen may be evacuated, replaced with an inert gas such as argon, or left intact. If atmospheric nitrogen is left intact, the calculation must take into consideration that a nonequilibrium mixture of gaseous nitrogen will result (Hauck *et al.*, 1958; see also Chapter 6 in this volume).

Among the earliest denitrification experiments using ^{15}N as a tracer, Jones (1951) maintained 25 g soil, 10 ml water and 5 mg NO_3^-–N under vacuum, and the evolved gases were conducted through a Dumas apparatus to measure N_2 manometrically. In one case, 0.5 % sucrose was added to the soil, and over 80 % of the added nitrogen was lost in 3 days. In other incubations without sucrose, 2.1, 4.1 and 4.2 mg of nitrogen were evolved after 53, 125 and 190 h, respectively, from the 5 mg NO_3^-–N additions. With the use of $Ca(^{15}NO_3)_2$ in the system it was ascertained that essentially all of the evolved nitrogen arose from the labeled source.

Wijler and Delwiche (1954) conducted a more elaborate denitrification experiment in an enclosed laboratory system in which the formation of N_2O, N_2 and NO was followed quantitatively, using $K^{15}NO_3$ as a tracer. In their experiments, N_2O was found to be the major product of denitrification under most soil conditions. Total denitrification rates were quite constant above pH 6, but the proportions of N_2O and N_2 were pH-dependent. Although N_2O could be reduced readily to N_2 above pH 7, the reduction was strongly inhibited below the neutral point. The formation of NO became more prominent below pH 6, and the rate of denitrification decreased. Oxygen tensions as low as 5 mm Hg decreased denitrification to rates almost one-tenth of the anaerobic rates. The reduction of NO_3^- to NH_4^+ was generally small and occurred chiefly when excessive amounts of organic matter were added under anaerobic conditions. The isotope content of all denitrification products showed them to be wholly derived from NO_3^-, indicating that non-enzymatic denitrification (van Slyke reaction) was not taking place.

An extensive investigation of denitrification in several Scandinavian soils was carried out by Nommik (1956), using $K^{15}NO_3$ as the tracer material. This study again showed the pronounced effect of pH on the relative proportions of N_2O and N_2 produced during denitrification in soil incubated at 100% water saturation (Fig. 4.6). At low pH values (3.8–4.1) only N_2O was produced, but at high pH values (7.8–8.0) essentially all of the gaseous phase was N_2. At intermediate pH levels the proportion of N_2O gradually decreased, while that of N_2 increased, as a function of soil pH. It appeared evident that N_2O was being converted to N_2 during the 12 day incubation at

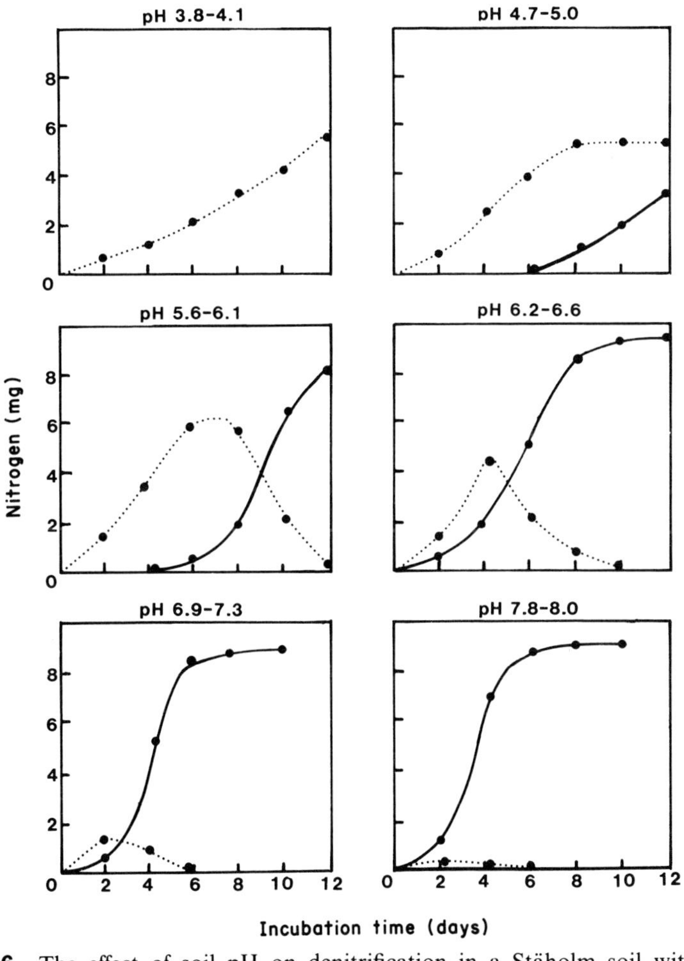

Fig. 4.6 The effect of soil pH on denitrification in a Stäholm soil with 100% water saturation and an initial nitrate content of 9.28 mg of nitrogen: ·····, N_2O; ———, N_2 (Nommik, 1956).

the intermediate pH levels. In an open system it would be expected that much of the N_2O would escape to the atmosphere before conversion to N_2 could take place.

Denitrification is markedly affected by temperature and proceeds very slowly below 10 °C (Wijler and Delwiche, 1954; Nommik, 1956). Nommik (1956) showed that the optimum temperature for the reaction in a Väring soil during a 2 day incubation was about 65–70 °C, as shown in Fig. 4.7. Rapid denitrification at elevated temperatures suggests an active thermophilic flora, and this suggestion is verified by the ease of obtaining enrichment cultures of thermophilic denitrifiers. Low temperature effects have significance since there is less possibility of denitrification in the colder part of the year when plants are not growing.

In early studies of denitrification in closed systems, the mixture of evolved gases was analyzed by mass spectrometry, and appropriate measurements and corrections were made to account for interfering fragment ions from such substances as CO_2 and N_2O. In later work, the interfering substances were separated from the sample before introduction into the mass spectrometer (Hauck *et al.*, 1958). As instrumentation developed, infra-red analysis was used in some cases for the identification of N_2O. With the advent of gas chromatography, rapid, unequivocal identification and quantification of gaseous components of soil became possible. Combined gas

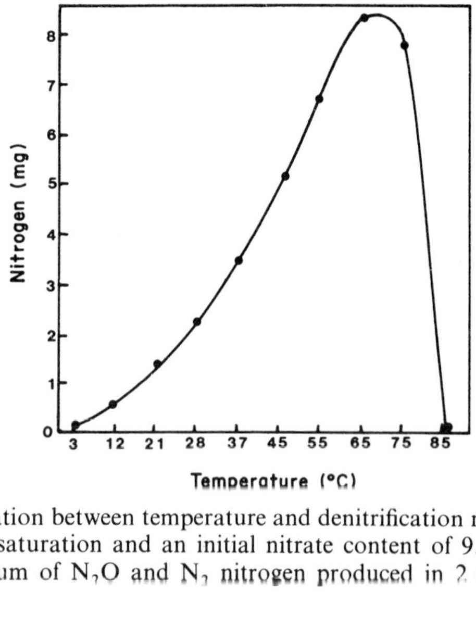

Fig. 4.7 The relation between temperature and denitrification rate in a Väring soil with 100% water saturation and an initial nitrate content of 9.21 mg of nitrogen, expressed as the sum of N_2O and N_2 nitrogen produced in 2 days of incubation (Nommik, 1956).

chromatography–mass spectrometry allows separation of gases before introduction to the mass spectrometer, and also enables one to discriminate between two or more compounds that have similar retention times by monitoring only the ion voltage produced by molecular ions of the desired compound. An excellent review of methods for analysis of denitrification has been provided by Focht (1978).

Most of the studies of denitrification by direct means have been carried out in the laboratory as in the examples given above. Other closed systems for use in the field have been described (e.g. McGarity and Rajaratnam, 1973), but they are generally more complex and require considerable effort to maintain uniform conditions. More recently, direct methods of measuring N_2 and N_2O gas fluxes from denitrification of ^{15}N-labeled nitrate in open field systems have been investigated by California workers (Rolston *et al.*, 1976; Rolston and Broadbent, 1977; Rolston, 1978). In a comparison of indirect and direct measures of denitrification, using gaseous diffusion theory, Rolston *et al.* (1976) applied 300 kg of nitrogen per hectare in the form of ^{15}N-labeled NO_3^- to soil columns in the laboratory and a field plot that were instrumented with tensiometers, soil solution samplers and gas samplers. Gaseous concentration gradients and measured diffusion coefficients were used to calculate the fluxes of $^{15}N_2$ and $^{15}N_2O$ from the soil. Residual soil nitrogen, plant uptake and leaching were measured to determine denitrification by difference (indirect). There was relatively good agreement between the indirect and direct methods for denitrification in the soil columns; however, there was less agreement between the methods in the field plot. This disparity was attributed to the uncertainty in the gaseous concentration near the soil surface and the spatial variability of the apparent diffusion coefficient near water saturation.

In a later study, Rolston *et al.* (1979) maintained field plots at different water, carbon and temperature levels. The flux of gases at the soil surface was measured from the accumulation of N_2 and N_2O beneath air-tight covers placed over the soil surface for 1 or 2 h per day. Denitrification measured directly was generally 0–30 kg of nitrogen per hectare less than that determined by difference for the water treatments, and 12–65 kg of nitrogen per hectare less for the drier treatments. In the latter case the difference between the two methods was attributed to the inability to continuously monitor and measure very small gas fluxes of $^{15}N_2$ over relatively long periods. The application of gaseous diffusion theory to the measurement of denitrification appears promising, and further refinements in methodology should provide a precise means of measuring denitrification losses directly. Presently, however, the indirect method of nitrogen balance in field systems predominates, especially where leaching losses are negligible or can be measured accurately (Legg and Meisinger, 1982).

C. The use of ^{13}N in denitrification studies

Use of the radioactive isotope, ^{13}N, with a high specific activity (1.88×10^{10} Ci per gram-atom) provides extremely high sensitivity during short-term experiments, and allows the direct measurement of N_2 losses in any atmosphere. In denitrification studies with ^{13}N, the NO_3^- level in the system is not significantly changed, and rates of denitrification in freshly collected samples can be measured before changes in enzyme concentrations can occur. The utility of ^{13}N for denitrification studies is enhanced by the direct production of $^{13}NO_3^-$ with H_2O as a target. Gersberg *et al.* (1976), for example, obtained 99.6% of the radiochemical form as $^{13}NO_3^-$ using this method, and found no measurable $^{13}NO_2^-$ or $^{13}NH_4^+$.

The short half-life (10 min) of ^{13}N limits its experimental use to laboratories located adjacent to production facilities and to short-term experiments of 1–2 h. For these reasons, few denitrification studies with ^{13}N have been reported (e.g. Gersberg *et al.*, 1976; Tiedje *et al.*, 1979), and those deal largely with methods for the production of ^{13}N and the apparatus required for the measurement of denitrification rates. The methodology that has been developed should enhance the possibilities of determining transient events in denitrification such as response to changes in soil moisture, O_2 content or carbon energy sources.

VII. Mineralization and immobilization of nitrogen

In soils a continuous cycling of nitrogen takes place between the organic and inorganic phases, as illustrated in Fig. 4.8. Much of the older soil organic nitrogen is resistant to decomposition by microorganisms; consequently, the most active interchange occurs after additions of fresh organic matter. General reviews of mineralization–immobilization reactions in soils (e.g. Bartholomew, 1965; Allison, 1966; Paul and Juma, 1981; Jansson and Persson, 1982) cover much of the pertinent literature in which isotopic nitrogen has been used. In this section an attempt will be made to summarize some of the major advances concerning mineralization–immobilization of nitrogen in soils obtained by the use of tracer technology.

Jansson and Persson (1982) used the abbreviation "MIT" for mineralization–immobilization turnover. For simplicity, the same terminology will be used in the following discussion.

A. Energy in relation to MIT

Rapid MIT requires a supply of readily utilizable carbonaceous material as an energy source for the soil microorganisms. Jansson (1958) in his early

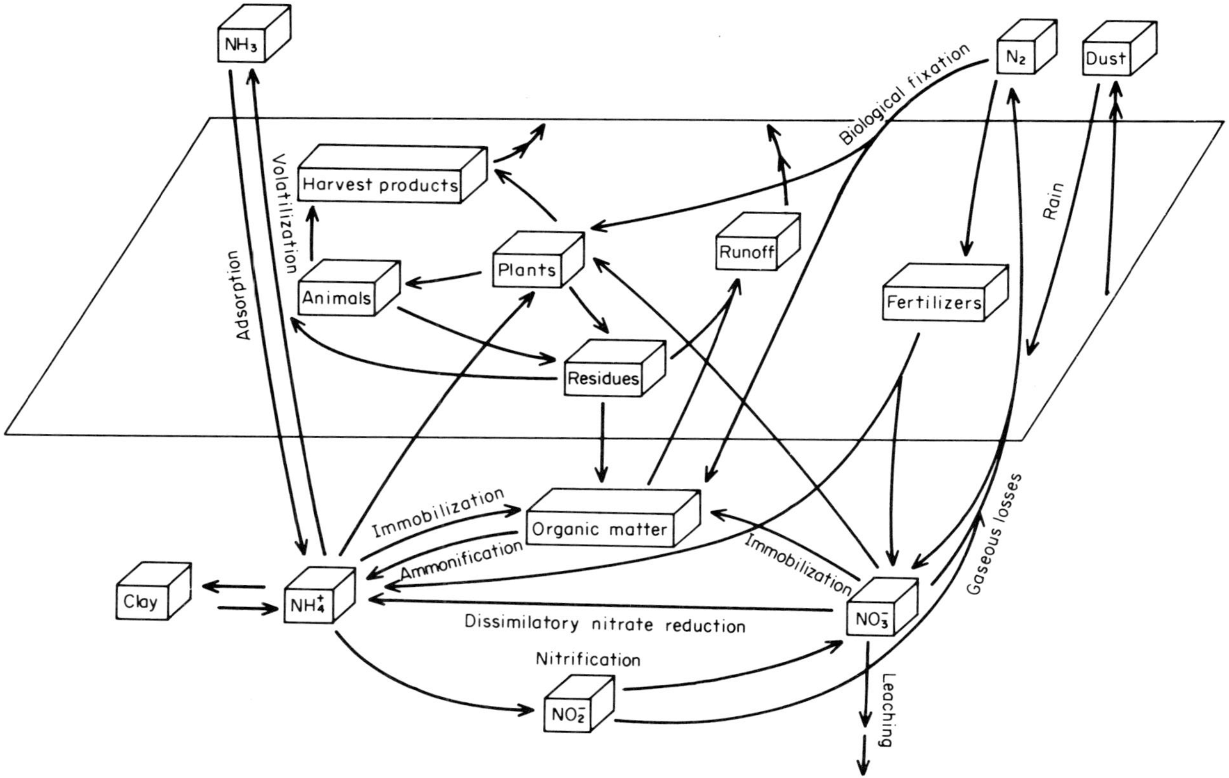

Fig. 4.8 Schematic representation of nitrogen transformations in a general agricultural nitrogen cycle (Legg and Meisinger, 1982).

studies of nitrogen transformations of soils, using ^{15}N as a tracer, emphasized the concept of energy supply as the driving and directing force in soil nitrogen metabolism. As noted above, the native soil organic matter is relatively resistant to decomposition, and microorganisms depend greatly upon fresh additions of carbonaceous materials for their energy requirements. Mineralization and immobilization occur simultaneously, and in unlabeled systems measures of either reaction are essentially net differences between the two reactions. When a state of equilibrium exists, little change will be observed in the inorganic nitrogen content, although microbiological activity continues. The microbial activity, and thus the relative energy state, can be ascertained by incubating the soil with labeled NH_4^+ or NO_3^- and determining the amounts immobilized or remaining in inorganic forms. For example, Broadbent (1965) found that about 10% of an addition of 200 ppm N in the form of $^{15}NH_4^+$ was immobilized in a soil after 30 days, while only 1% of a similar addition of $^{15}NO_3^-$ was immobilized. These results demonstrate that there was sufficient available carbon in the soil for immobilization to proceed, that NH_4^+ was preferentially utilized and that NO_3^- remained almost completely in a passive pool as discussed by Jansson (1958). If the energy level is high enough, however, NO_3^- as well as NH_4^+ will be utilized in the immobilization process (Broadbent, 1966).

B. Influence of plants on MIT

In soils with low available energy levels and in which added NO_3^- remains in an inorganic form in the absence of plants, one might expect that plants would be able to absorb all of the NO_3^- to the extent of its absorption capacity. Numerous experiments, with and without ^{15}N-labeled fertilizer, have shown that this total absorption does not occur, and about 20% of the added nitrogen is commonly immobilized in organic forms. Working with 12 soils fertilized with 0, 50, 100 and 200 ppm N as $Na^{15}NO_3$, Legg and Allison (1967) and Legg and Stanford (1967) found that plant recovery in greenhouse tests ranged from 46 to 76%, averaging 64% for all soils and treatments. With exhaustive cropping by two succeeding crops, only an additional 6.6 and 1.4% of the ^{15}N-labeled fertilizer was removed. After removal of roots, containing an average of 1.7% of the ^{15}N, 21% of the ^{15}N applied remained in the soil in organic forms. The low residual value of the ^{15}N additions is indicative of the low mineralization rates of nitrogen recently incorporated into soil organic nitrogen.

In the above studies, 98% or more of the $^{15}NO_3^-$ added to the 12 soils was recovered as inorganic nitrogen during a 40 day incubation test. The amount of ^{15}N immobilized during cropping results from the energy material supplied by the root system and the large number of microorganisms in the rhizosphere. The energy material is derived chiefly from the ever-changing root hair tissues, sloughed off root material and root excretions. The amount of nitrogen so immobilized varies widely in relation to such influences as crop species, stage of growth, soil environment and nutritional conditions. The amount of added ^{15}N immobilized has ranged as high as 60% (Allison, 1966).

C. Effect of nitrogen additions on soil nitrogen mineralization and nitrogen utilization by plants

Mineralization of soil organic nitrogen during the growing season usually supplies a major portion of the nitrogen needs of crops. In early work with ^{15}N-labeled fertilizers, it was found that increased uptake of soil nitrogen occurred with increased ^{15}N fertilizer rates (e.g. Legg and Allison, 1960). Since the tests were conducted in greenhouse pots, the volume of soil exploited by the roots was the same for all nitrogen treatments.

Some workers have considered the effect of nitrogen rates on increased nitrogen uptake by plants as a "priming" effect of the fertilizer nitrogen on soil nitrogen mineralization (e.g. Hauck and Bremner, 1976). Allison (1966), however, provided calculations indicating that soil nitrogen mineralization remained essentially constant with ^{15}N-labeled nitrogen additions ranging from 37.5 to 600 mg of nitrogen per pot in which more than threefold increases in soil nitrogen uptake occurred. Legg and Allison (1960) attributed the observed differences to greater immobilization of soil nitrogen at lower fertilizer nitrogen levels than at higher nitrogen levels. It is apparent that one of the main effects of increasing nitrogen additions is a reduction in the C:N ratios of the plant material produced. This, in turn, decreases the need of additional nitrogen by microorganisms decomposing the root residues and excretions. As plant growth proceeds, the supply of mineral fertilizer nitrogen in the soil is depleted, and any additional nitrogen needed by rhizosphere microorganisms must be obtained largely from mineralized soil nitrogen.

The increased plant uptake of soil nitrogen induced by fertilizer nitrogen applications is mainly of academic interest when it is considered that the additional uptake is usually more than balanced by fertilizer nitrogen immobilization. The immobilization process is important in nitrogen conservation, particularly in soils with low mineralization capacities. Legg and Stanford (1967) showed that six soils, having low mineralization rates, immobilized more fertilizer nitrogen at the higher rates than was removed by plants from soil sources. Six other soils, having high mineralization rates, showed net losses of nitrogen, since the uptake of mineralized soil nitrogen exceeded the amounts of fertilizer nitrogen immobilized.

D. Stability of immobilized ^{15}N-labeled nitrogen in soils

Since significant quantities of fertilizer nitrogen are usually immobilized in the soil, it is important to know the availability of the recently immobilized nitrogen to subsequent crops. The composition of recently incorporated soil organic nitrogen has not been clearly elucidated, but tracer data have provided some important comparisons between indigenous soil nitrogen and organically incorporated labeled nitrogen.

In early mineralization studies of recently incorporated ^{15}N, Jansson (1958) concluded that the active phase of soil organic nitrogen including the biomass, accounted for about 10–15% of the total nitrogen of mineral soils. The remainder was considered a biologically passive organic nitrogen pool undergoing slow interchange with the active phase. To estimate the active phase using tracer data, the general equation for calculating quantities involved in isotopic equilibria was employed:

$$A = B(1 - y)/y, \tag{4.1}$$

where A is the amount of organic ^{15}N in the active phase, B is the amount of newly immobilized labeled nitrogen in the organic state, and y is the net mineralization outflow of labeled nitrogen.

In long-term experiments, increasing A values are observed. This appears to be a reflection of the stabilization of the labeled nitrogen and its equilibration with increasing amounts of soil nitrogen in a passive form.

Broadbent and Nakashima (1967) studied the reversion of labeled inorganic nitrogen to soil organic nitrogen forms and subsequent re-mineralization over periods of about 500–600 days in laboratory and greenhouse experiments. Straw was incorporated in the soil to intensify immobilization in some cases. They calculated a quantity designated as "availability ratio" according to the equation

$$\text{Availability ratio} = \frac{Tc/(Uc + Tc)}{Ts/(Us + Ts)}, \tag{4.2}$$

where Tc represents the tagged nitrogen in plant material (or in mineralized nitrogen from incubation), Uc is the untagged nitrogen in the corresponding material, Ts is the tagged nitrogen in soil at the beginning of the vegetative or incubation period, and Us is the untagged nitrogen in soil at the same time. In this equation the availability ratio will be equal to unity if the labeled nitrogen in the soil has the same availability to plants or microorganisms as the soil nitrogen. If the relative availability of the labeled nitrogen is greater than that of the soil nitrogen, the ratio will be greater than one. Broadbent and Nakashima (1967) found that in successive cuttings of sudangrass (*Sorghum sudanense*) availability ratios ranged from 11.9 in the second cutting to 1.2 in the seventh cutting of plants grown on Yolo loam. In exhaustive incubation tests the availability ratios were usually between 1 and 2 during the latter part of the experiment.

Similar studies were carried out by Legg *et al.* (1971), in which labeled nitrogen was incorporated into the soil organic matter by successive cropping with oats in the greenhouse. The plant material was returned to the soil, except for pots that were removed from the experiment at the end of each cropping period. Plant material and soil from each set of triplicate pots were analyzed for total and ^{15}N-labeled nitrogen. Availability ratios declined from an initial value of 10.4 to about 2 at the end of the plant growth experiment. Soil samples obtained after two, five and eight cropping periods were incubated at 35 °C and leached with 0.01 M $CaCl_2$ (followed by –N solution) biweekly for 40 weeks. The availability ratios, calculated for the mineralized nitrogen at each incubation period, gradually declined for the second-crop soil until it was about 2; thereafter it remained essentially constant. For the fifth- and eighth-crop soils, little change was noted in availability ratios, which remained about 2 throughout incubation. The data suggest that inorganic nitrogen applied to soils may be converted rather quickly to stable organic forms similar to indigenous organic nitrogen. They speculated that with a soil having a stable availability ratio of 2, one-half of the indigenous soil organic nitrogen is essentially inert, owing to physical or biochemical inaccessibility to microorganisms.

Stanford *et al.* (1970) subjected samples of the fifth-crop soil mentioned above to repeated extractions with boiling 0.01 M $CaCl_2$ (100 °C), or autoclaving (121 °C) or both, to study the behavior of the immobilized ^{15}N relative to soil organic nitrogen. They found the distribution of ^{15}N among chemical fractions of the hot water extracts, acid hydrolyzate and acid-insoluble portion was similar to the distribution of indigenous soil nitrogen. Since appreciable equilibration between added ^{15}N and the more resistant forms of soil nitrogen appeared unlikely, it was postulated that the fertilizer nitrogen reacted to form new compounds, similar in composition to pre-existing forms of soil nitrogen. This postulate was supported by the observation that the percentage removals of total and ^{15}N-labeled nitrogen by repeated extraction were almost identical.

Using the same soil but different fractionation and extraction procedures, Chichester (1970) found further support for the above postulate. He initially divided the soil components into separate size fractions by wet sieving and centrifugation without using chemical dispersants. The resulting silt-size separate was then subjected to repeated 15 min ultrasonic vibration treatments, followed by centrifugation to separate the clay-size material that resulted from the physical dispersion. The separates were extracted with 0.5 N sodium pyrophosphate at 100 °C, and the extracts were fractionated into distillable, acid-hydrolyzable, ninhydrin-reactive and residual nitrogen components. The results obtained in this detailed study gave further evidence of the similarity in composition of the labeled and unlabeled fractions for the different treatments. It appears highly significant that biological, chemical and physical methods lead to the same conclusion concerning the fate of immobilized fertilizer nitrogen in soil, and it is likely that findings of this nature will lead to a greater understanding of the basic mineralization–immobilization processes.

VIII. Biological nitrogen fixation

Knowledge about nitrogen fixation processes in agricultural systems has expanded in recent years, and interest in biological nitrogen fixation has been stimulated by increased costs of energy and nitrogen fertilizer. In the long term, genetic engineering may be the best means of improving nitrogen fixation; the possibilities are intriguing. In the meantime, however, considerable effort is being expended on the selection and breeding of rhizobia–legume combinations for increased nitrogen fixation. Although nodulated legumes are the primary nitrogen-fixing plants in agriculture, plant–algal associations are highly important in rice (*Oryza sativa*) culture, and most of the nitrogen required for woodlands and water habitats is provided by a variety of nonleguminous nitrogen-fixing associations and free-living microorganisms.

The development and use of various methods for studying nitrogen fixation have been reviewed in detail (Bergersen, 1980). In earlier work, gains in nitrogen content of the system under study was the primary method; however, this technique was inadequate where the amounts of nitrogen fixed were relatively small, and the method allowed little insight into the fixation process.

The stable isotopes, ^{15}N and ^{14}N (^{15}N-depleted nitrogen), and the radioactive isotope, ^{13}N, have all been used to some extent in nitrogen fixation studies. As isotopic nitrogen first became available for research, it was possible to conduct nitrogen fixation experiments in which ^{13}N or ^{15}N was contained in an enclosed atmosphere. This provided a direct sensitive method of detection in which the fixed nitrogen could be detected as an enrichment of the total nitrogen or chemical constituents of the system or in specific components of the biomass.

A. Use of ^{15}N in enclosed atmospheres to determine nitrogen fixation

Early nitrogen fixation work with the stable isotope ^{15}N was pioneered at the University of Wisconsin (Burris and Miller, 1941; Burris, 1942; Burris *et al.*, 1943) and involved free-living organisms and excised nodulated roots which were rather easily maintained in small enclosed atmospheres. These early studies have led to a greater understanding of nitrogen fixation processes, especially with roots and microorganisms.

The ^{15}N tracer studies of nitrogen fixation involved exposure of the nitrogen-fixing system to $^{15}N_2$ in gas-tight containers. Bergersen (1980) and Burris (1972, 1974) have given detailed descriptions of the apparatus required to generate and purify gaseous ^{15}N, and the techniques involved in providing the nitrogen-fixing system with a ^{15}N atmosphere. Although ^{15}N can be purchased in the gaseous form, it is usually more economical to generate $^{15}N_2$ in the laboratory when a number of experiments are involved. In general, the gas is obtained by treating a labeled ammonium salt with 40% NaOH in an evacuated system in which the NH_3 released upon heating can be circulated over hot CuO (600 °C) by means of a Toepler pump. The gas is finally circulated through a liquid nitrogen trap to remove any remaining impurities and then pumped to a storage container.

Other closed atmosphere systems used to determine nitrogen fixation have included Warburg vessels, vacuum desiccators, gas lysimeters and specially constructed systems to fit the particular needs (e.g. Delwiche and Wijler, 1956; Ross *et al.*, 1968; Burris, 1974; Bergersen, 1980). One consideration for small systems is that the gas space must be sufficient to contain adequate O_2 for aerobic systems.

Closed atmosphere systems have also been used in evaluations of non-symbiotic fixation. Many microorganisms reportedly are capable of fixing small amounts of nitrogen, and the closed ^{15}N atmosphere system provides a very sensitive, direct means of confirming or refuting such reports. Sensitivity is increased by increasing the atom % of ^{15}N in the system when fixation is very low. In most cases, the closed ^{15}N atmosphere system is the best method of studying fixation mechanisms, intermediate products, fixation in cell-free preparations, influence of inorganic nitrogen on fixation, and the effect of various nutrient and environmental conditions on fixation. It is especially useful for short-term experiments (hours or days), but may also be used for periods of several weeks if the air-tight system can be maintained successfully with the desired atmosphere and growth conditions.

Although most nitrogen fixation studies with enclosed atmosphere have been concerned with free-living microorganisms or excised nodules, higher plants have been included in such studies to a certain extent. It is difficult to maintain plants in a confined atmosphere for long periods, but several early nitrogen fixation studies with legumes were conducted chiefly to determine the influence of ^{15}N-labeled fertilizer on nitrogen uptake and the effect of the fertilizer on symbiotic nitrogen fixation (e.g. Norman and Krampitz, 1945; Thornton, 1947). Ross *et al.* (1968) proposed the application of gas lysimeters to nitrogen-fixing systems, while Loginov (1966) grew lupins (*Lupinus* spp.) with only the roots and lower stems enclosed in the ^{15}N atmosphere. In Loginov's study, a special sealing compound was used around the stems of the plants and the top of the container to allow the tops of the plants to grow freely in the natural atmosphere and, thus, reduce problems with temperature and humidity control.

B. Use of ^{15}N in isotope dilution studies

The closed atmosphere system is impractical for field measurements which are required to determine improvements in nitrogen fixation in agricultural practice. An alternative method for determining nitrogen fixation depends on labeling the substrate (soil) and measuring nitrogen fixation by dilution of the labeled nitrogen from the soil by the fixed nonlabeled atmospheric nitrogen. The feasibility of measuring nitrogen fixation by isotopic dilution methodology was demonstrated in the greenhouse by McAuliffe *et al.* (1958). In their studies, fixing and nonfixing plants were grown in association. Fixation was determined from differences in the ^{15}N composition of the respective plants. In the past decade interest has been revived in this method. This revival in interest was stimulated by increased costs of nitrogen fertilizer, and lower costs of isotopic nitrogen that made field-scale studies possible.

The isotope dilution method, as it is generally used, requires that both nonfixing and fixing plants be grown on the labeled soil. The nonfixing plant provides a measure of the atom % of ^{15}N derived from the soil, and, using the atom % of ^{15}N in the fixing plant, the fraction of total nitrogen in the fixing plant derived from the atmosphere by fixation is calculated as follows:

$$N_f = 1 - \frac{A}{B}, \tag{4.3}$$

where N_f is the fraction of nitrogen fixed, A is the atom % of ^{15}N excess in fixing plants, and B is the atom % of ^{15}N excess in nonfixing plants. In this method there is one basic assumption: the atom % of ^{15}N derived from the soil is similar for both the fixing and nonfixing plants.

Labeled nitrogen may be incorporated into the soil in either inorganic or organic forms. If the inorganic form is used, there is a possibility that it will affect the amount of nitrogen fixed, but this can be minimized by using low amounts of total nitrogen with a high atom % ^{15}N content. Legg and Sloger (1975) proposed a method in which inorganic ^{15}N was incorporated in the soil organic matter of microplots through the addition of sucrose plus cellulose together with inorganic ^{15}N; an incubation period followed before plants were grown. The method was based on evidence that inorganic ^{15}N-labeled compounds can be incorporated into stable organic forms similar in nature to indigenous soil nitrogen in a relatively short time (Stanford *et al.*, 1970; Legg *et al.*, 1971; Chichester *et al.*, 1975). The ^{15}N can be incorporated by growing one or more crops after application and incorporating the plant material, by incorporating labeled plant material from other experiments, or by incubating inorganic ^{15}N with carbonaceous material as indicated above. Main advantages are the low level of inorganic nitrogen in the soil and the fact that the labeled soil can be used for several years for fixation studies.

The inorganic system of labeling the soil has the advantage of simplicity of application, and the ^{15}N can be applied directly to existing plants (e.g. legume–grass combinations in pastures and meadows). When it is applied directly to plants, the ^{15}N should be washed from the foliage into the soil (Edmeades and Goh, 1979). Fried and Broeshart (1975) introduced the *A* value concept into calculations of nitrogen fixation, but this was later modified (Fried and Middelboe, 1977). Theoretically, the *A* value concept allows fertilization of non-fixing plants at a higher nitrogen rate than fixing plants to reduce nitrogen deficiency conditions. Under field conditions, however, the *A* value often does not remain constant at different nitrogen rates, and this can lead to inflated values for nitrogen fixation. For example, Vasilas (1981) found that in his studies with soybeans, nitrogen fixation values determined by *A* values exceeded those obtained by the standard isotope dilution equation by an average of 23 %. In the same study, nitrogen fixation by isotope dilution and by difference were in close agreement. Fixation by difference was determined by subtracting the total nitrogen in non-nodulating soybeans from the total nitrogen in the nodulating plants. In other comparisons of nitrogen fixation involving nodulated and non-nodulated soybeans, there has generally been good agreement between values obtained by the isotope dilution and difference methods (e.g. Legg and Sloger, 1975; Talbott, 1980).

For best results with the isotope dilution system, it is necessary to label field plots as uniformly as possible. This can be accomplished in microplots with thorough mixing. In large-scale field plots lack of uniform labeling may result from variability in soil nitrogen availability over the plot and/or failure to apply the ^{15}N uniformly. Precision of measurements is improved by growing the paired fixing and non-fixing plants in the same subplot area (Talbott, 1980).

Use of ^{15}N isotope dilution methodology in field tests has resulted in improved estimates of nitrogen fixation under a variety of conditions (Vallis *et al.*, 1967, 1977; Henzell *et al.*, 1968; Legg and Sloger, 1975; Goh *et al.*, 1978; Edmeades and Goh, 1979; Vasilas, 1981; Broadbent *et al.*, 1982; Wagner and Zapata, 1982). In recent years the trend has been to use this method for the selection and testing of improved legume cultivars with high nitrogen fixation capability (e.g. Talbott, 1980; Heichel *et al.*, 1981; Stangarone, 1981; Rennie *et al.*, 1982). Continued work along these lines should help to determine the role of biological nitrogen fixation in agricultural practice.

C. Use of ^{15}N-depleted nitrogen in fixation studies

Since ^{15}N-depleted nitrogen contains about 99.99% ^{14}N, it is equivalent to about 0.35 atom% ^{15}N excess (Hauck, 1973). Although it is less expensive than ^{15}N, the ^{15}N-depleted materials can be used only where the dilution is not excessive. Little work has been conducted to test the possible use of ^{15}N-depleted nitrogen as a tracer in nitrogen fixation since the dilution can be large. In studies of nitrogen fixation by isotope dilution methods, it is difficult to add sufficient ^{15}N-depleted nitrogen in an inorganic form to the soil for adequate detection without influencing the fixation process. In soils that are not excessively high in indigenous nitrogen, however, ^{15}N-depleted materials can be incorporated in organic forms in sufficient quantity to obtain precise measurements, comparable to those obtained with ^{15}N (Legg, 1978). Where plant material is available from ^{15}N-depleted fertilizer experiments, such material can be used very effectively in nitrogen fixation studies, or the inorganic material may be incorporated into organic forms as described in the previous section. Broadbent *et al.* (1982) used a field soil which had been fertilized with ^{15}N-depleted $(NH_4)_2SO_4$ for five consecutive years during a previous cropping experiment with corn to obtain estimates of nitrogen fixation. Use of ^{15}N-depleted nitrogen in fixation studies depends on a careful balancing of dilution factors from both the available soil nitrogen and the expected atmospheric nitrogen fixation with the amounts of tracer that can be incorporated organically without increasing mineralization rates to a point that adversely affects fixation. Analytical precision of measurements and sample variation must also be achieved.

D. Use of ^{15}N natural abundance to study nitrogen fixation

Small variations in the nitrogen isotope compositions of natural substances occur over time during biological and chemical reactions involving nitrogenous substances. This generally leads to a small positive ^{15}N enrichment of the organic nitrogen of surface soils that may provide a very weakly labeled substrate for nitrogen fixation studies. The variations in natural abundance are usually within ten $\Delta^{15}N$ units of atmospheric nitrogen (one $\Delta^{15}N$ unit $= 0.000\,37$ atom % ^{15}N). This low level of enrichment, high dilution by nitrogen fixation and the spatial variability in measurements of the soil's natural abundance reduce the probability of attaining nitrogen fixation values that are comparable to those attained with more highly enriched systems. Problems in interpretation of data obtained by this method have been discussed by Hauck (1973), Hauck and Bremner (1976) and Bremner (1977). With the lower cost of ^{15}N-enriched materials and the recent developments in ^{15}N tracer techniques in field studies of nitrogen fixation, it is difficult to justify studies involving natural abundance measurements.

E. Nitrogen fixation methods using ^{13}N

A study involving radioactive isotope ^{13}N (half-life 10 min) was reported as early as 1940 by Ruben *et al.*, but this isotope has not been used extensively. Use of this short-lived radioactive isotope in nitrogen fixation studies has been limited since ^{13}N must be produced at the time and site of the experiment; thus, experiments must necessarily be short, and expensive, sophisticated equipment is required for the preparation and purification of ^{13}N gas for fixation research. It is produced by a variety of nuclear reactions using a proton or deuteron beam from cyclotrons or higher energy Van de Graaff accelerators. Tiedje *et al.* (1979) have summarized the most useful reactions, and have given the techniques involved in the production and use of ^{13}N obtained from irradiations carried out with proton beams from a sector-focused cyclotron. The production and use of ^{13}N in biological experiments have been reviewed by Straatman (1977).

Although the short half-life of ^{13}N allows an experimental time of only 1–2 h, the high specific activity (1.88×10^{10} Ci per gram-atom) provides extremely high sensitivity. With optimal conditions the minimum detectable amount of ^{13}N is less than 3.3×10^{-21} mol, compared to 7×10^{-9} mol for ^{15}N at 99 atom % excess (Tiedje *et al.*, 1979).

Ruben *et al.* (1940) reported the use of ^{13}N in nitrogen fixation studies by nonleguminous plants even before Burris and Miller (1941) described their studies of nitrogen fixation by the bacterium *Azotobacter* sp., using ^{15}N. Since 1940 there have been few published reports of nitrogen fixation involving ^{13}N. Nicholas *et al.* (1961) used ^{13}N to study nitrogen fixation by bacterial cells and cell extracts, and Campbell *et al.* (1967) used ^{13}N to assess the nitrogen-fixing potential of microorganisms from sub-Arctic soil. The products of fixation in blue-green algae have been traced with ^{13}N by Wolk *et al.* (1976) and Meeks *et al.* (1978), who also described rapid, high-voltage electrophoretic and chromatographic methods, using thin layers of cellulose on glass plates for the identification of nitrogen fixation products following short exposures of 1–120 s to ^{13}N. Recent improvements in techniques and instrumentation for both radioactive and stable nitrogen isotopes should provide refinements in measurements of nitrogen fixation, as well as other biological reactions.

F. Comparison of nitrogen tracer methods with acetylene reduction methods

In recent years the development of the acetylene reduction method has had a major influence on the proliferation of nitrogen fixation research. This indirect, nonisotopic method developed from the observation that the enzyme nitrogenase can reduce acetylene to ethylene (Dilworth, 1966; Schollhorn and Burris, 1967) and, thus, provides a measure of nitrogenase activity. Both gases can be measured easily at low concentrations by gas chromatography. With its high sensitivity, low equipment costs, simplicity and rapidity of use, the acetylene reduction assay has become a standard procedure in many laboratories. As a measure of nitrogen fixation in natural habitats, however, the acetylene reduction method suffers from the disadvantage of using short-term assays to assess overall activity. In field systems it is often difficult to obtain by acetylene reduction reliable estimates of nitrogen fixation comparable to those obtained by ^{15}N or total nitrogen methods (e.g. Legg and Sloger, 1975). Bergersen (1970) and Turner and Gibson (1980) have pointed out some of the difficulties in deriving quantitative nitrogen fixation data from acetylene reduction assays. The method has the disadvantages of any system that uses short-term assays to assess overall activity. It is difficult to achieve meaningful integration of a series of short-term assays with the variations in nitrogenase activity that occur with changes in light, temperature, moisture and other conditions. This limitation is especially true in field systems, where it is very difficult to obtain reliable estimates of nitrogen fixation comparable to those obtained by tracer methods.

IX. Conclusions

Many studies on the decomposition of organic compounds have used ^{14}C-labeled materials and measured $^{14}CO_2$ evolution. Future studies will most likely follow a similar pattern, but greater attention will be focused on such areas as microbial biomass and efficiency.

Use of ^{15}N has been common in nitrogen transformation studies in soil and many future research studies will undoubtedly include the use of ^{15}N. Studies to define parameters which will increase the efficiency of nitrogen utilization by crops will be of great importance.

Radioactive and stable isotopes have played a major role in the study of soil microbiological and biochemical reactions in the past. Future research will undoubtedly involve even greater utilization of isotopes.

References

Alexander, M. (1977). "Introduction to Soil Microbiology", p. 451, John Wiley and Sons, New York.

Allison, F. E. (1955). *Adv. Agron.* **7**, 213–250.

Allison, F. E. (1966). *Adv. Agron.* **18**, 219–258.

Atlas, R. M. and Bartha, R. (1972). *Biotechnol. Bioeng.* **14**, 297–308.

Bartha, R. and Pramer, D. (1965). *Soil Sci.* **100**, 68–70.

Bartholomew, W. V. (1965). *In* "Soil Nitrogen" (W. V. Bartholomew and F. E. Clark, eds), pp. 285–306, American Society of Agronomy, Madison, Wisc.

Behera, B. and Wagner, G. H. (1974). *Soil Sci. Soc. Am. Proc.* **38**, 591–594.

Bergersen, F. J. (1970). *Aust. J. Biol. Sci.* **23**, 1015–1025.

Bergersen, F. J. (1980). *In* "Methods for Evaluating Biological Nitrogen Fixation" (F. J. Bergersen, ed.), pp. 65–110, John Wiley and Sons, New York.

Billes, G. (1978). *Ecol. Plant.* **13**, 339–353.

Bondietti, E., Martin, J. P. and Haider, K. (1972). *Soil Sci. Soc. Am. Proc.* **36**, 597–602.

Bremner, J. M. (1977). *In* "Biological Nitrogen Fixation in Farming Systems of the Tropics" (A. Ayanaba and P. J. Dart, eds), pp. 335–352, John Wiley and Sons, New York.

Broadbent, F. E. (1965). *Soil Sci. Soc. Am. Proc.* **29**, 692–696.

Broadbent, F. E. (1966). *Hilgardia* **37**, 165–180.

Broadbent, F. E. and Clark, F. E. (1965). *In* "Soil Nitrogen" (W. V. Bartholomew and F. E. Clark, eds), pp. 344–359, American Society of Agronomy, Madison, Wisc.

Broadbent, F. E. and Nakashima, T. (1967). *Soil Sci. Soc. Am. Proc.* **31**, 648–652.

Broadbent, F. E. and Stojanovic, B. F. (1952). *Soil Sci. Soc. Am. Proc.* **16**, 359–363.

Broadbent, F. E., Nakashima, T. and Chang, G. Y. (1982). *Agron. J.* **74**, 625–628.

Burris, R. H. (1942). *J. Biol. Chem.* **143**, 509–517.

Burris, R. H. (1972). *Methods Enzymol.* **24**, 415–431.

Burris, R. H. (1974). *In* "The Biology of Nitrogen Fixation" (A. Quispel, ed.), pp. 9–33, North Holland, Amsterdam.

Burris, R. H. and Miller, C. E. (1941). *Science* **93**, 114–115.

Burris, R. H., Eppling, F. J., Wahlin, H. B. and Wilson, P. W. (1943). *J. Biol. Chem.* **148**, 349–357.

Cady, F. B. and Bartholomew, W. V. (1960). *Soil Sci. Soc. Am. Proc.* **24**, 477–482.

Campbell, N. E. R. and Lees, H. (1967). *In* "Soil Biochemistry" (A. D. McLaren and G. H. Peterson, eds), pp. 194–215, Marcel Dekker, New York.

Campbell, N. E. R., Dular, R., Lees, H. and Standing, K. G. (1967). *Can. J. Microbiol.* **13**, 587–599.

Carter, J. N., Bennett, O. L. and Pearson, R. W. (1967). *Soil Sci. Soc. Am. Proc.* **31**, 50–56.
Chahal, K. S. and Wagner, G. H. (1965). *Soil Sci.* **100**, 96–103.
Chiariello, N., Hickman, J. C. and Mooney, H. A. (1982). *Science* **217**, 941–943.
Chichester, F. W. (1970). *Plant Soil* **33**, 437–456.
Chichester, F. W. and Smith, S. J. (1978). *J. Environ. Qual.* **7**, 227–233.
Chichester, F. W., Legg, J. O. and Stanford, G. (1975). *Soil Sci.* **120**, 455–460.
Cortez, J. (1977). *Geoderma* **18**, 177–192.
Delwiche, C. C. and Wijler, J. (1956). *Plant Soil* **7**, 113–129.
Dilworth, M. J. (1966). *Biochim. Biophys. Acta* **127**, 285–294.
Edmeades, D. C. and Goh, K. M. (1979). *Commun. Soil Sci. Plant Anal.* **10**, 513–520.
Focht, D. D. (1978). *In* "Nitrogen in the Environment", Vol. II (D. R. Nielsen and J. G. MacDonald, eds), pp. 433–490. Academic Press, New York and London.
Fried, M. and Broeshart, H. (1975). *Plant Soil* **43**, 707–711.
Fried, M. and Middelboe, V. (1977). *Plant Soil* **47**, 713–715.
Gersberg, R., Krohn, K., Peek, N. and Goldman, C. R. (1976). *Science* **192**, 1229–1231.
Goh, K. M., Edmeades, D. C. and Robinson, B. W. (1978). *Soil Biol. Biochem.* **10**, 13–20.
Haider, K. and Martin, J. P. (1975). *Soil Sci. Soc. Am. Proc.* **39**, 657–662.
Haider, K., Martin, J. P. and Rietz, E. (1977). *Soil Sci. Soc. Am. J.* **41**, 556–562.
Hattingh, M. J., Gray, L. E. and Gerdemann, J. W. (1973). *Soil Sci.* **116**, 383–387.
Hauck, R. D. (1971). *In* "Nitrogen-15 in Soil Plant Studies", pp. 65–80, IAEA, Vienna.
Hauck, R. D. (1973). *J. Environ. Qual.* **2**, 317–327.
Hauck, R. D. and Bremner, J. M. (1976). *Adv. Agron.* **28**, 219–266.
Hauck, R. D. and Tanji, K. K. (1982). *In* "Nitrogen in Agricultural Soils" (F. J. Stevenson, ed.), pp. 891–925, American Society of Agronomy, Madison, Wisc.
Hauck, R. D., Melsted, S. W. and Yankwich, P. E. (1958). *Soil Sci.* **86**, 287–291.
Heichel, G. H., Barnes, D. K. and Vance, C. P. (1981). *Crop Sci.* **21**, 330–335.
Henzell, E. F., Martin, A. E., Ross, P. J. and Haydock, K. P. (1968). *Aust. J. Agric. Res.* **19**, 65–77.
Hsu, T. S. and Bartha, R. (1979). *Appl. Environ. Microbial.* **37**, 36–41.
Hurst, H. M. and Wagner, G. H. (1969). *Soil Sci. Soc. Am. Proc.* **33**, 707–711.
Jansson, S. L. (1958). *Kung. Lantbruks-högsk Ann.* **24**, 101–361.
Jansson, S. L. and Persson, J. (1982). *In* "Nitrogen in Agricultural Soils" (F. J. Stevenson, ed.), pp. 229–252, American Society of Agronomy, Madison, Wisc.
Jenkinson, D. S. (1977). *J. Soil Sci.* **28**, 209–213.
Jenkinson, D. S. and Ayanaba, A. (1977). *Soil Sci. Soc. Am. J.* **41**, 912–915.
Jenkinson, D. S. and Ladd, J. N. (1981). *In* "Soil Biochemistry", Vol. V (E. A. Paul and J. N. Ladd, eds), pp. 415–471, Marcel Dekker, New York.
Jenkinson, D. S. and Powlson, D. S. (1976). *Soil Biol. Biochem.* **8**, 209–213.
Jenkinson, D. S. and Rayner, J. H. (1977). *Soil Sci.* **123**, 298–305.
Jones, E. J. (1951). *Soil Sci.* **71**, 193–196.
Kassim, G., Martin, J. P. and Haider, K. (1981). *Soil Sci. Soc. Am. J.* **45**, 1106–1112.
Kassim, G., Stott, D. E., Martin, J. P. and Haider, K. (1982). *Soil Sci. Soc. Am. J.* **46**, 305–309.
Kaufman, D. D. (1974). *In* "Pesticides in Soil and Water" (W. D. Guenzi, ed.) pp. 133–202, Soil Science Society of America, Madison, Wisc.

Kearney, D. C. and Kaufman, D. D. (1975). "Herbicides", Marcel Dekker, New York.
Legg, J. O. (1978). *Proc. 11th Int. Congr. Soil Sci., Edmonton* **1**, 329.
Legg, J. O. and Allison, F. E. (1960). *Trans. 7th Int. Congr. Soil Sci., Madison* **3**, 545–550.
Legg, J. O. and Allison, F. E. (1967). *Soil Sci. Soc. Am. Proc.* **31**, 403–406.
Legg, J. O. and Meisinger, J. J. (1982). *In* "Nitrogen in Agricultural Soils" (F. J. Stevenson, ed.), pp. 503–566, American Society of Agronomy, Madison, Wisc.
Legg, J. O. and Sloger, C. (1975). *In* "Proceedings of the 2nd International Conference on Stable Isotopes" (E. R. Klein and P. D. Klein, eds), pp. 661–665, Argonne National Laboratory, Argonne, Ill.
Legg, J. O. and Stanford, G. (1967). *Soil Sci. Soc. Am. Proc.* **31**, 215–219.
Legg, J. O., Chichester, F. W., Stanford, G. and De Mar, W. H. (1971). *Soil Sci. Soc. Am. Proc.* **35**, 273–276.
Loginov, Yu. M. (1966). *Agrokhimiya* No. 11, 21–28.
Loos, M. A., Kontson, A. and Kearney, P. C. (1980). *Soil Biol. Biochem.* **12**, 583–585.
McAuliffe, C., Chamblee, D. S., Uribe-Arango, H. and Woodhouse, W. W., Jr (1958). *Agron. J.* **50**, 334–337.
McGarity, J. W. and Rajaratnam, J. A. (1973). *Soil Biol. Biochem.* **5**, 121–131.
MacRae, I. C., Ancajas, R. R. and Salandanan, S. (1968). *Soil Sci.* **105**, 327–334.
MacVicar, R., Garman, W. L. and Wall, R. (1950). *Soil Sci. Soc. Am. Proc.* **15**, 265–268.
Martens, R. (1982). *Soil Biol. Biochem.* **14**, 315–317.
Martin, J. P. and Focht, D. D. (1977). *In* "Soils for Management of Organic Wastes and Waste Waters" (L. F. Elliott and F. J. Stevenson, eds), pp. 115–169, Soil Science Society of America, Madison, Wisc.
Martin, J. P. and Haider, K. (1971). *Soil Sci.* **111**, 54–63.
Martin, J. P. and Haider, K. (1976). *Soil Sci. Soc. Am. J.* **40**, 377–380.
Martin, J. P., Haider, K. and Wolf, D. C. (1972). *Soil Sci. Soc. Am. Proc.* **36**, 311–315.
Mayaudon, J. (1971). *In* "Soil Biochemistry", Vol. II (A. D. McLaren and J. Skujins, eds), pp. 202–256, Marcel Dekker, New York.
Meeks, J. C., Wolk, C. P., Lackau, W., Schilling, N., Shaffer, P. W. and Chien, W. S. (1978). *J. Bacteriol.* **134**, 125–130.
Moreale, A. and Van Bladel, R. (1980). *J. Environ. Qual.* **9**, 627–633.
Nicholas, D. J. D., Silvester, D. J. and Fowler, J. F. (1961). *Nature* (*Lond.*) **189**, 634–636.
Nommik, H. (1956). *Acta Agr. Scand.* **6**, 195–228.
Norman, A. G. and Krampitz, L. O. (1945). *Soil Sci. Soc. Am. Proc.* **10**, 191–196.
Paul, E. A. and Juma, N. G. (1981). *Ecol. Bull.* (*Stockholm*) **33**, 179–199.
Paul, E. A. and Van Veen, J. A. (1978). *11th Int. Congr. Soil Sci., Symp. Session Papers* **3**, 61–102.
Rennie, R. J., Dubetz, S., Bole, J. B. and Muendel, H. H. (1982). *Agron. J.* **74**, 725–730.
Rhodes, L. H. and Gerdemann, J. W. (1978). *Soil Biol. Biochem.* **10**, 355–360.
Rolston, D. E. (1978). *In* "Nitrogen in the Environment", Vol. I (D. R. Nielsen and J. G. MacDonald, eds), pp. 309–335, Academic Press, New York and London.
Rolston, D. E. and Broadbent, F. E. (1977). U.S. Environmental Protection Agency Research Report Series EPA-600/2-77-23, Ada, Okla.

Rolston, D. E., Broadbent, F. E. and Goldhamer, D. A. (1979). *Soil Sci. Soc. Am. J.* **43**, 703–708.

Rolston, D. E., Fried, M. and Goldhamer, D. A. (1976). *Soil Sci. Soc. Am. J.* **40**, 259–266.

Ross, P. J., Martin, A. E. and Henzell, E. F. (1968). *Trans. 9th Int. Congr. Soil Sci., Adelaide* **2**, 487–494.

Rovira, A. D. and McDougall, B. M. (1967). *In* "Soil Biochemistry", Vol. I (A. D. McLaren and G. H. Peterson, eds), pp. 417–463, Marcel Dekker, New York.

Ruben, S., Hassid, W. Z. and Kamen, M. D. (1940). *Science* **91**, 578–579.

Schollhorn, R. and Burris, R. H. (1967). *Proc. Nat. Acad. Sci. USA* **58**, 213–216.

Scott, H. D., Wolf, D. C. and Lavy, T. L. (1982). *J. Environ. Qual.* **11**, 107–112.

Sissons, C. H. (1976). *Anal. Biochem.* **70**, 454–462.

Sommers, L. E., Gilmour, C. M., Wildung, R. E. and Beck, S. M. (1981). *In* "Water Potential Relations in Soil Microbiology" (J. F. Parr, W. R. Gardner and L. F. Elliott, eds), pp. 97–117, Soil Science Society of America, Madison, Wisc.

Stanford, G., Legg, J. O. and Chichester, F. W. (1970). *Plant Soil* **33**, 425–435.

Stanford, G., Legg, J. O., Dzienia, S. and Simpson, E. C., Jr. (1975). *Soil Sci.* **120**, 147–152.

Stangarone, J. H. (1981). MS thesis, University of Maryland, College Park, Md.

Stotzky, G. (1965). *In* "Methods of Soil Analysis" (C. A. Black, ed.), pp. 1550–1572, American Society of Agronomy, Madison, Wisc.

Straatman, M. G. (1977). *Int. J. Appl. Radiat. Isotopes* **28**, 13–20.

Stutte, C. A., Weiland, R. T. and Blem, A. R. (1979). *Agron. J.* **71**, 95–97.

Talbott, H. J. (1980). MS thesis, University of Maryland, College Park, Md.

Thorton, G. D. (1947). *Soil Sci. Soc. Am. Proc.* **11**, 249–251.

Tiedje, J. M., Firestone, R. B., Firestone, M. K., Betlach, M. R., Smith, M. S. and Caskey, W. H. (1979). *Soil Sci. Soc. Am. J.* **43**, 709–715.

Turner, G. L. and Gibson, A. H. (1980). *In* "Methods for Evaluating Biological Nitrogen Fixation" (F. J. Bergersen, ed.), pp. 111–138, John Wiley and Sons, New York.

Vallis, I., Haydock, K. P., Ross, P. J. and Henzell, E. F. (1967). *Aust. J. Agric. Res.* **18**, 865–877.

Vallis, I., Henzell, E. F. and Evans, T. R. (1977). *Aust. J. Agric. Res.* **28**, 413–425.

Vasilas, B. J. (1981). PhD thesis, University of Minnesota, St Paul, Minn.

Verma, L. and Martin, J. P. (1976). *Soil Biol. Biochem.* **8**, 85–90.

Verma, L., Martin, J. P. and Haider, K. (1975). *Soil Sci. Soc. Am. Proc.* **39**, 279–284.

Wagner, G. H. and Krzywicka, A. M. (1975). *In* "Biodegradation et Humification" (G. Kilbertus, O. Reisinger, A. Mourey and J. A. Cancela da Fonseca, eds), pp. 202–207, Pierron, Sarreguemines, France.

Wagner, G. H. and Medynska-Rabinska, H. (1971). *Humus Planta* **5**, 27–32.

Wagner, G. H. and Zapata, F. (1982). *Agron. J.* **74**, 607–612.

Warembourg, F. R. (1975). *Rev. Ecol. Biol. Sol.* **12**, 261–272.

Webley, D. M. and Jones, D. (1971). *In* "Soil Biochemistry", Vol. II (A. D. McLaren and J. Skujins, eds), pp. 446–485, Marcel Dekker, New York.

Wijler, J. and Delwiche, C. C. (1954). *Plant Soil* **5**, 155–169.

Wilson, R. G., Jr. and Cheng, H. H. (1978). *J. Environ. Qual.* **7**, 281–286.

Wolf, D. C. and Martin, J. P. (1974). *Soil Sci. Soc. Am. Proc.* **38**, 921–925.

Wolf, D. C. and Martin, J. P. (1976). *Soil Sci. Soc. Am. J.* **40**, 700–704.

Wolk, C. P., Thomas, J., Shaffer, P. W., Austin, S. M. and Galonsky, A. (1976). *J. Biol. Chem.* **251**, 5027–5034.

Zeller, A., Oberlander, H. E., Roth, K. and Stadler, I. (1966). *In* "The Use of Isotopes in Soil Organic Matter Studies", pp. 275–279, FAO/IAEA Technical Meeting Brunswick-Volkenrode, Pergamon Press, Oxford.

Zwarun, A. A. (1972). *Soil Sci. Soc. Am. Proc.* **36**, 968–969.

5

The Detection and Measurement of Radionuclides

Michael F. L'Annunziata

I. Introduction

This chapter deals with current techniques used for the detection and quantitative assay of radionuclides used as tracers in agricultural research. Emphasis is placed on the radioassay of beta-particle- and gamma-ray-emitting nuclides as it is these which are most commonly used as tracers in the agricultural sciences. Only passing reference is made to alpha-particle emitters and their assay by ionization or liquid scintillation counting because alpha-particle-emitting nuclides, such as the transuranium elements, have limited application in agricultural research. As neutrons are not emitted from nuclides used as tracers, the production and detection of neutrons are not covered in this chapter.

Among the radioassay techniques presented herein, major emphasis is placed on liquid scintillation and Cherenkov counting and semiconductor detectors as significant advances have been made in these fields in the past decade. The use of gas ionization counters is presented in a condensed form. The reasoning here is that the gas ionization counting of radiotracers has become, for the most part, outdated in agricultural research.

A thorough understanding of the physical and chemical mechanisms of the detection and assay of radionuclides requires a background knowledge of the physics of nuclear radiation and its interaction with matter, as well as familiarity with the concepts of radioisotope decay and half-life. For those unfamiliar with these concepts, an introduction can be obtained from Chapters 1 and 2 of a previous book (L'Annunziata, 1979).

II. Gas ionization counters

Ionization counting is based upon one of the principal effects of the interaction of nuclear radiation with matter – that of ionization. The ion pairs formed in a gas-filled counting chamber by incident radiation are collected at their respective electrodes (anode and cathode) and the collected charges are registered as counts.

Proportional and Geiger–Müller (GM) counters are constructed to permit the detection of nuclear radiation which is capable of producing ionization in a gas. The use of controlled gas amplification and a pulse height discriminator gives the proportional counter the unique ability to distinguish between alpha and beta radiation. Both gas amplification and the use of a discriminator are discussed below.

A basic problem in the detection of ionizing radiation in a simple ionization chamber is that a very small electrical pulse is produced by a single ionizing particle when passing through gas within the chamber. To alleviate this problem a higher potential gradient is applied across the electrodes in the chamber, greater than that required to collect the ions at the electrodes before ion recombination can occur. The higher potential draws the negative ion towards the anode at an ever-increasing velocity, resulting in the production of more ion pairs. An individual electron approaching the anode at an increasing velocity produces additional ion pairs and consequently additional electrons. These newly produced electrons can in turn produce more ion pairs, resulting in an eventual avalanche of ions collected at the electrodes. The process of producing an avalanche of ions, known as gas amplification, may occur within a time period of only 2 μs and produces a pulse sufficiently large to be registered by an external scalar circuit. With the proportional counter the gas amplification is such that the number of ions collected is directly proportional to the number of ions initially produced by the nuclear radiation.

Because of the large mass and double charge of the alpha particle, more ion pairs per path length are produced by an alpha particle than by the smaller, lighter and single-charged beta particle (L'Annunziata, 1979). For this reason the charge resulting from an incident alpha particle is greater than that from a beta particle at any given applied potential. The higher charge produced by alpha particles over that produced by beta particles, in addition to the proportionality of the total ions collected to the primary ions formed, establishes the basis of proportional counting which enables an investigator to distinguish between alpha and beta radiations.

By means of an electronic pulse height discriminator employed in proportional counters, pulses of a greater magnitude are permitted to pass on to the scaling circuit and be registered. Thus, if a given sample emits both alpha and beta particles, the alpha particles which produce larger pulses would be detected first at lower voltages. Only after sufficiently increasing the high voltage will the beta particles produce a charge sufficient to pass the discriminator and be registered.

An example of alpha–beta discrimination is illustrated in Fig. 5.1, the curve of which was obtained from a ^{210}Pb (^{210}Bi) sample by plotting the counts registered with a proportional counter at various high voltage

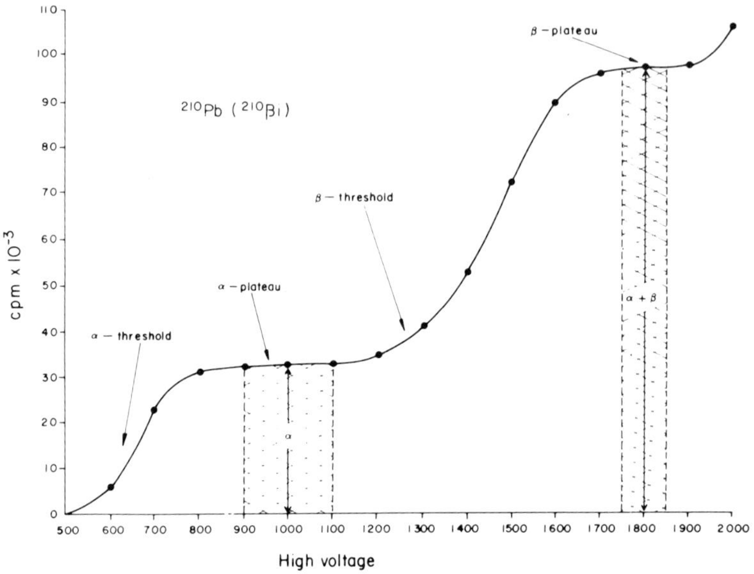

Fig. 5.1 Proportional plateau curve for ^{210}Pb (^{210}Bi) determined with a proportional counter. The hatched areas indicate the range of voltages which would not produce any significant change in the sample count rate for the particular counter tested.

settings. At the lower voltage settings from 500 to 1100 V, designated the alpha-threshold and alpha-plateau, there is the first sign of counts followed by a rise in count rate due to alpha particles. In this region the alpha particles produce sufficient ionization to make the low amplification obtained at these lower voltage settings adequate to yield pulses which can pass the discriminator and be registered. The comparatively weakly ionizing beta particles are not detected until greater amplification is obtained at the higher voltage levels. The beta-threshold begins at ~1200 V, where amplification is high enough to produce pulses from beta particle ionization which can pass the discriminator. The maximum count rate obtained via amplification of both alpha and beta particle ionization is indicated by the beta-plateau, between 1700 and 1900 V for the particular detector used here. The rise in count rate at voltages above 1900 V is attributable to continuous discharge from the ionization detector. Prolonged operation of the counter in this region is avoided so as not to shorten the life of the detector.

Count rate due to alpha radiation alone is obtained directly from the count rate of the alpha plateau. To determine the count rate due to beta radiation it is necessary to subtract the count rate of the alpha-plateau from that at the beta-plateau since the beta-plateau represents both alpha and beta radiation events.

The basic difference between the proportional and GM counters is in gas amplification. In GM counters the gas amplification is higher, and thus there is no proportionality between the number of ions collected and those initially produced by the incident radiation. Consequently, any type of incident ionizing radiation will produce the same pulse size, and thus alpha and beta radiation cannot be distinguished with a GM counter. The relationship between count rate and high voltage of a GM counter is that of a single plateau curve for either alpha or beta radiation. This is in contrast to that of a double-plateau curve obtained with proportional counters, as illustrated in Fig. 5.1.

It should be noted that the plateau curves of proportional and GM counters are not identical and vary from counter to counter. Consequently, a plateau curve is characteristic of a particular counter and must be plotted in order to determine the proper operational voltage of the instrument.

Proportional and GM gas ionization counting techniques are not normally used to assay gamma radiation because of the very low ionization produced by secondary electrons originating from the interaction of electromagnetic radiation with gases. This rests principally upon the low density and consequently low gamma-ray "stopping power" of gases. Furthermore, gas ionization counters are seldom employed currently to assay for radionuclides used as tracers in agricultural research because of the relatively low counting efficiencies obtained. Consequently, only a very brief discussion of this counting technique is given here.

III. Semiconductor detectors

Semiconductor radiation detectors are essentially solid ionization detectors. The ionization produced by either particulate or non-particulate nuclear radiation in a semiconductor crystal is collected as an electrical pulse, amplified and registered or counted by a scaler or other suitable quantifying device. The semiconductor detector may be compared in some ways with the gas ionization detector in that both devices register the ionization produced by nuclear radiation in the detection medium – gas or solid; however, the solid detector medium offers certain advantages over a gaseous one.

One major advantage of the semiconductor over the gas ionization detector is the much higher density of the solid in comparison to the gaseous medium, which attributes a greater "stopping power" of nuclear radiation to the former. Because semiconductors are small devices (a few micrometres to $\sim$7.5 cm of detection medium), this "stopping power" advantage is particularly relevant in the analysis of charged particles such as alpha and beta (positron and negatron) radiations among other particles which have short ranges in solids. The width of the detection medium of a semiconductor detector may be chosen to stop the particulate nuclear radiation completely and obtain a complete energy transfer from the nuclear particle to the semiconductor detector.

Another important property of the semiconductor detector is the higher energy resolution obtainable, which may be as much as 20 times that of a NaI(Tl) solid scintillation crystal detector. Semiconductor detectors are highly popular as spectrometers because of their superior energy resolution capabilities for both particulate and electromagnetic radiation.

In addition to the utility of semiconductor detectors in spectrometry, an ever-increasing number of applications are being found in the agricultural and biological sciences. The small sizes of these detectors permit their use as probes where other detector types cannot be located, e.g. in contact with animal organs, at different depths of a soil profile or within various parts of a tree or other vegetation.

Prior to discussing the physical and chemical mechanisms of a semiconductor radiation detector, a semiconductor is defined and the components of semiconductor detectors are briefly discussed.

A. Insulators, semiconductors and conductors

Insulators and semiconductors are similar in that their highest filled energy band of electrons bound to the atoms (valence band) is separated from the energy band of free electrons (conduction band) by an energy gap known as the forbidden energy band (Fig. 5.2). The difference between insulators and semiconductors is in the magnitude of the forbidden energy band. In insulators the electrons are all occupied as bonds between the atoms, i.e. there are no electrons which are free to travel throughout the insulator structure to allow conduction to take place, and a relatively excessive amount of energy (such as high temperature) would be required to remove an electron from its bonds (valence band) and permit its free movement in the conduction band. However, in semiconductors the energy required to remove an electron from its valence band to the conduction band is much lower. In semiconductors a small amount of energy (<3 eV), from light, moderate temperatures or nuclear radiation, imparted to a valence electron, is sufficient to displace it from the valence band into the conduction band where it may travel freely and enable conduction to occur. Conductors such as metals have overlapping valence and conduction bands; i.e. there is no forbidden energy band and consequently high concentrations of free electrons are encountered in the conduction band.

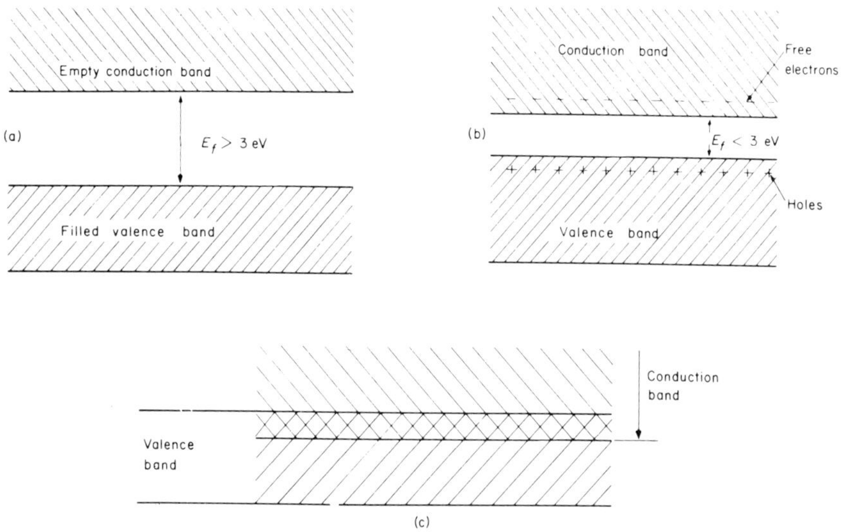

Fig. 5.2 Band structure of (a) insulators, (b) semiconductors and (c) metals (Deme, 1967).

If, in any given material, impurities are present which have more or fewer valence electrons than the atomic species of that material, the height of the forbidden energy band may be lowered and the conduction potential of the material increased. This results because neither an excess nor a deficiency of electrons (i.e. electron holes) is occupied in structural bonds; thus, the electrons have more freedom to move in the conduction band of a given material than the valence band electrons of the same pure material. This is illustrated and discussed in more detail in the following section.

B. n- and p-type semiconductor materials

Consider the crystal structure of pure silicon or pure germanium. Each silicon or germanium atom with its four valance electrons is combined to four other silicon or germanium atoms in a tetrahedral structure by covalent bonds formed with the valence electrons of the other atoms. All of the electrons take part in the covalent bonds between the atoms; i.e. there are no free electrons. All of the electrons are thus in the valence band and there are none in the conduction band; the crystal as such is an insulator.

Impurities may be incorporated into the silicon or germanium crystal intentionally or by chance. If the impurity atom has five valence electrons, it is considered as a donor atom. Examples of donor atoms are phosphorus, antimony and arsenic. Such a case is shown in Fig. 5.3(a). This represents a magnified portion of a silicon crystal in which one of the donor atoms can be seen (black spot). The crystal is connected to an electric power supply. Notice that four of the five valence electrons of the donor atom are shared in covalent bonds with silicon atoms. These fixed electrons are in the valence energy band and consequently cannot carry conductive charge. The fifth valence electron is an excess electron. It cannot share in covalent bonds with silicon atoms since all of the bond locations are occupied. The excess electron is located near the conduction band and the energy supplied by the electric field is sufficient to permit it to enter into this band. In the conduction band

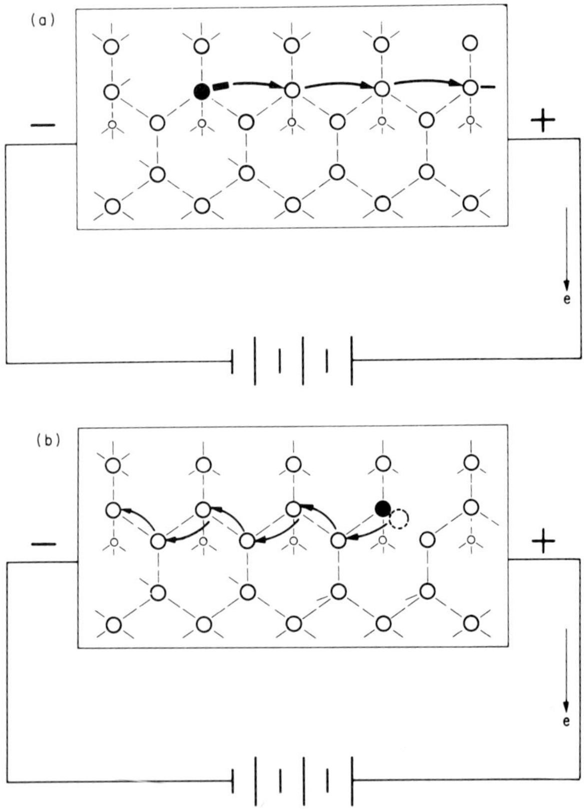

Fig. 5.3 Magnified portion of a silicon crystal in an electric field in which is seen (a) a donor atom impurity, its excess valence electron and the travel of the surplus electron towards the positive pole while in the conduction energy band; (b) an acceptor atom impurity, its adjacent electron hole (absent electron) and the travel of the hole towards the negative pole while in the conduction energy band.

the electrons may leave the donor atom and travel from atom to atom towards the positive pole. The donor atom which loses the excess valence electron then becomes positively charged. Semiconductor crystals which contain donor atoms contain negative charge carriers and are thus considered as n-type materials.

Now, if the impurity incorporated into the silicon or germanium crystal has three valence electrons, it is considered as an acceptor atom. Examples of acceptor atoms are boron, aluminium, indium and gallium. Figure 5.3(b) shows a magnified portion of a silicon crystal in which an acceptor atom is present (black spot). The three valence electrons of the acceptor atom form covalent bonds with three electrons of neighbouring silicon atoms. A deficient fourth valence electron, necessary for bonding of the acceptor atom with a fourth silicon atom in the tetrahedron, is represented in the figure as an electron hole (dotted circle). This electron hole may be looked upon as a positively charged entity since it represents the absence of a necessary electron. The positive electron hole may easily trap an electron from an adjacent silicon atom leaving the hole with the silicon atom which donated the electron. The electron entrapment and new electron hole position (electron hole travel) can occur from silicon atom to silicon atom, as indicated in Fig. 5.3(b), particularly under the influence of an electromotive force. In the electric field illustrated, the electron hole is shown as travelling towards the negative pole. The acceptor atom becomes negatively charged once its electron hole entraps a neighbouring electron; however, the semiconductor crystal maintains an overall neutral charge since the number of electron holes and negative ions in the crystal are the same. Semiconductor crystals which contain acceptor atoms contain positive charge carriers and are thus termed p-type materials.

C. Junction semiconductor detectors

Consider the case where a p-type material is joined to an n-type material by a neutral region, referred to as the depletion zone, which may consist of silicon or germanium free of electron donor or electron acceptor atoms. If an electrical potential or bias is applied to this junction, as illustrated in Fig. 5.4(a), the electron holes of the p-type material migrate towards the negative pole and the electrons towards the positive. Thus current flows and the depletion zone disappears.

If a reverse bias is applied to the p–n junction (Fig. 5.4(b)), the holes and electrons of the p- and n-type materials migrate towards the negative and positive poles, respectively, and the depletion zone increases in depth. In this case only noise current flows unless electron–hole pairs are formed in the depletion region by the absorption of nuclear radiation. The application of a

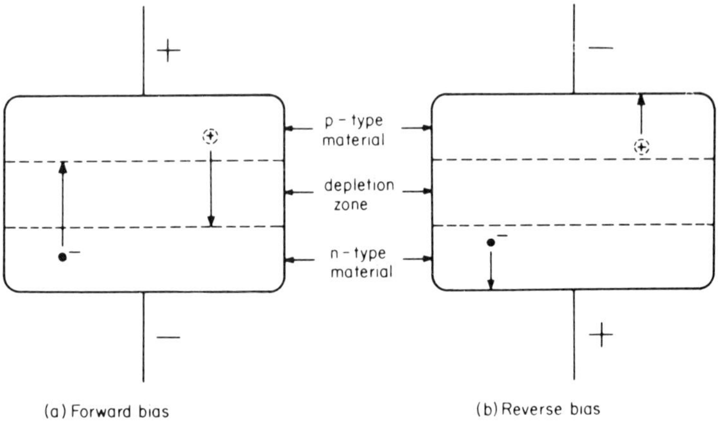

Fig. 5.4 Electron and hole migrations in p–n-junction devices with forward and reverse bias (electrical potential).

reverse bias, with the subsequent formation of a depletion region with a depth dependent upon the reverse bias potential, is the basis of operation for both non-compensated and compensated semiconductor radiation detectors.

D. Non-compensated surface barrier detectors

The non-compensated semiconductor detector consists of a union of p- and n-type materials with the application of a reversed bias voltage. The depletion region is produced by the migration of the holes and electrons toward their respective electrodes, and a zero current results. The depletion region produced becomes free of charge carriers and consequently acts as an insulator preventing the flow of current. Such a non-compensated p–n-junction detector is shown in Fig. 5.5.

This type of instrument usually consists of p- and n-type silicon materials and is used for the detection of charged particle radiation such as alpha and beta particles. Upon bombardment of the thin ($\sim 40\ \mu g\ cm^{-2}$) gold window and p-type silicon the charged particle is absorbed by the thicker depletion region. The width of the depletion zone is proportional to the square root of the bias voltage and resistance of the silicon semiconductor material.

The depth of the depletion region may be controlled by the bias voltage so that it may be superior to the range of the charged particle, thereby assuring its complete absorption in this region, and may range from a few micrometres to a few millimetres in width. The presence of impurities in the silicon semiconductor will cause a reduction in the resistance and consequently a reduction in the maximum possible depletion zone width

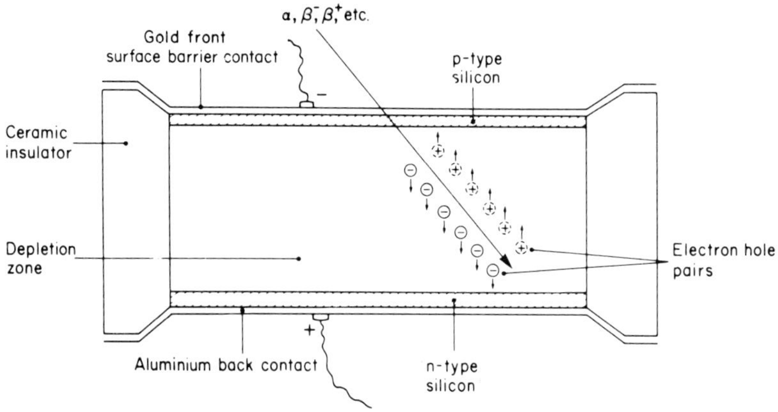

Fig. 5.5 Cross-section of a non-compensated p–n-junction detector.

obtainable. This harmful effect of impurities is corrected for in the so-called compensated semiconductor detectors discussed later.

The absorption of energy of the charged particle in the depletion zone produces ionization in which some electrons of the valence band of the silicon are elevated to the conduction band, leaving holes in the former. This results in the formation of electron–hole pairs as illustrated in Fig. 5.5. The electrons and holes produced may then migrate to the positive and negative poles according to the mechanism described previously, with the resultant production of current with a pulse height dependent upon the number of electron–hole pairs produced or the energy of the incident radiation.

Silicon surface barrier detectors are used almost exclusively for the analysis of particulate radiation or low-energy X-radiation. This is because the low atomic number (14) and density (2.33 g cm^{-3}) of the silicon crystal in comparison to germanium (32 and 5.32 g cm^{-3}, respectively) and the shallow depletion zone depths of these crystals make them efficient only in stopping particulate radiation. Gamma radiation and high-energy X-radiation are most efficiently analysed or detected by compensated germanium semiconductor crystals which have superior atomic number, density and depleted zone depths, resulting in a greater "stopping power" for these penetrating radiations. The compensated semiconductor detectors are described below.

E. Compensated p–i–n detectors

In the compensated semiconductor detectors, the harmful effect of impurities in silicon and germanium is greatly reduced by diffusing Li^+ ions through the p-type material at elevated temperatures and reverse bias voltage. The Li^+ ions are electron donors and their diffusion through the p-type material compensates for the holes in this material. The number of holes is reduced by an equal number of Li^+ donor ions. With this technique it is possible to attain full compensation, i.e. a layer without holes or free electrons. This compensated region is more often referred to as the intrinsic region and is located between the p- and n-type materials. These detectors are consequently called p–i–n-junction devices and are designated Ge(Li) or Si(Li) detectors. The Li^+ ions have a greater mobility in germanium than in silicon, which results in greater possible depths of the intrinsic region in germanium ($\sim$10 mm) than in silicon ($\sim$5 mm). Because of the elevated temperatures employed for the Li^+ diffusion and the higher mobility of Li^+ in germanium, the Li^+ concentration in germanium at room temperature is exceeded. Consequently, to maintain the intrinsic layer width and detector properties constant, the Ge(Li) detectors must be kept at very cold temperatures. Liquid nitrogen, boiling at a temperature of 77 °K, is contained in a large (15–30 litre) Dewar flask and is used to cool the Ge(Li) semiconductor detector by means of a copper cold finger contact which leads from the liquid nitrogen and maintains a thermic contact with the semiconductor detector. The Si(Li) p–i–n-junction detector may be operated at room temperature without harm to the units, although they are usually maintained at reduced temperatures to minimize thermal noise and attain optimum resolution.

F. Choice of the most suitable detector

In choosing the most suitable detector one must consider the properties of the radiation, the detector types and the experimental application. The radiation properties of most concern are the type and energy which determine the range of the radiation in the detector. The detector type and possible depletion thickness must be considered. The Ge(Li) devices are used for the detection of X- or gamma radiation because of the higher atomic number and density of germanium and because of the greater intrinsic region depths obtainable with Ge(Li). The coaxial design of the Ge(Li) detector is the most appropriate for high-energy X- or gamma radiation because of its wider depletion thickness. The Si(Li) devices are used mostly for the detection of low-energy gamma radiation, soft X-radiation or charged particles such as beta radiation. The non-compensated Si(p–n) surface barrier detector is most useful where room or body temperatures are necessary, e.g. in the location of detectors near body organs or plant parts for *in vivo* experimentation. A Ge(Li) detector cannot be used in such cases because of the low liquid nitrogen temperatures required to maintain detector stability.

IV. Semiconductor detector applications

The importance of semiconductor detectors in biological experiments is ever-increasing. Due to the usual small size of these solid-state devices, they can be implanted via surgery or injected via needle probes into the organs of animals or plant parts for *in vivo* experimentation. The dynamics of radiotracers in living animal organs, tissues and blood vessels can be more easily determined with solid-state detectors. The difficulty of *in vivo* experimentation with miniature gas ionization counters is now being eliminated with the recently developed windowless or needle probe semiconductor detector. A comprehensive discussion of these detectors is not given here; however, some recently developed probes and their uses are described briefly. Details of their construction and operation may be obtained from the references.

A. Particulate radiation counting

The particulate radiations most commonly encountered in radiotracer studies in the biological sciences are alpha and, especially, beta particles. These particles are counted efficiently with the silicon semiconductor detector. For the counting of such charged particle radiation, the silicon semiconductor detector has certain advantages over the gas ionization counter. One of these is that the window thickness is less in the semiconductor detector. This is an important factor for low-energy particles which can be absorbed by the window material. The much smaller detector size, freedom from cumbersome gas attachments, and low background counts without shielding are important advantages over the gas ionization counting technique. For large-area detectors, typical beta detection efficiencies can range from less than one-half the efficiency of a gas flow counter for low-energy beta-particle emitters such as ^{14}C to more than that of a thin-window gas-flow counter for high-energy beta particles. The beta particle counting efficiencies are not as high as those obtained by liquid scintillation counting, but the compactness of the silicon detector and its room or body temperature operation make the silicon detector preferable in certain *in vivo* and *in vitro* animal experiments.

The sample counting arrangement is similar to that of the gas ionization counting system. The current pulse produced by the colletion of electron–hole pairs in the semiconductor detector travels on to the preamplifier and amplifier to be received by the electronic discriminator. Those pulses with heights above that of the discriminator level go on to be counted by the scaler.

B. Gamma-ray spectrometry

The mechanisms of interaction of gamma radiation with the semiconductor detector are similar to those which occur with solid scintillation detectors discussed subsequently. The gamma rays dissipate their energy within the semiconductor depletion zone principally via the photoelectric effect, Compton effect and pair production, with the photon energy appearing as the kinetic energy of the photoelectron, Compton electron and electron–positron pair, respectively. The kinetic energy of these electrons is then absorbed by the semiconductor to produce the electron–hole pairs necessary for charge collection and counting.

The semiconductor detector offers superior energy resolution to that of the NaI(Tl) solid scintillation crystal detector. This is due in part to the smaller amount of nuclear radiation energy required to produce an ion pair in a silicon (3.6 eV per ion pair) or germanium (2.8 eV per ion pair (semiconductor as compared to the energy required by a gas ionization or solid scintillation detector ($\sim$30 eV per ion pair). The lower energy requirement of semiconductor detectors consequently results in a greater production of ion pairs per given energy quantity absorbed by the detection medium. The ion pairs are also more efficiently collected by semiconductor detectors, which results in the production of signal pulses with magnitudes of smaller statistical variance (greater resolution).

In addition to the greater energy resolution, the semiconductor detector also produces optimum peak-to-Compton ratios, i.e. the optimum ratio of the full energy peak height to the Compton peak height of a gamma-ray spectrum. The full energy peak results when monoenergetic gamma rays dissipate all of their energy in the detector material via the photoelectric effect, Compton scattering effects and pair production. The pulses which result from full energy dissipation of the gamma rays appear in the full energy channel of the radiation analyser. In addition to full energy dissipation, gamma rays may lose only a fraction of their energy in the detector material via the Compton scattering effect. For example, a gamma ray may enter the detector, deposit part of its energy via the Compton effect, and then escape from the detector. This partial energy dissipation via the Compton effect produces a rather broad peak (because the Compton scatter electrons are not monoenergetic) at energy channels inferior to the full energy channel. The peak-to-Compton ratio of solid scintillation counters is rather low ($\sim$5:1) in comparison to the semiconductor detector ($\sim$32:1), as illustrated in Fig. 5.6.

The efficiency of a Ge(Li) semiconductor detector is conventionally defined as the ratio of the percentage of ^{60}Co gamma rays (1.33 MeV) which lose all of their energy in the semiconductor detector (counts in the 1.33 MeV peak/total number of gamma rays emitted) to the percentage of ^{60}Co gamma

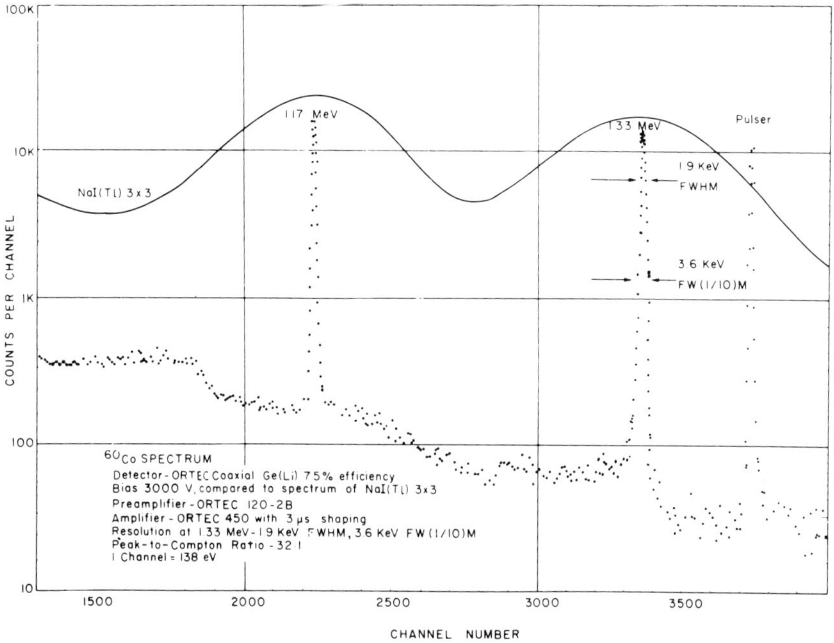

Fig. 5.6 Cobalt-60 spectra obtained by NaI(Tl) (upper) and Ge(Li) (lower) detectors. (Courtesy of EG & G Ortec, Oak Ridge, Tenn.)

rays which lose all of their energy in a 7.62 cm × 7.62 cm (3 in × 3 in) NaI(Tl) solid scintillation detector. The two measurements are made with the same ^{60}Co source (National Bureau of Standards) at a distance of 25 cm from the detector.† The photopeak efficiency of the Ge(Li) detector is inferior to that of the NaI(Tl) detector because this measurement is a function of the active volume of the detector, e.g. an efficiency of 7.5% is shown in Fig. 5.6. Consequently, in experimental applications which involve routine sample counting techniques of gamma-ray-emitting tracers, solid scintillation counting with a NaI(Tl) crystal is preferable. However, in applications involving radiation analysis via gamma-ray spectrometry such as neutron activation analysis (see Chapter 1), the superior energy resolutions and peak-to-Compton ratios offered by the semiconductor detector are most advantageous.

In a review of semiconductor detector applications, Pehl (1978) points out how the germanium detector has been very useful in the analysis of complex gamma-ray spectra encountered in neutron activation analysis. Neutron

† IEEE Test Procedures for Germanium Detectors for Ionizing Radiation, ANSI/IEEE 325 – 1971 (Reaff, 1977).

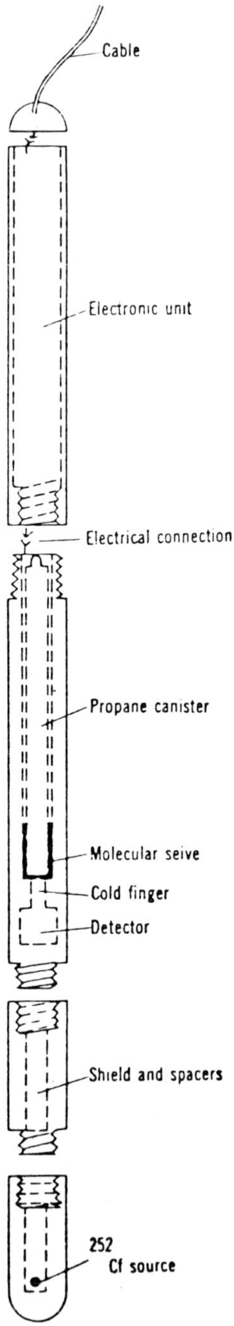

Fig. 5.7 Sketch of a typical soil borehole probe using a canister cryostat. The ^{252}Cf in the soil probe is a neutron source used for the neutron activation analysis of soil elements. The probe may be used without the neutron source if only an assay of natural, fallout and tracer gamma-ray-emitting nuclides is desired. The probe is 3.7 cm in diameter and 57 cm long. The molecular sieve may consist of zeolite (metallic aluminium silicate) to maintain vacuum conditions via the absorption of molecules (Boynton, 1975).

activation analysis involves firstly the neutron irradiation of the material to be analysed. After capturing slow neutrons, elements within the material are analysed in terms of the gamma rays emitted by excited nuclides in the sample. The use of neutron activation analysis with germanium semiconductor detectors for the determination of trace elements in all sorts of materials is now widespread. Its applications in entomological research are discussed in Chapter 2 in Volume 2.

The *in-situ* Ge(Li) gamma-ray spectrometric analysis of naturally occurring radioactive nuclides such as ^{40}K, ^{236}U and ^{232}Th, as well as fallout nuclides such as ^{137}Cs in the soil, is reported by Sakai *et al.* (1976). The Ge(Li) detector, cryostat, preamplifier, high voltage supply and 7.5 litres of liquid nitrogen coolant are suspended by a tripod 25 cm above the soil surface. This experimental arrangement gives results over a wide soil surface area and volume with a knowledge of the depth distribution of soil radioactivities (discussed below); and it is a very promising method for distinguishing exposure rates of man-made radionuclides from those occurring naturally in the soil. The depth distribution of soil radioactivity originating from natural, fallout or tracer nuclides which emit gamma radiation can be determined with the borehole probe. Because the high resolution Ge(Li) semiconductor detector requires cooling at temperatures of 77–90 °K to maintain the intrinsic layer width and detector properties constant, the large Dewar flask required to house the liquid nitrogen coolant has posed a serious impediment to the use of these detectors in soil probes. The solution to this bulky problem has been the use of a canister of prefrozen propane or fluorinated hydrocarbon refrigerant (e.g. Freon 13 or Freon 22), as illustrated in Fig. 5.7. The detector, as reported by Boynton (1975), is cooled at the melting point of a solid cryogen rather than at the boiling point of a liquid cryogen. In practice, the soil probe is lifted from the borehole just prior to the melting of the last trace of solid cryogen. Once back on the soil surface the probe is disconnected at the cryostat and the warmed canister is removed and replaced by a prefrozen canister of cryogen. The probe is then reassembled for further use. The canisters can be changed within 10 min, leaving the probe ready for use with a detector temperature stabilized at 86 °K for about 6 h. Schnerr *et al.* (1980) report the use of a similar probe for the neutron activation analysis of soil elements and the potential application of the probe in waste product monitoring in soils.

C. Miniature probe applications

1. *Semiconductor needle probes*

A semiconductor needle probe is a small (millimetre or submillimetre in diameter) needle-shaped semiconductor detector used to determine regional radioisotope activity under *in vivo* or *in vitro* experimental conditions. The needle semiconductor probe was first devised by Lauber and Rosencrantz (1964) and several modifications have since been made by Grangsjo *et al.* (1966), Lauber (1970, 1972), Lauber and Malmsten (1970) and Caine *et al.* (1978a,b) to improve its sensitivity to beta and gamma radiation by increasing the detector size or reducing the probe wall (dead layer) thickness. The needle probe developed by Lauber and coworkers was a silicon p–n-junction semiconductor detector. It has been applied to the *in vivo* measurement of beta particle tracer activity in animal tissue by Grangsjo *et al.* (1966), Wolgast (1968), Pinter (1969) and Lauber and Wolgast (1972) in regional blood flow studies in animals.

The needle probe designed by Caine *et al.* (1978a,b) contains a HgI_2 detector. The HgI_2 detector has been gaining in popularity because of its high atomic number (80 for Hg as compared to 32 for Ge and 14 for Si) and its large band gap of 2.1 eV, whereas the band gaps for Ge and Si are 0.66 and 1.1 eV, respectively. As reported in a review by Miller (1977) a band gap of $\sim$1.4 eV or greater gives rise to thermally generated currents sufficiently small to permit a high resolution detector at room temperature operation. Also the higher atomic number increases the detectors' efficiency for trapping the gamma-ray energy through photoelectric, Compton and pair production conversion processes. The history, fabrication, theory of operation and possible future use of HgI_2 detectors are given in a recent review by Whited and Schieber (1979). A cross-sectional diagram of the HgI_2 needle probe designed by Caine *et al.* (1978a,b) is given in Fig. 5.8. The detector is fabricated by cleavage of a HgI_2 single crystal grown from the vapour phase. The cleaved wafer is sawn to size, palladium wires are attached and the detector unit is encapsulated in Humiseal.

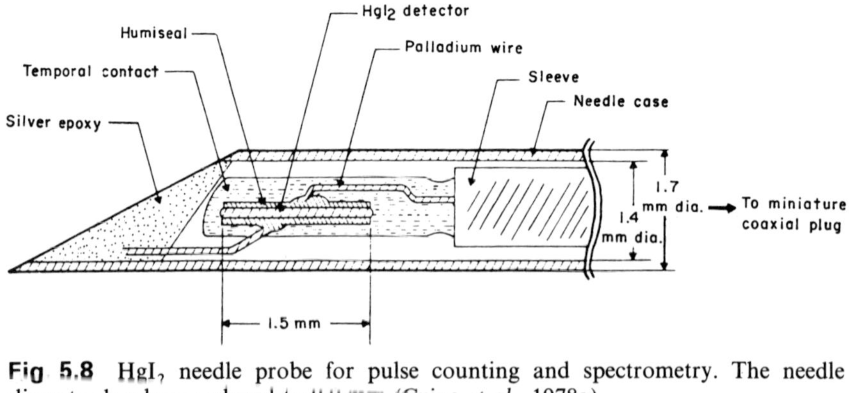

Fig. 5.8 HgI_2 needle probe for pulse counting and spectrometry. The needle diameter has been reduced to 0.9 mm (Caine *et al.*, 1978a).

The tube walls of the Lauber needle probe (Lauber and Wolgast, 1972) may be of 0.05 mm thickness, or extra thin (0.015 mm). The wall thickness greatly influences the absorption of beta particles and the mechanical strength of the probe. Probes with a 0.05 mm wall thickness are directly introduced into animal tissue; however, the low mechanical strength of the thin-walled probe requires that an incision be made before insertion. A compromise is made between mechanical strength and beta- or gamma-ray absorption by the wall. The absorption of beta radiation by the probe wall is highly significant for weak beta-particle emitters such as ^{14}C (E_{max} = 0.15 MeV) which for most applications would require the thin-walled probe. Background count rates vary from a few tenths of a count per minute to a few counts per minute as controlled by the discriminator setting. The background noise counts are dependent on the discriminator setting, as selected for the optimum signal-to-noise ratio for given experimental conditions. The count rates from signal pulses may vary from 50 to 10^4 cpm for 0.15–0.50 MeV maximum beta-particle energy sources of 1 $\mu Ci\ ml^{-1}$ activity. Much higher count rates are obtained with higher-energy beta particles such as the 1.7 MeV maximum energy particles emitted from ^{32}P. The following conditions are responsible for the rapid reduction in count rate (sensitivity) with decreasing beta-particle energies:

(i) The active volume of the tissue or liquid viewed by the probe decreases with decreasing beta-particle energies due to shorter beta-particle ranges at lower energies.
(ii) The absorption of beta particles by the probe wall increases with decreasing beta-particle energy.
(iii) Of the total pulses produced in the detector by the low-energy beta-particle-emitting nuclides, more are rejected by the electronic discriminator than of those pulses produced by high-energy beta-particle-emitting nuclides. This is because a greater portion of the pulses produced by the lower-energy beta-particle emitters are of the same height as the noise pulses. For average beta-particle energies >0.3 MeV, the influence of the discriminator is reported to be negligible with the Lauber needle probe. Consequently, high-energy beta-particle emitters such as ^{32}P (E_{max} = 1.7 MeV) may be counted with high signal-to-noise ratios.

The needle probes work most efficiently with gamma-ray-emitting isotope tracers. An example may be taken from the work of Sasaki and Pitt (1976) employing ^{85}Kr to measure regional myocardial blood flow in dogs.

2. *Implantable and surface-located encapsulated detectors*
Semiconductor detectors of surface area larger than those discussed in

Section IV.C.1 above may yield higher radiation detection efficiencies; however, they cannot be inserted into tissues and organs as needle probes. These detectors may be encapsulated within an inert material, such as stainless steel, and implanted or placed on the surface of animal organs or plant tissue. Such a detector consisting of Si(Li) encapsulated in stainless steel is described by Khouri *et al.* (1977) for the determination of myocardial blood flow in experimental animals using ^{85}Kr. A novel experimental application with plants has been made by Petit and van de Geijn (1978) and van de Geijn and Petit (1978). They report the determination of internal distribution processes and migration kinetics of ^{115m}Cd in the stems of tomato (*Lycoperscion esculentum* Mill.) by placing two or three silicon surface barrier detectors (ORTEC, 25 mm^2 × 300 μm thick) at selected positions along the plant stem. The beta particles emitted from ^{115m}Cd have an $E_{max} = 1.63$ MeV and are therefore comparable to those emitted from ^{32}P with an $E_{max} = 1.71$ MeV. Thus, this experimental approach could also be attempted with ^{32}P in plants as the beta particles of ^{32}P would have a slightly higher penetration power and should be detected more efficiently, if only slightly, than those emitted from ^{115m}Cd.

Some radiobiological applications of the semiconductor radiation probes described previously are reported by Martini (1973). Among these may be cited (i) the use of ^{32}P to measure renal and cerebral blood flow in animals; (ii) the estimation of hepatic blood flow with colloidal ^{198}Au; (iii) the measurement of retinal blood flow in animals employing ^{85}Kr tracer; (iv) the detection of corneal lesions in animals by measuring the disappearance rate of [^{75}Se]sodium selinite; and (v) the use of ^{85}Kr to measure blood flow in intact dental pulp.

3. *Windowless detectors*

In addition to those previously mentioned, numerous types of semiconductor detector have been developed in the early work of Appelgren *et al.* (1965), Brock *et al.* (1967), Morley and Parker (1967), Abson *et al.* (1968), Takayanagi *et al.* (1968) and Ueda (1969) for *in vivo* tracer studies with animals. These detectors usually have a thick metal or plastic window which greatly restricts their efficient use as detectors of low-energy beta-particle-emitting nuclides such as ^{14}C ($E_{max} = 0.15$ MeV). The previously mentioned needle probes may employ a very thin metal wall; however, significant absorption of weak (low-energy) beta particles still occurs.

A windowless semiconductor radiation detector was developed by Kobayashi *et al.* (1972). They applied this detector to the *in vitro* continuous measurement of $^{14}CO_2$ released by the microbial oxidation of ^{14}C-labelled organic compounds in culture vials. The detector consists of a silicon p–n junction with a thin film of silicon dioxide ~300 nm thick. For *in vivo* applications the detectors are mounted in headers described by Kobayashi *et al.* (1973). Detectors with window areas of up to 1 cm^2 are designed for insertion into living body cavities or implantation into living body tissue. For *in vitro* applications, detectors ranging from 4 to 200 mm^2 in area are prepared from p–n junction diodes. The diode is affixed to the can of an ordinary transistor with an aperture at the top which exposes the sensitive area of the diode, and is connected to the headers which contain electrodes.

A discriminator level of 15 keV produces background counting rates of ~1 cpm. The counting efficiency with a discriminator level of 15 keV is reported as 200 cpm cm^{-2} of a ^{14}C solid source. A 20 mm^2 area detector may be used for the *in vivo* measurement of low-energy beta particles such as those emitted from ^{14}C and ^{35}S.

Kobayashi *et al.* (1974) report the application of this windowless detector for the telemetric measurement of the dynamics of radioisotope tracers in the organs of free-ranging animals. The system consists of an implantable windowless silicon radiation detector protected with a SiO_2 film, a transmitter package, an FM receiver and a data handling unit. The advantage offered by this new technique is that a cable between the implanted radiation detector and the radiation-measuring electronics is not necessary. In addition, more reliable results of *in vivo* studies are obtained when the animal is permitted to roam free rather than restricted to cage confinement. Improvements in the manufacture and design of these silicon semiconductor detectors have been reported by Kobayashi *et al.* (1980).

4. *Other semiconductor detectors and their applications*

Cadmium telluride (CdTe) is being investigated as a practical semiconductor detector in experimental animal probes. Meyer *et al.* (1975) and Taguchi *et al.* (1975) have demonstrated its usefulness, which is due to the greater gamma-ray absorption obtained because its elemental constituents possess a higher atomic number than silicon and germanium (Cd = 48, Te = 52). It has been applied by Cerretelli *et al.* (1978) to the measurement of muscle blood flow from the clearance rate of ^{133}Xe. Bojsen and Rossing (1978) report the application of CdTe detectors in the biotelemetric determination of kidney function, i.e. glomerular filtration rate, from the disappearance rate of intravenously injected chelated ^{99m}Tc. The CdTe detector has been reviewed by Whited and Schieber (1979).

The semiconductor detector is now being applied to multi-label tracer experiments. The high resolution obtainable with semiconductor detectors permits the analysis of more than one and even several gamma- and X-ray-emitting radionuclides in the same sample with the detector connected to a multi-channel analyser. Some reports are (i) the simultaneous measurement of ^{99m}Tc and ^{131}I in the rat (Meyer *et al.*, 1972); (ii) the determination of potassium and sodium transport in the guinea-pig using ^{42}K and ^{22}Na (Konishi *et al.*, 1978); (iii) blood-iron studies by the simultaneous measurement of ^{59}Fe and ^{55}Fe even in the presence of ^{57}Co and ^{51}Cr (Kumpulainen and Saukkonen, 1979); and (iv) glomerular filtration rates by external measurements of ^{51}Cr-EDTA clearance (Bojsen *et al.*, 1981).

Some reviews on semiconductor radiation detectors are given by Siffert (1978), Ewan (1979) and McKenzie (1979).

V. Liquid scintillation counting

Liquid scintillation counting in agricultural research is the most popular experimental tool for the quantification of radiotracers, at least where beta-particle-emitting nuclides are concerned. The technique has been expanded to the analysis of alpha-, weak gamma- and X-ray-emitting nuclides. The popularity of liquid scintillation counting is a consequence of numerous advantages, among which are the high efficiency of detection, the improvements in sample preparation techniques, automation and the spectrometer capability permitting the simultaneous assay of more than one different nuclide.

A. General theory

Liquid scintillation counting has its origin in the independent discovery by Kallman (1950) and by Reynolds *et al.* (1950) that certain organic chemicals emit fluorescence upon being bombarded by nuclear radiation. Thus, the fluorescence or photons emitted by the organic compounds (fluors) as a result of excitation can be readily converted to a burst of electrons with the use of photocathode and photomultiplier (PM) tube, and subsequently measured as an electronic pulse.

Liquid scintillation counting normally entails the use of 20 ml capacity glass or plastic vials to contain the radioactive sample and fluor system. The fluor consists of a solution of the organic fluorescent compounds in a suitable organic solvent. Numerous fluors may be used, and these are decribed in more detail later in this chapter.

The sample vial is placed mechanically between two PM tubes, as illustrated in Fig. 5.9, which can simultaneously detect photon emissions from the vial. The simultaneous detection of a given fluorescence emission produces separate and simultaneous electrical pulses in the two PM tubes which pass on to a coincidence circuit after passing through preamplifier, linear amplifier and discriminator circuits. The coincidence circuit offers the advantage that only the electrical pulses produced simultaneously by the two

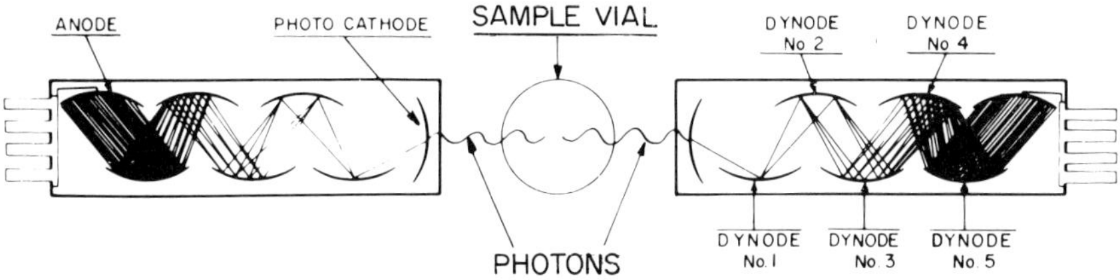

Fig. 5.9 Photomultiplier tubes with sample vial.

PM tubes can continue on and be registered by the scalers. This prevents the registering of pulses produced by the PM tubes which are a result of electronic noise, since the noise in the two independent PM tubes is not normally produced in coincidence. In this manner, the background count resulting from electronic noise is reduced to a minimum.

In older PM tubes with caesium–antimony photocathodes, thermal noise (electronic noise produced by elevated temperatures) is very significant. Consequently, with such tubes refrigeration is utilized in addition to the above-mentioned coincidence circuitry to reduce background counts produced by this source of noise. In modern instruments, which employ PM tubes with the bialkaline photocathode containing potassium, refrigeration is not necessary to reduce noise, although it is sometimes employed to maintain constant sample temperatures needed for reproducible counting efficiencies in regions where large variations $>15\,°C$ in room temperature are encountered. Temperature effects in liquid scintillation counting are reported by Stanley (1980) and reviewed in detail by ten Haaf and Kuipers (1977).

The photoelectrons produced at the photocathode are then accelerated towards a positively charged dynode (Fig. 5.9). The acceleration of the photoelectrons produces numerous secondary electrons upon impact with the positive dynode. Each subsequent dynode possesses an increased positive potential to further accelerate the new electrons produced so that upon impact each new electron previously produced will "knock out" even more electrons. In this manner, after having been accelerated through several dynodes, one photoelectron produces an avalanche of secondary electrons which are collected at the anode. Thus, the current output from the photoelectrons may be amplified by a factor of 10^6–10^9 after acceleration from dynode to dynode, with the final current produced being proportional to the intensity of the photons emitted from the vial.

The proportionality between the pulse intensity (height) produced and the initial photon intensity is of great use with the liquid scintillation counter as a spectrometer. After passing through the preamplifier and linear amplifier circuits, the voltage pulse must pass through a discriminator circuit. The discriminator circuitry may be controlled so that the pulses below a selected minimum height and above a selected maximum height are rejected. The pulses which pass through this window continue on to the coincidence circuitry and scaler. If the discriminator circuitry is such that two or three minimum and maximum discriminator settings (windows or channels) can be established by the investigator, there are a corresponding number of scalers to register the number of pulses which pass each respective window. A common number of channels and corresponding scalers is three, although more may be utilized, as discussed later. This permits the simultaneous assay of three beta-particle-emitting nuclides which differ significantly in their maximum beta-particle energies.

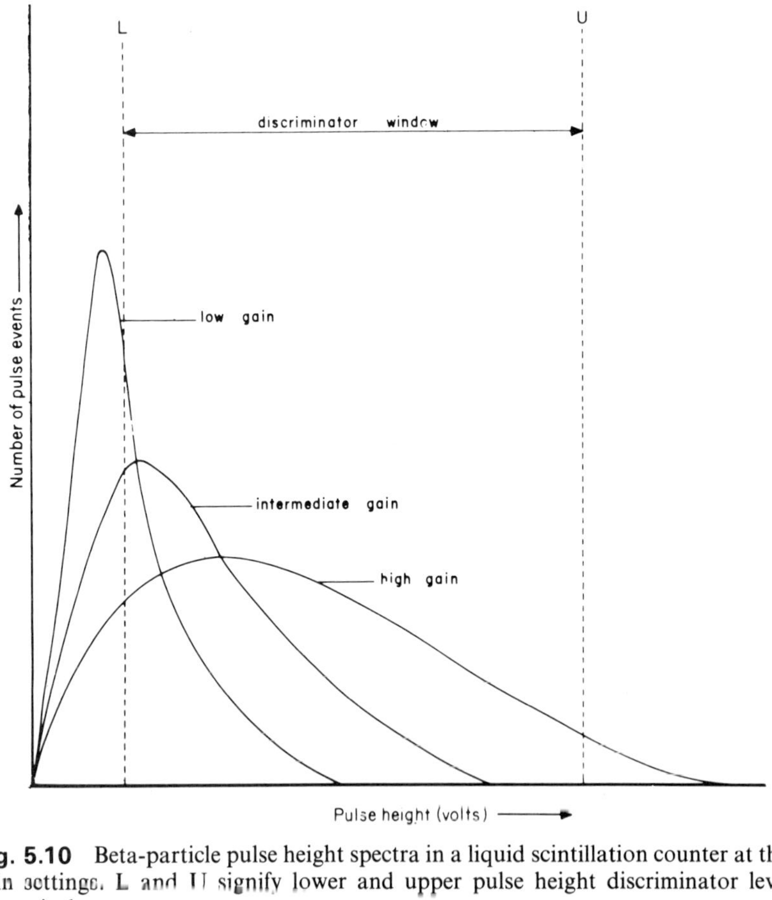

Fig. 5.10 Beta-particle pulse height spectra in a liquid scintillation counter at three gain settings. L and U signify lower and upper pulse height discriminator levels, respectively.

Prior to passing on to the pulse height discriminator circuitry, the pulse is amplified to a level within the limited range of voltage values (e.g. 0.5–9.9 V) which may be received by the channel analyser or pulse height discriminators. The range of voltage values may be controlled by an instrument setting most often referred to as the amplifier gain or attenuator, which permits energy discrimination over the range of the beta-particle pulse height spectrum (Fig. 5.10). This is so because if one wishes to count a specific range of voltage values which pass the amplifier, the lower and upper discriminators are set so that only those voltages above the lower discriminator and below the upper discriminator pass on to the scalers to be counted.

B. Liquid scintillation fluors

The fluors previously mentioned usually consist of two organic, fluorescing compounds referred to as the primary and secondary fluors. These fluors are dissolved in an organic solvent such as toluene, *p*-dioxane or pseudocumene. The mode of action and criteria used in the selection of fluors and fluor solvent are discussed further on in this section.

The solvents are toxic and very flammable with relatively low flash points. The less hazardous solvent pseudocumene (1,2,4-trimethylbenzene) has a much higher flash point (49 °C) compared to toluene (7 °C) and *p*-dioxane (12 °C); and it has been found very useful as a solvent compatible with water, as reported by Nibeck *et al.* (1980). A new and promising non-volatile solvent is silicon oil, reported by Kato and Hatagami (1980). It is comparable in efficiency to toluene and *p*-dioxane as a fluor solvent and is an excellent absorber of radioactive gases in air. Toluene is not miscible with aqueous samples and consequently can only be used for assay of radionuclides dissolved in organic solvents.

Many primary and secondary fluors are available. A commonly used fluor solution for the radioassay of non-aqueous samples consists of 3.0 g of 2,5-diphenyloxazole (PPO) as primary fluor and 0.1 g of 2,2′-*p*-phenylene-bis-5-phenyloxazole (POPOP) as secondary fluor dissolved in 1 litre of toluene. Two common fluor cocktails for the assay of aqueous samples are (i) 7.0 g of PPO, 0.05 g of POPOP and 50 g of naphthalene dissolved in 1 litre of *p*-dioxane, and (ii) Bray's solution as reported by Nibeck *et al.* (1980), consisting of 4.0 g of PPO, 0.20 g of POPOP, 20 ml of ethylene glycol, 60 g of naphthalene and 1 litre of *p*-dioxane. The formulation of other fluor cocktails, as well as several ready-made and commercially available cocktails, are described by L'Annunziata (1979) and Nibeck *et al.* (1980).

The fluors are low in concentration, with the secondary fluor being approximately less than one-tenth the concentration of the primary fluor. Due to the low concentration of fluor, the energy dissipated by the nuclear radiation of the sample is absorbed to a greater extent by the solvent molecules than by the fluor molecules. A portion of the absorbed energy results in the excitation of the π-electrons of the solvent molecules. This energy is then transferred to the less-abundant fluor molecules, particularly to the primary fluor molecules, which upon excitation emit this energy as photons in the visible or near-u.v. region. The photons may not be of the wavelength detected most efficiently by the photocathodes of the PM tubes. If this is the case, secondary fluor molecules are employed which can absorb the photon energy emitted by the primary fluor molecules and re-emit this energy as photons of a longer wavelength more efficiently detected by the photocathodes. A good review of the theoretical aspects of such solvent–fluor energy transfers in liquid scintillation counting is given by Horrocks (1976b). Secondary fluor wavelength shifters often are not required with instruments employing the new bialkaline photocathodes.

C. Optimal instrument settings

It was mentioned previously that the pulse height produced in a liquid scintillation counter is proportional to the photon intensity emitted from the fluor. The photon intensity, in turn, is proportional to the nuclear particle energy absorbed by the solvent and fluor molecules. The magnitude of the proportionality factor between the pulse height and the nuclear particle energy is governed by the high voltage applied to the photomultiplier and the amplifier gain (attenuator). In most present-day liquid scintillation spectrometers the photomultiplier high voltage is usually fixed. Consequently, the only instrument setting of concern in the control of pulse height is the amplifier gain.

To visualize the effect of amplifier gain on the pulse height spectrum produced by beta-particle radiation, consider the example shown in Fig. 5.10. The pulses within the discriminator window defined by the lower (L) and upper (U) level discriminators pass on to the scaler to be registered as counts. Those pulses outside of these discriminator levels are not registered. Because beta radiation from a given radionuclide is not monoenergetic, but rather consists of a continuous spectrum of energies up to a given maximum, the pulse heights produced in a liquid scintillation spectrometer consist of a continuous spectrum. At a low gain setting the amplification of pulses is a minimum and consequently the pulse height spectrum produced is characterized as that predominantly of low voltage levels. At higher gain settings pulses are amplified to higher voltage levels which result in a continuous spectrum of pulse heights predominantly of greater magnitude than those produced at lower gain settings.

In practice, the lower and upper discriminator levels are selected and fixed so that the majority of pulses produced by the radionuclide–fluor solution are counted, i.e. they fall within the window selected. This ensures the highest counting efficiency possible, with all other factors affecting the counting efficiency at an optimum. This is referred to as a "wide-open" counting window. With many modern liquid scintillation counters the counting employed is "completely open", that is, L at its lowest possible level and U at its highest. However, with the relatively older models, samples are seldom counted in such wide-open windows because small pulses originating from phototube thermal noise are appreciable and are eliminated by setting the lower discriminator at a level sufficient to reject them. Likewise, large pulses resulting from cosmic radiation can be rejected by setting the upper discriminator at a relatively low level. This technique produces minimal background counts. An example is illustrated in Fig. 5.11. A ^{14}C standard

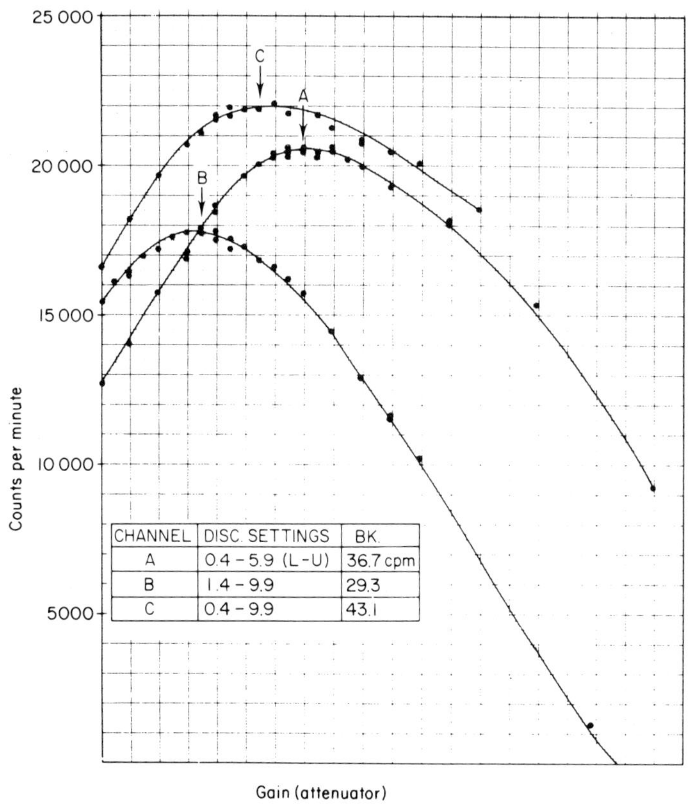

Fig. 5.11 Typical balance point curves for a liquid scintillation counter having variable gain controls. The balance points for ^{14}C in the discriminator settings chosen with each channel are indicated by arrows. BK represents background count rates.

with an absolute activity of 24 700 dpm in a PPO–POPOP–toluene fluor was counted in three channels designated A, B and C, respectively, with the discriminator levels set as noted in Fig. 5.11 for an Amersham/Searle Model 6860 (Mark I) liquid scintillation spectrometer which has variable gain controls. With this instrument the discriminator levels may be adjusted from 0.0 to 9.9 divisions from the lower to the upper end of the energy scale. Other scintillation spectrometers may have different window divisions, e.g. 0–1000 or 0–100 %.

In all of the channels that have variable gain settings it can be seen that as the amplifier gain is increased for each counting channel, the count rate rises to a maximum and then decreases. This effect can be explained by the fact that at low gain settings the pulse heights are predominantly of low magnitude, many of which are not counted because they fall below the lower discriminator level (Fig. 5.11). As the gain is increased, some pulse heights which were below that of the lower discriminator level may rise to a height within the range fixed by the counting (discriminator) window. Some pulses of high magnitude are also amplified to a height above that of the upper discriminator level. Thus, the count rate reaches a maximum when the pulses are amplified so that a majority fall within the preselected window, and then falls to a minimum when the gain is so high that the majority of the resultant amplified pulses are of a height superior to that of the upper discriminator level. For given lower and upper discriminator levels (settings), the optimum gain settings are those which correspond to the curve peaks or balance points (e.g. as indicated by arrows in Fig. 5.11). The balance point is thus the gain setting at which the number of pulses amplified out of the counting window is equal to (balanced by) the number of pulses amplified into the counting window.

It should be pointed out from the example of Fig. 5.11 that the highest count rate was observed in channel C where the discriminator window was "wide open". A higher background count rate was also observed in this channel. At relatively elevated sample counts, a background count rate such as that observed in channel C is considered low. A much lower sample count rate was observed in channel B because of the elevated lower discriminator (1.4); moreover, a much lower background count rate was observed. Notice that in channel A of this example fewer pulses were rejected by lowering the upper level discriminator than by raising the lower level discriminator. A higher background count rate was observed in channel A as compared to channel B, which indicates that pulses which produced background counts in this instrument were mostly of low magnitude.

In the operation of spectrometers with variable gain settings, the counting of samples at the balance point offers a twofold advantage. Firstly, the highest sample count rates and figure-of-merit are obtained at the balance point. Secondly, balance point operation offers the greatest stability of count rate because it is at this point that minimal changes in count rate occur as a function of gain setting.

D. Quenching and quench correction

The popularity of the liquid scintillation counting technique for the assay of beta-particle-emitting radiotracers rests largely on the high counting efficiencies attainable with this technique. These high counting efficiencies are mostly due to efficient beta-particle–solvent–fluor energy transfer processes, the overall high luminescence yield obtainable from solvent–fluor mixtures and the approximate 4π (360°) counting geometry obtained when the scintillation fluor encompasses the radioisotope atoms.

The principal problem affecting this relatively efficient detection method is that of the presence of quenching substances; these are chemical constituents of the sample which lower the overall quantum efficiency of the energy transfer–light output scintillation process and thereby reduce the counting efficiency. Quenching substances of principal concern to the experimenter may be classified, for the most part, as chemical quenchers and colour quenchers. Chemical quenchers interact "chemically" with the solvent–fluor system in such a way that they compete with the energy transfer processes of the solvent or fluor molecules. Colour quenchers absorb light in the visible region of the spectrum, and consequently reduce the intensity of photons reaching the photocathodes. Some common organic quenching substances are listed in Table 5.1. The classification is divided into strong quenchers, mild quenchers and diluters. The diluters, as described by Yang and Lee (1969), are compounds which absorb far u.v. radiation and merely lower the concentration of the excited solvent molecules by competing for the beta-particle energy.

Table 5.1 Classification of quenching substances[a] (Kerr *et al.*, 1957)

Strong quenchers	Mild quenchers	Diluters
RCOR	RCOOH	RCOOR
RCHO	RCH=CHR	ROH
R_3N	RNH_2	ROR
RSH	RSR	$(RO)_3PO$
RI	RBr	RCl
RNO_2		RF
		RH

[a] Dissolved O_2, not listed, is also a significant inorganic chemical quencher.

The effect of quenching on the sample counting efficiency may be determined and corrections made by (i) internal standardization, (ii) sample channels ratio (SCR) determinations, (iii) external standard channels ratio (ESCR) determinations, (iv) external standardization, (v) H-number ($H\#$), (vi) spectral index of the sample (SIS) and external standard (SIE) and (vii) relative pulse height (RPH) or inverse relative pulse height (RPH) or inverse relative pulse height (ESP). Each of these techniques is described below.

1. *Internal standardization*

In the internal standardization or "spiking" technique the sample is counted before and after the addition of a labelled standard to the sample, and the counting efficiency (E) is then calculated according to the equation

$$E = \frac{(\text{cpm})_{\text{s+i}} - (\text{cpm})_{\text{s}}}{(\text{dpm})_{\text{i}}}, \tag{5.1}$$

where $(\text{cpm})_{\text{s+i}}$ is the count rate of the sample after the addition of internal standard, $(\text{cpm})_{\text{s}}$ is the count rate of the sample before the addition of internal standard and $(\text{dpm})_{\text{i}}$ is the disintegration rate of the added aliquot of internal standard. The disintegration rate of the sample $(\text{dpm})_{\text{s}}$ may then be calculated as follows:

$$(\text{dpm})_{\text{s}} = (\text{cpm})_{\text{s}}/E. \tag{5.2}$$

The internal standard added to the sample must be labelled with the same radionuclide being counted in the sample, e.g. a tritium-labelled standard must be used with samples containing tritium. Hendee *et al.* (1972) showed that [^{3}H]toluene or [^{3}H]hexadecane are good internal standards for toluene-based fluors; and [^{3}H]toluene, [^{3}H]water or [^{3}H]hexadecane serve well for dioxane-based fluors when assaying for tritium. These compounds, labelled with ^{14}C, are good internal standards for counting efficiency determinations of samples containing ^{14}C.

Other important rules which must be followed with regard to internal standardization are (i) the standard must not quench significantly and (ii) the volume of the standard aliquot added to the sample vial must be so small that it has a negligible effect on the counting efficiency. Dobbs (1965) and Thomas *et al.* (1965) have investigated syringe dispensing techniques for the addition of internal standards to samples in scintillation vials. The method is good; however, it is more tedious than the techniques discussed below.

2. *Sample channels ratio* (*SCR*)

When quenching occurs there is a reduction in the energy transfer efficiency in the beta-particle–solvent–fluor energy transfer steps. This results in a change in the pulse height spectrum such as that illustrated in Fig. 5.12. The

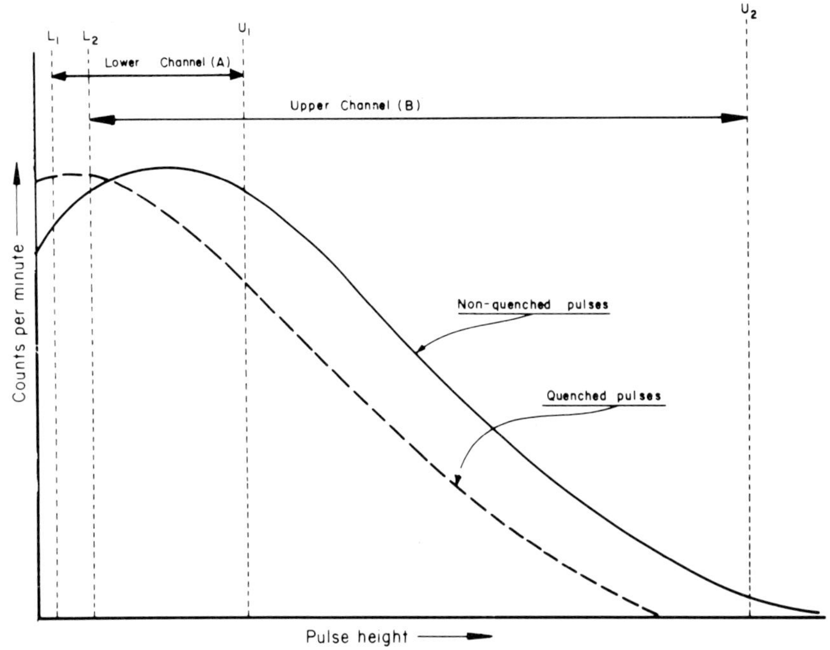

Fig. 5.12 Quenched and non-quenched pulse height spectra in relation to two overlapping channels (windows). The lower and upper channels are specified by the discriminators L_1, U_1 and L_2, U_2, respectively. The broken curve illustrates how quenching produces a change in the ratio of the pulses received by the two channels.

quenched pulses are of lower magnitude (height) than the non-quenched pulses, and consequently the pulse heights shift towards the lower end of the pulse size distribution spectrum. If the discriminator levels are selected so as to provide one channel which can receive pulses from only the lower end of the pulse spectrum and another which can receive most of the pulses (those of both low and high magnitude), this shift in pulse height due to quenching produces a change in the ratio of the pulses received by the two channels. The degree of spectral shift and magnitude of change occurring in the channels ratio are dependent upon the severity of quenching. Consequently, if a series of scintillation vials each containing the same amount of radioactivity but increasing amounts of quenching agent were counted, they would show a variation in the channels ratio and counting efficiency such as that illustrated in Fig. 5.13. The quench correction curve, once prepared for a given radiotracer, may be used as a standard curve in determining the counting efficiency of a sample from the channels ratio obtained. Although this

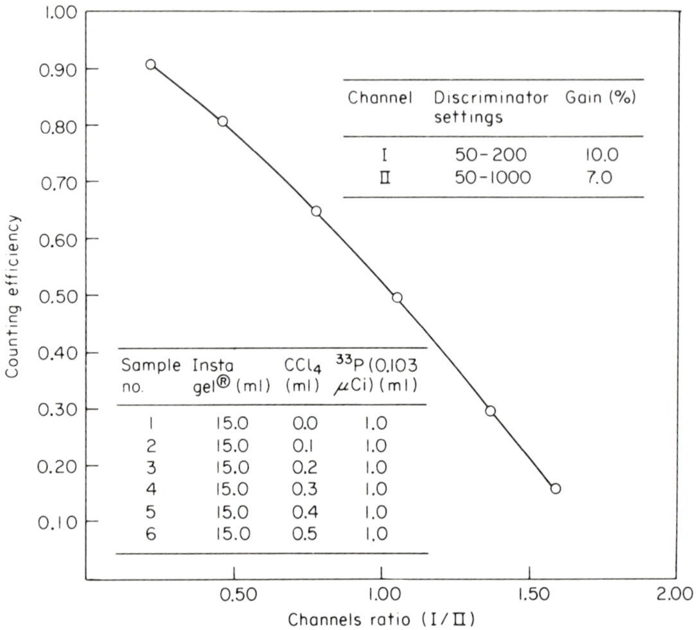

Fig. 5.13 Typical channels ratio quench correction curve. The channels ratio (I/II) represents the count rate of ^{33}P from channel I divided by the count rate from channel II. The discriminator and gain settings for each channel are given. The data were obtained from six samples each containing 15 ml of commercial fluor (Insta Gel) and 1.0 ml of [^{33}P]orthophosphate of known activity (0.103 μCi). Each sample contained increasing amounts of quenching agent (CCl_4) as described in the table inset.

method is reported to be reasonably independent of volume (Gibson, 1980), different quench correction curves are obtained for colour and chemical quenching methods, as demonstrated by ten Haaf and Verheijke (1977). A detailed procedure which may be followed for the preparation of SCR quench correction curves is given by L'Annunziata (1979).

The SCR quench correction curves may or may not be linear. If linear, fewer points are needed and a simple linear regression analysis can give the appropriate straight line. If non-linear, more calibration points may be required, and a second-order quadratic function approximation or curve fitting may be applied, as described by Parmentier and ten Haaf (1969). Barrows *et al.* (1976) report best fits to non-linear quench correction curves by using second-order polynomial regression. The major disadvantage of the SCR technique is that with samples of low count rate or high quench the counts in one or both of the channels may be so low that a channels ratio becomes meaningless, or long periods of counting time are required to achieve acceptable levels of statistical accuracy (Kolb, 1978).

3. *External standard channels ratio* (*ESCR*)
The external standard channels ratio technique for the determination of counting efficiencies is similar to the SCR method described above. The principal difference is that the channels ratio of an external rather than internal radioactive standard is obtained. Many liquid scintillation spectrometers contain installed gamma-ray-emitting radioactive standards which may be mechanically positioned external to a sample counting vial. That is, within the liquid scintillation counter an elevator can automatically position a radioactive standard or standards to one side of a liquid scintillation vial. Some gamma-ray sources used as external standards are ^{241}Am ($t_{1/2}$ = 458 y), ^{226}Ra ($t_{1/2}$ = 1620 y), ^{137}Cs ($t_{1/2}$ = 30 y) and ^{133}Ba ($t_{1/2}$ = 7.5 y). The relatively long half-lives obviate the necessity of making adjustments for radionuclide decay. The ESCR technique offers the advantage that the optimum channel widths and gains for the channels ratio determinations are often factory set to monitor quenching of scintillation events produced by Compton electrons which result from the interaction of gamma rays of the external standard with the scintillation solution and vial wall. In the previously described SCR technique the channel widths and gain settings which produce the best (most linear) quench correction curve must be determined experimentally and will differ from radionuclide to radionuclide.

In practice, the ESCR quench correction curve is prepared firstly by counting a series of variably quenched radionuclide samples of known activity in a preselected counting channel from which counting efficiency values are obtained. After each of the above sample counts are obtained, further counts are then made with the samples exposed to the external gamma-ray source. The external standard counts are made in two other preselected channels and the net count rate due to the external standard in these two channels is computed by subtracting from both channels those pulses contributed by the sample nuclides. The ESCR is obtained from the net count rates of the external standard in these two channels. A plot of counting efficiency versus ESCR is then made. A detailed treatment on this method is given by Stanley (1976) and by Cummins and Horne (1977).

4. *Automatic external standardization* (*AES*)
This method involves the counting of an external standard(s), such as ^{226}Ra, in one channel and the determination of the relationship between the drop in count rate or counting efficiency of the external standard and the drop in the counting efficiency of the sample. A series of variably quenched nuclide samples of known activity are prepared and counted in one channel from

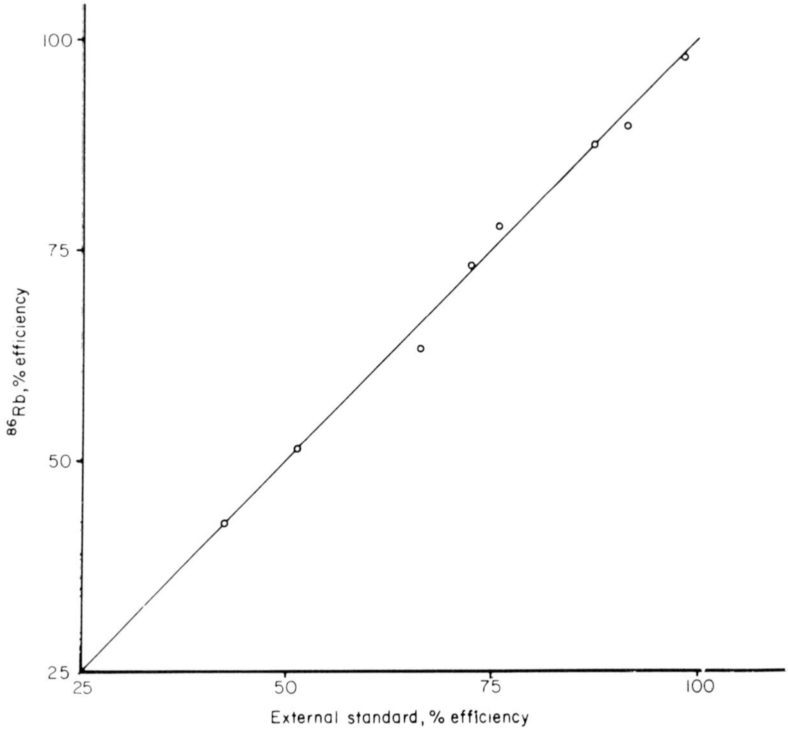

Fig. 5.14 External standard quench correction curve for the Cherenkov counting of ^{86}Rb in plant material (Brownell and Lauchli, 1969).

which values of sample counting efficiency are calculated. An external standard count is made in another channel with a discriminator level set to reject pulses/counts originating from the sample. The effect of quenching on the external standard count rate or counting efficiency is obtained from this second count. A graph can then be plotted as illustrated in Fig. 5.14. Parmentier and ten Haaf (1969) conclude that this method has drawbacks because the quench correction curve depends on the volume of the sample and the type of quenching.

5. *H-number* ($H\#$)

The H-number ($H\#$) concept, first reported by Horrocks (1976a,b, 1978a,b), is a popular method for quench correction in liquid scintillation counting. The technique involves the irradiation of different samples, varying in their degree of quench, with an external radionuclide source (e.g. ^{137}Cs—^{137m}Ba) emitting monoenergetic gamma radiation. The radiation reaching the

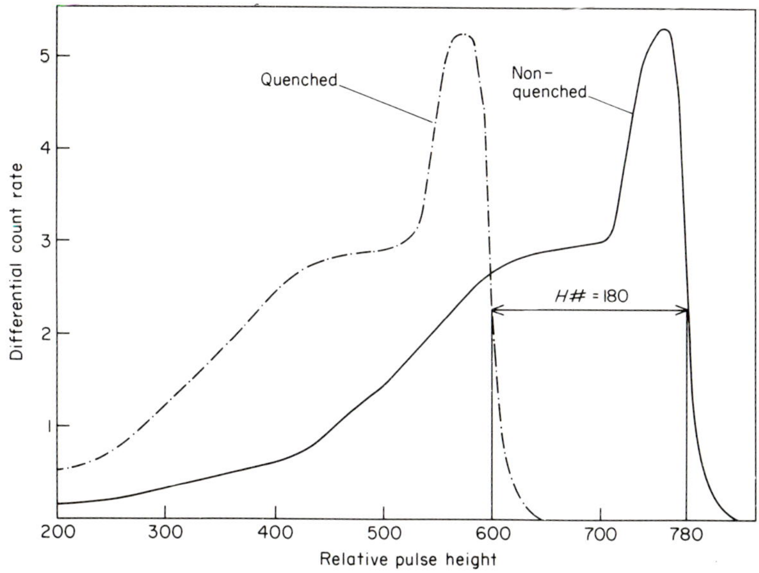

Fig. 5.15 Hypothetical pulse height spectra for Compton scatter electrons produced in quenched and non-quenched liquid scintillation fluor by the 0.662 MeV gamma rays from a ^{137}Cs–^{137m}Ba external standard.

samples consists exclusively of 0.662 MeV gamma rays as the 0.032 MeV X-rays from ^{137m}Ba are absorbed by the source container. Via the Compton effect (see L'Annunziata, 1979), the gamma radiation produces a spectrum of Compton scatter electrons of varying energies between zero and E_{max} in the scintillation fluor medium. The energies of the electrons are constant from sample to sample; however, the pulse heights produced by the electrons through scintillation can vary and are a function of the amount and type of quenching agent in each sample. The Compton scatter electrons produce a spectrum of pulses as illustrated in Fig. 5.15. If we consider only those Compton scatter electrons of E_{max}, these will produce pulse heights of maximum magnitude as a peak referred to as the Compton edge. The magnitude of the pulse height at the Compton edge is maximum for a sample free of quenching agents and saturated with nitrogen gas (non-quenched sample). The Compton edge for quenched samples is encountered at lower pulse heights than those for the non-quenched sample; and the degree of spectral shift is a function of the amount of quench in the sample. A measure of the degree of spectral shift or difference in E_{max} pulse response is called the $H\#$ and is a measure of the amount of quench in a sample.

In practice, the $H\#$ concept is applied to the quench correction of samples counted with liquid scintillation spectrometers equipped with logarithmic pulse height conversion. Such systems convert the initial pulse height response to the logarithm of the pulse height. Thus, initial pulse responses which may differ by a factor of 1000, may be handled by a single amplifier and pulse height analyser and, as reported by Horrocks (1978c), there is a constant logarithmic difference between response relationships at different quench levels. For example, a 50 % reduction in photon yield or an increase in quench by a factor of 2 represents a constant difference of 0.301 or the logarithm of 2 between the logarithmic response relationships of different quench levels. This is illustrated in Fig. 5.16.

With logarithmic response relationships, the measured pulse height (PH) may be defined, using the notation of Horrocks (1976a,b, 1978b,c), as

$$\mathrm{PH} = a + b \log E, \tag{5.3}$$

where PH is commonly expressed in discriminator divisions, a is the pulse height response for a 1 keV electron and b is the slope of the energy response

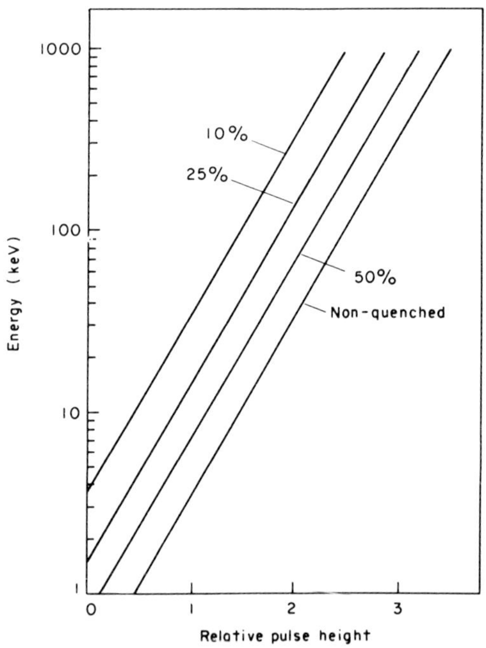

Fig. 5.16 Relative quench effect on the relative logarithmic response for different electron energies. Curves are marked to indicate logarithmic response curves at quench levels of 50, 25 and 10 % compared to the pulse height response from non-quenched scintillation media. Curves marked 25 %, 50 % and non-quenched are separated by a constant value of 0.301 relative pulse height (Horrocks, 1978c).

curve. For different levels of quench the slope b will remain constant but the valuc of a will differ. For a non-quenched sample the measured pulse height (PH_0) may be defined as

$$PH_0 = a_0 + b \log E. \tag{5.4}$$

Likewise, for a quenched system the measured pulse height (PH_q) may be written as

$$PH_q = a_q + b \log E. \tag{5.5}$$

The difference between the measured pulse height responses for electrons of the same energy (e.g. E_{max} from ^{137}Cs Compton edge) in non-quenched and quenched systems is defined as the $H\#$ or

$$H\# = PH_0 - PH_q \tag{5.6}$$

$$= a_0 - a_q + b \log E - b \log E \tag{5.7}$$

$$= a_0 - a_q. \tag{5.8}$$

The $H\#$ is determined by taking the difference between the relative pulse heights at the inflection points of ^{137}Cs Compton spectra of non-quenched and quenched samples as illustrated in Fig. 5.15. In the hypothetical case drawn up in Fig. 5.15 an $H\#$ of 180 is illustrated. Greater and lesser degrees of quench will result in a corresponding variation in the $H\#$. The inflection points of the quenched and non-quenched samples are determined automatically by multichannel analysis while exposing the samples to an external ^{137}Cs—^{137m}Ba source as reported by M. T. O'Neill in Horrocks (1978b). The multichannel analyser divides the pulse height scale into narrow channels and accumulates the counts in each channel over a given period of time. A microprocessor then compares the counts in each channel to define the Compton spectra and precisely locate the inflection points.

With certain liquid scintillation spectrometers of Beckman Instruments, Inc., the inflection point of the Compton edge produced by a non-quenched fluor is factory set at 780 discriminator units, as reported by M. T. O'Neill in Horrocks (1978b). Quenched samples produce Compton edges with inflection points at lower discriminator levels and the magnitude of the difference defines the $H\#$ (Fig. 5.15).

In practice, a standard curve is prepared to relate counting efficiency to $H\#$. This requires the preparation of a set of standards in liquid scintillation vials containing the same and known activity (dpm) of radionuclide and increasing amounts of quenching agent in scintillation fluor. These standards are then counted at optimal window settings and the percentage counting efficiency ((cpm/dpm) $\times$ 100) for each standard is plotted against the $H\#$ as illustrated in Fig. 5.17. The counting efficiencies of unknown samples are determined from their $H\#$ and the standard curve.

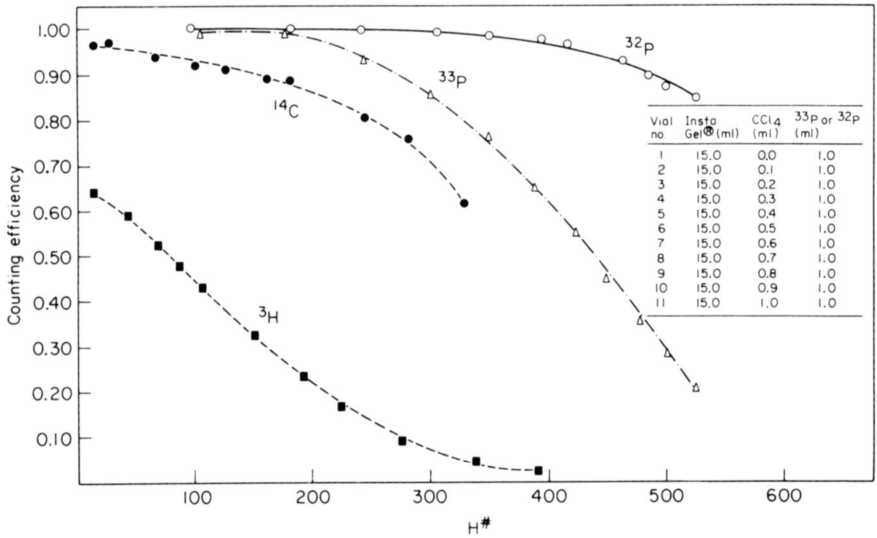

Fig. 5.17 Plots of counting efficiencies of ^3H, ^{14}C, ^{33}P and ^{32}P in scintillation fluors of different quench levels as a function of the $H\#$. The ^3H and ^{14}C plots were obtained from commercially prepared quenched standards; those of ^{33}P and ^{32}P were obtained from standards prepared with Insta Gel® fluor and CCl_4 quenching agent according to the table inset.

The $H\#$ technique offers certain advantages over other conventional quench correction methods such as SCR and ESCR, as reported by Long (1977), Kolb (1978) and Kolb and Fujimoto (1978, 1979). These advantages are as follows: (i) any sample can have only one $H\#$ value, contrary to channels ratio techniques where one sample may produce many different channels ratios depending on the channel widths and gains selected; (ii) the $H\#$ technique results in less variable quench correction curves over a wider range of counting efficiency; (iii) proper calibration with the $H\#$ of a given sample is constant from instrument to instrument, although the counting efficiency may not necessarily be constant; and (iv) quench correction curves using $H\#$ are identical for all fluor cocktails, and all chemical and colour quenching agents.

6. *Relative pulse height* (*RPH*) *and external standard pulse* (*ESP*)

The relative pulse height (RPH) and external standard pulse (ESP) quench correction techniques are similar in concept to the previously described $H\#$ quench monitor procedure. In the ESP technique reported by Laney (1976, 1977) and evaluated by McQuarrie *et al.* (1980) the liquid scintillation spectrometer measures the degree of quench in a sample by the shift in the

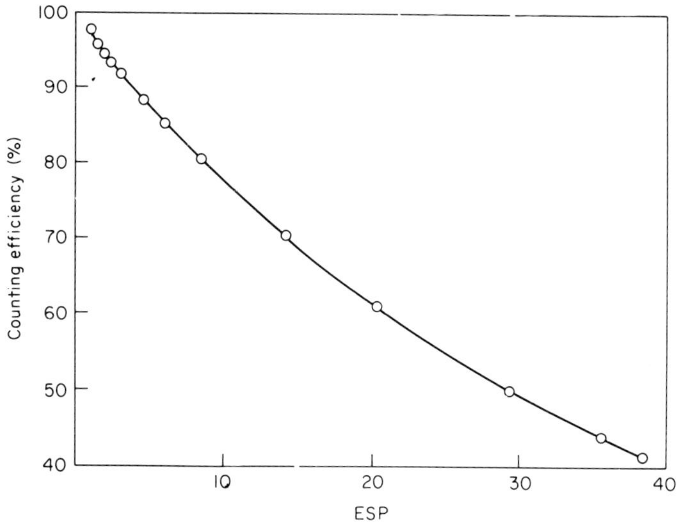

Fig. 5.18 Variations in counting efficiency with ESP, using quenched ^{14}C standards (plotted from data of McQuarrie *et al.*, 1980).

average pulse height (P_s) due to Compton electrons produced by an external ^{133}Ba gamma-ray source as compared to the average pulse height (P_r) produced in a sealed non-quenched reference vial stored in the elevator mechanism of the counter. The shift in the average pulse heights is defined by the ratio

$$\text{ESP} = \frac{P_r - P_\infty}{P_s - P_\infty}, \tag{5.9}$$

where P_∞ as defined by Laney (1976) and McQuarrie *et al.* (1980) is a correction term corresponding to Compton electrons produced by the external gamma-ray source under "infinite" quenching. In ESP determinations the entire external standard pulse height spectrum of quenched and non-quenched samples is stored in a multichannel analyser from which the average pulse height is determined. As previously described in the case of $H\#$, a quench correction curve is prepared relating counting efficiency with ESP, using a set of quenched standards containing the same activity of radionuclide and increasing amounts of quenching agent in fluor solution. A linear curve is obtained by plotting counting efficiency against ESP as illustrated in Fig. 5.18. Laney (1976) defines ESP as the reciprocal of the relative pulse height of the external standard, $[\text{RPH}]^{-1}$, as more linear quench correction curves are obtained by plotting efficiency versus reciprocal RPH.

The real difference between the ESP and *H*# quench monitor techniques lies in the method of measuring the pulse heights in quenched and non-quenched samples. *H*# determinations are made from a small portion of the pulse height spectra, that is, the inflection point of the upper edge of the Compton spectra. McQuarrie *et al.* (1980) indicate that greater accuracy might be expected in ESP quench monitoring because a larger number of pulses are used to measure the average pulse height; however, they also indicate that the *H*# can be precisely determined because it involves locating only the steep upper edge of the Compton spectrum.

7. *Spectral index of the sample (SIS) and spectral index of the external standard (SIE)*

The automated quench correction techniques previously described entail the use of channels ratios (e.g. SCR and ESCR) or the determination of differences in magnitude of the average pulse heights of external standard Compton spectra (e.g. RPH and ESP) or inflection points at the Compton edge (e.g. *H*#) of quenched and non-quenched samples. Other quench monitor methods have recently been reported by Everett *et al.* (1980) and Ring *et al.* (1980) under the designation of spectral index of the sample (SIS) and spectral index of the external standard (SIE). The SIS and SIE methods are different in that channels ratios are not used and more complex analysis of the "entire" pulse height spectra is performed as reported by Ring *et al.* (1980).

In counting systems employing SIS and SIE the pulses produced from photon events are linearly amplified, digitalized and stored to produce the complete pulse height spectra in a region from 0 to 2000 keV. Some examples of pulse height spectra as stored in instrument memory are illustrated in Fig. 5.19. The ultimate objective in spectral analysis is to characterize pulse height spectra to the extent of quantifying the various features of the pulse height distribution and any changes in these features which can occur as a result of quenching.

Ring *et al.* (1980) classify features of pulse height spectra as "local" and "overall" features. Features considered as local are (i) the spectral peak or most frequent pulse height, (ii) the maximum pulse height, (iii) the slope at various points, and (iv) the Compton edge or its inflection point when considering pulse height spectra produced by external standards. Overall features are (i) the total counts collected over defined regions of the spectra, (ii) moments of the pulse height distribution, and (iii) mathematical functions defining the pulse height distribution. Local features describe only a small portion of a spectrum while overall features define a very large portion, if not all of it.

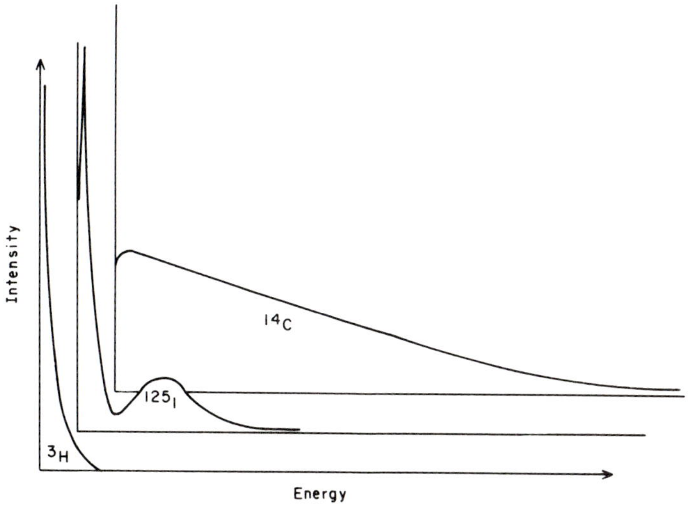

Fig. 5.19 Spectra of ^{3}H, ^{125}I and ^{14}C as stored in a Packard Model Tri-Carb 460C (Everett *et al.*, 1980).

Because quenching causes changes in the features of pulse spectra, a feature or a number of features of pulse spectra, if defined by mathematical relations, can serve as indices to monitor for quench. Such indices as coined by Ring *et al.* (1980) are the SIS (or spectral index obtained from the sample spectrum) and the SIE (or spectral index obtained from the external standard spectrum). As described by Ring *et al.* (1980), the starting point for either index is the first moment $M(1)$ of the pulse height distribution, which is approximated by

$$M(1) = \sum_{L}^{U} X \cdot Y(X) / \sum_{L}^{U} Y(X), \tag{5.10}$$

where $Y(X)$ represents the number of counts with pulse heights between X and $X + \Delta X$, and L and U are the lower and upper limits of the summation along the pulse height axis. When SIS determinations are made, the value of L is set as low as possible. For SIE, the lower limit is set just above the "wall effect" spectrum, which are pulses of very low magnitude produced by a weak scintillation effect occurring when toluene-based liquid scintillation fluor penetrates the walls of plastic vials exposed to the external standard gamma radiation. The upper limit, U, for both indices is defined by the end of the spectrum.

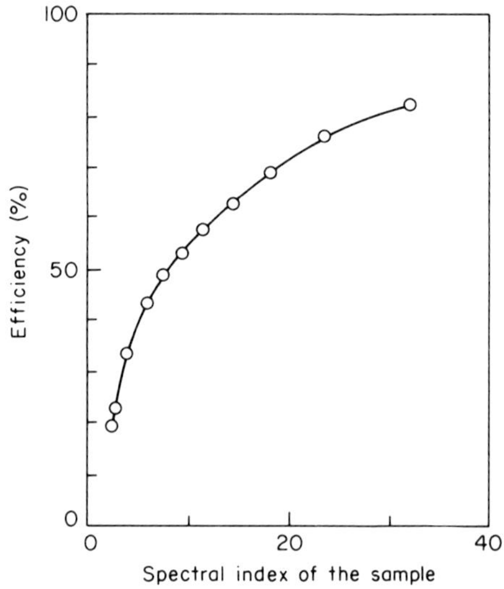

Fig. 5.20 Iodine-125 efficiency correlation curve based on the spectral index of the sample (SIS) (Ring *et al.*, 1980).

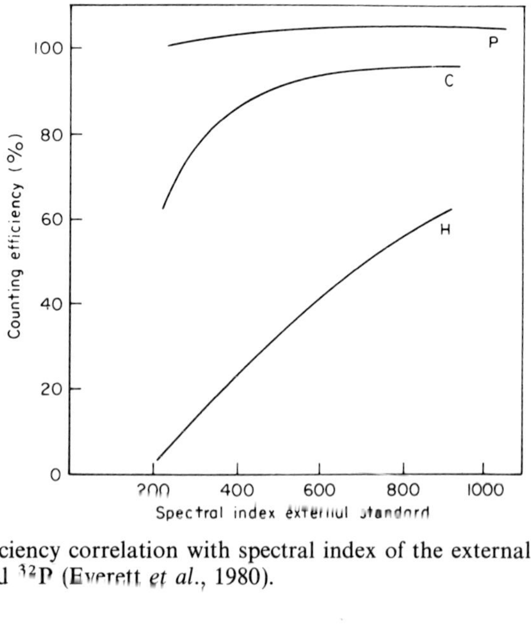

Fig. 5.21 Efficiency correlation with spectral index of the external standard (SIE) for ^{3}H, ^{14}C and ^{32}P (Everett *et al.*, 1980).

In practice, whether SIS or SIE is used as an index of quench, the preparation of quenched standards is necessary as in previously described quench correction methods; and a plot of counting efficiency versus spectral index is made as a quench correction curve. Such a plot for the quench correction of ^{125}I is illustrated in Fig. 5.20.

Everett *et al.* (1980) report that the values of SIE may be very accurately reproduced for any liquid scintillation cocktail. For SIE determinations ^{226}Ra is used as the external standard gamma-ray source. The radium pellet is placed mechanically next to the counting vial for 15 s for the collection of data on the external standard spectrum. The pellet is then removed and the sample's contribution to the external standard spectrum is determined and deducted from the numerator and denominator of equation (5.10). The actual SIE value is multiplied by a factor to give an SIE of 1000 for a non-quenched standard. Examples of typical counting efficiency correlation curves with SIE are illustrated in Fig. 5.21. Like all other quench correction factors, the values of SIS and SIE are unitless.

Reviews on liquid scintillation counting and related quench correction techniques are given by Gibson (1980) and Stanley (1980).

E. Sample preparation techniques

In the preparation of samples for liquid scintillation counting, a researcher should attempt to obtain the following sample–fluor properties which are important in optimizing the sample counting efficiency: (i) a fluor cocktail free of insoluble materials whereby the radionuclide is in intimate contact with the scintillation fluor, (ii) a homogeneous or single-phased fluor cocktail, and (iii) a clear translucent cocktail free of chemical and colour quenching agents and chemiluminescence.

Depending on the sample, nuclide and sample preparation technique employed, all of the above sample–fluor properties may not always be attainable. However, we must be aware that the above sample–fluor properties are very important in maintaining an optimum energy transfer of the nuclear radiation to the solvent and fluor molecules, and optimum photon emission yields. These are criteria which must be met to attain reliable count rates and maximum counting efficiencies.

The techniques used for the preparation of a wide variety of samples encountered in biological and agricultural research, such as animal and plant matter, trapped $^{14}CO_2$, soil extracts and aqueous samples, including samples high in metal salts, are presented and discussed in detail in a book by L'Annunziata (1979). The procedures differ depending on the type of material to be counted; and the limitations of space in this chapter do not enable any detailed treatment of this topic. Despite this variation in their preparation, all samples usually are prepared to contain the radiotracer dissolved in an appropriate fluor cocktail. However, if insoluble, the sample is usually digested into soluble form, suspended as a fine powder in a gel matrix containing the fluors and solvent, or burnt and the combustion products trapped by an appropriate solvent. The preparation of samples by dissolution (digestion), suspension and combustion is treated in detail by L'Annunziata (1979).

F. Counting of X-ray, gamma-ray and Auger electron emitters

Liquid scintillation counting is used principally for the radioassay of beta-particle emitters. However, this radioassay technique is becoming increasingly popular for the counting of certain gamma-ray emitters and nuclides decaying by electron capture and emitting X-rays and Auger electrons. In certain cases, the counting efficiencies may be higher with the liquid scintillation technique than those attainable with the thin-walled NaI(Tl) solid scintillation crystal detector discussed further on in this chapter. The liquid scintillation counter with an automatic sample changer and microprocessor is more commonly encountered in most biochemical tracer laboratories than its solid scintillation counterpart. This of course reflects the importance of finding broader ranges of application for liquid scintillation counting.

The interaction of X- or gamma rays with liquid scintillators is principally the result of the Compton effect whereby part of the energy of the electromagnetic radiation is imparted to orbital electrons. An ejected orbital electron (Compton electron) imparts its energy in material in a fashion similar to that of a beta particle; the absorption of its energy by the scintillation fluor results in fluorescence or the emission of photons of visible light. In liquid scintillators, the photoelectric effect usually does not occur above 30 eV. However, Bransome (1973) reports that it may be evident at higher gamma-ray energies in the glass vial walls or if the scintillator is loaded with heavy elements. Cherenkov photons may also be produced in liquid scintillators if the gamma-ray energies are high enough to produce Compton electrons of sufficient energy to cause the Cherenkov effect. The Cherenkov effect is discussed further on in this chapter.

Numerous studies concerning the liquid scintillation counting of ^{55}Fe, which decays exclusively by electron capture, have been undertaken by Dern and Hart (1961a,b), Perry and Warner (1963), Eakins and Brown (1966), Cosolito *et al.* (1968), Miller *et al.* (1969), Cramer *et al.* (1971), Horrocks (1971) and Grau-Malonda (1982). An optimum counting efficiency of 36% is reported by Grau-Malonda (1982) in a study of quench correction for ^{55}Fe determinations in various scintillators.

A maximum counting efficiency of 56% for ^{125}I by standard liquid scintillation counting techniques is reported by Rhodes (1965). Rhodes compared the liquid scintillation counting of ^{125}I and ^{129}I to that of a standard solid scintillation counter. The comparison showed similar counting efficiencies for ^{125}I in the two counters; however, liquid scintillation was highly superior to solid scintillation in the counting of ^{129}I. Liquid scintillation counting efficiencies of ~100% for ^{129}I are reported (Rhodes, 1965). A liquid scintillation counting efficiency of 100% for ^{131}I is also reported by Rhodes. The higher counting efficiencies for ^{129}I and ^{131}I over ^{125}I may be attributed to the fact that ^{129}I and ^{131}I are beta-particle and gamma-ray emitters, whereas ^{125}I decays exclusively by electron capture.

Tothill (1977) gives a detailed treatment on the decay schemes of ^{125}I and the liquid scintillation assay for this nuclide, together with ^{3}H as a double label. The work of Jordan *et al.* (1974) serves as a good example of the practical application of liquid scintillation counting for the assay of ^{125}I.

Ashcroft (1970) suggests the incorporation of a heavy atom into the liquid scintillator solution to increase the effective atomic number of the solution and thus enhance the gamma-ray absorption coefficient. If this is achieved without quenching, the counting efficiency can be increased. This technique is suitable for nuclides which emit weak X- and gamma radiation and was tested for ^{125}I which emits photons in the energy range 27.4–35.3 keV. An eightfold increase in the liquid scintillation counting efficiency of ^{125}I was reported by Ashcroft after incorporating tetrabutyltin into the liquid scintillator as a source of heavy atoms. This is a colourless stable liquid and is completely miscible with dioxane- or toluene-based scintillation fluors. A suitable solution for the detection of ^{125}I in toluene is 5 g PPO per litre, 0.1 g POPOP per litre and 35% w:v tetrabutyltin.

Higher liquid scintillation counting efficiencies of 70% for ^{125}I are reported by Ting (1973). The ^{125}I samples are counted in microcentrifuge tubes suspended in a liquid scintillation cocktail containing 5% tetrabutyllead as a heavy atom carrier. Tetrabutyllead is used to increase the effective atomic number and consequent X- or gamma-ray "stopping power" of liquid scintillation fluors. Helman and Spiehler (1974) suggest the use of lead acetate, however, since it is less expensive and easier to handle. Lead- and thallium-doped fluors, discussed by Lundqvist *et al.* (1976), improve the separation of the pulse height spectra of a ^{3}H–^{125}I double label employing lead(II) nitrate and thallium(I) formate and malonate as heavy metal sources. The Pb and Tl obviously affect the shape of the pulse height spectrum produced by the weak X- and gamma rays emitted by ^{125}I without causing any change in the ^{3}H pulse height spectrum.

Gutcho *et al.* (1973) report the use of a liquid scintillation counter for the measurement of ^{57}Co in vitamin B_{12} assays. A polycrystalline sodium iodide vial for the counting of ^{125}I, ^{57}Co and other X-ray- and gamma-ray-emitting nuclides in a liquid scintillation counting system has been devised by Noakes and Menefee (1975) and Noakes (1976). These vials are reported by Prokop *et al.* (1980) to give very high counting efficiencies of 74 % and 54 % for ^{125}I and ^{57}Co, respectively. Even higher counting efficiencies (> 100 %) were obtained when pulse summation rather than coincidence counting was used in the liquid scintillation counter. Chase (1980) reviews the application of liquid scintillation counting for the assay of some gamma- and X-ray-emitting nuclides used in radioimmunoassays.

Borosi *et al.* (1980) report the liquid scintillation assay of the ^{59}Fe–^{51}Cr double label in red blood cell kinetic studies. ^{51}Cr decays by electron capture in 91 % of its disintegrations and by gamma-ray emission in the remaining 9 %. If the double label is assayed by solid scintillation counting of the gamma-ray photopeaks, optimum counting efficiencies of 15 % and 3 % are reported for ^{59}Fe and ^{51}Cr, respectively. However, if liquid scintillation counting is used, optimum counting efficiencies of 20 % and 15 % are reported for the ^{59}Fe and ^{51}Cr double label, respectively. The fivefold increase in the counting efficiency of ^{51}Cr is due to the liquid scintillation detection of its X-rays.

G. Counting of alpha-particle emitters

In agricultural and biological research beta-particle-emitting radiotracers such as ^{3}H, ^{14}C, ^{35}S, ^{36}Cl, ^{32}P, ^{33}P and ^{45}Ca, among others, are of most concern. However, increasing interest today is being directed towards the use of uranium, plutonium, americium, curium, polonium and other alpha-particle-emitting nuclides in soil, plant and animal studies. This interest stems from growing concern over the biological effects of nuclear wastes. As reported by Gibson (1976), the increasing levels of ^{211}Pu in man due to increased burn-up of reactor fuel is a problem.

Alpha particles and other charged particles such as protons produce similar scintillation effects in fluors to that of beta particles. Because of the short range of alpha particles in solid and liquid materials, self-absorption of the alpha particles in samples within the counting planchets employed in conventional gas ionization detection techniques is very significant and disadvantageous. The liquid scintillation counting technique in which the detector (scintillation fluor) is in intimate contact with the alpha-particle-emitting radionuclides overcomes this problem.

Studies by Flynn *et al.* (1964) and Horrocks (1966) demonstrate the advantage of liquid scintillation counting for the determination of the absolute activities as well as the energy spectra of alpha particles. Numerous techniques for the preparation of alpha-particle-emitting radionuclide samples for liquid scintillation counting, such as (i) organic complex formation with the radionuclide, (ii) dissolving the radionuclide as a radiocolloid, and (iii) counting of the sample–scintillator mixture as a gel, are described by Ihle *et al.* (1965). A very simple method for the preparation of biological samples containing ^{239}Pu, ^{241}Am and ^{242}Cm for liquid scintillation counting is given by Seidel and Volf (1972), with average nuclide recovery and counting efficiencies of 98%. Similar recoveries for ^{241}Am, ^{243}Cm and ^{244}Cm in biological samples are reported by Guilmette and Bay (1981). Counting efficiencies of ~100% for ^{236}Pu, ^{239}Pu and other alpha-particle emitters in ashed or digested animal organs and bone are reported by Lindenbaum and Lund (1969), Keough and Powers (1970) and McDowell and Weiss (1976a). Energy and pulse shape discrimination of alpha particles with liquid scintillation counters is currently being considered. Initial studies in this field are discussed by Cross and McBeth (1976) and McKlveen (1976). McDowell and Weiss (1976b) also report the testing and evaluation of the current techniques used for the preparation of samples for the liquid scintillation counting of alpha-particle emitters in bone and tissue samples. A review of alpha-particle liquid scintillation counting is given by McDowell (1980).

H. Double-label counting

The proportionality between the beta-particle energy and the pulse height produced in a liquid scintillation counter was previously discussed. This proportionality forms the basis for the combined spectrometric and mathematical separation of two different beta-particle-emitting nuclides in the same sample. Nuclide pairs which have important applications in agricultural, biological and chemical studies are ^{3}H–^{14}C, ^{14}C–^{32}P, ^{3}H–^{35}S, ^{3}H–^{36}Cl and ^{33}P–^{32}P, among others. The minimum requirements for the counting of double-label samples are a dual channel liquid scintillation counter, with its respective pulse height discriminators, and a significant E_{max} difference between the two beta-particle-emitting nuclides of the mixture. The technique involved in the counting of these nuclide pairs is not difficult and is basically the same for most dual labels. Various methods may be used; these are discussed in more detail in a book by L'Annunziata (1979) and a paper by Gringorten (1983).

With the advent of the use of a microprocessor as a standard accessory to liquid scintillation counters, the most popular method of analysing dual tracer labels is to count both nuclides of the pair in two different channels. Although pulse events from both nuclides appear in the two counting channels, the channel discriminator levels are set so that one nuclide is counted predominately in one channel and the second nuclide predominately in the other channel. The proper channel discriminator settings are easily determined by a printout or cathode ray tube (CRT) display of the pulse height spectra of the dual nuclide mixture, as illustrated in Fig. 5.22, taking ^{3}H and ^{14}C as an example.

Because pulse events (counts) from both nuclides of the dual label appear in channel 1 as well as in channel 2, the following two equations are used to describe the count rates in the two channels:

$$N_1 = Hh_1 + Cc_1, \tag{5.11}$$

$$N_2 = Hh_2 + Cc_2, \tag{5.12}$$

where N_1 and N_2 are the net observed count rates (cpm) in channels 1 and 2, respectively; H and C are the activities or distintegration rates (dpm) of ^{3}H and ^{14}C, respectively; h_1 and h_2 are the counting efficiencies of ^{3}H in channels 1 and 2, respectively; and c_1 and c_2 are the counting efficiencies of

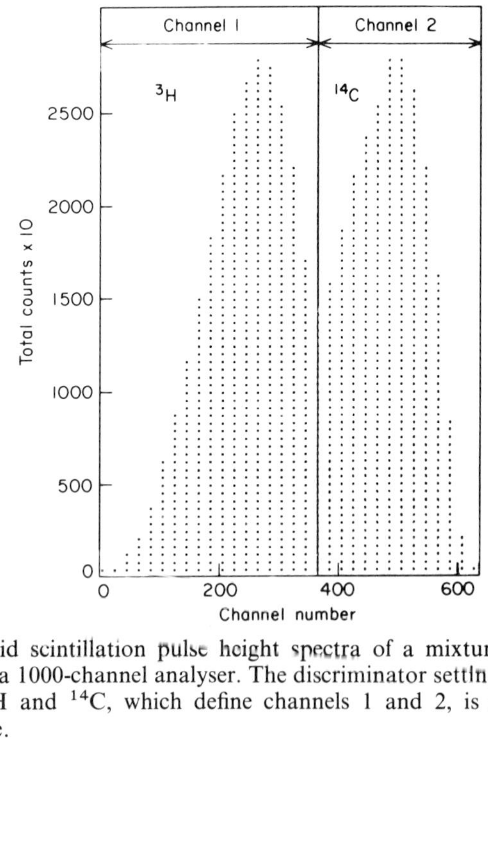

Fig. 5.22 Liquid scintillation pulse height spectra of a mixture of ^{3}H and ^{14}C determined with a 1000-channel analyser. The discriminator setting for the optimum separation of ^{3}H and ^{14}C, which define channels 1 and 2, is illustrated with a solid vertical line.

^{14}C in channels 1 and 2, respectively. The two unknowns, H and C, of equations (5.11) and (5.12) are solved to give

$$H = \frac{N_1 c_2 - N_2 c_1}{h_1 c_2 - h_2 c_1}, \tag{5.13}$$

$$C = \frac{N_1 h_2 - N_2 h_1}{c_1 h_2 - c_2 h_1}. \tag{5.14}$$

Equations (5.13) and (5.14) are used to calculate the ^3H and ^{14}C disintegration rates of unknown mixtures when a significant portion of the beta-particle pulse height spectra of the two nuclides are found in the two channels.

It should be pointed out that the above method requires the determination of four counting efficiencies, those of ^3H and ^{14}C in the two channels, h_1, h_2 and c_1, c_2, respectively. The four counting efficiency factors are easily obtained from quench correction curves as illustrated in Fig. 5.23. This may

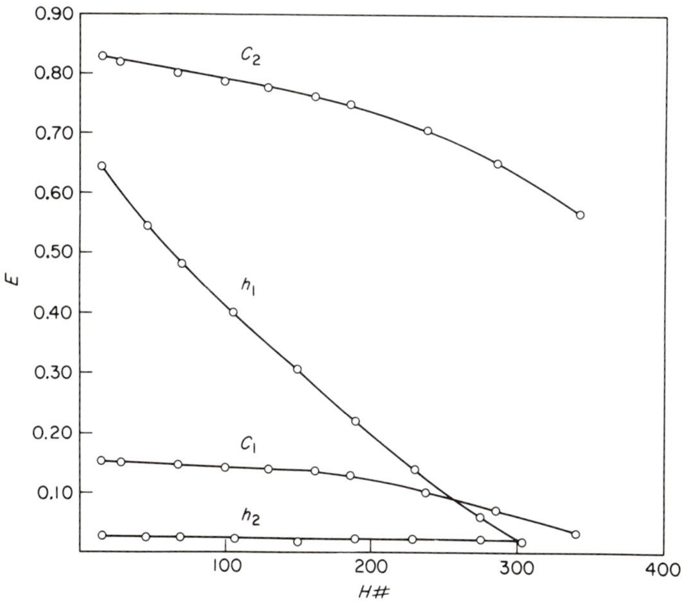

Fig. 5.23 Plots of counting efficiencies (E) in two channels for ^3H and ^{14}C in scintillation fluors at different quench levels as a function of $H\#$. The counting efficiencies of ^3H in channels 1 and 2 are obtained from curves marked h_1 and h_2, respectively; and the ^{14}C counting efficiencies in channels 1 and 2 are found from curves c_1 and c_2, respectively.

appear to be a tedious procedure at first sight; however, the four quench correction curves are easily prepared with present-day liquid scintillation counters equipped with microprocessors. With two sets of quenched standards, one containing known amounts of ^{3}H and another known amounts of ^{14}C in sample vials containing increasing amounts of quenched fluors, the counting efficiencies are determined as a function of a quench monitor factor such as $H\#$. With the aid of microprocessors the quench correction curves as illustrated in Fig. 5.23 are often not plotted; rather, the multilinear regression coefficients to the best fit equation for each plot are kept in the microprocessor memory as, for example,

$$\log E = A + Bx + Cx^2 + Dx^3, \tag{5.15}$$

where E is the counting efficiency (e.g. h_1, h_2, c_1 or c_2), A, B, C and D are numerical coefficients and x is the $H\#$. Four equations are determined to define the relationships between h_1, h_2, c_1 and c_2 as a function of the quench monitor factor (e.g. $H\#$). From these four equations and the value of the quench monitor of unknown experimental samples, the microprocessor can calculate the values of the counting efficiencies h_1, h_2, c_1 and c_2. The four counting efficiency factors are then used by the microprocessor to calculate the values of the ^{3}H and ^{14}C disintegration rates according to equations (5.13) and (5.14).

I. Chemiluminescence, phosphorescence (photoluminescence) and "static"

The energy liberated from chemical reactions can be absorbed by certain molecules which, as a result, reach an excited state. Upon de-excitation these molecules can emit photons of light. This phenomenon, whereby the energy of chemical reactions results in light emission, is known as chemiluminescence. It has particular relevance in liquid scintillation counting since chemiluminescence can occur in some of the chemical procedures used to prepare samples for counting; and the chemiluminescence photons can produce spurious counts during the radioassay of experimental samples. An awareness of the causes, means of prevention and detection of chemiluminescence is important to eliminate error, which can be introduced by such spurious counts.

A thorough review of chemiluminescence in liquid scintillation counting is given by Peng (1976) and a current assessment of the problem has been reported by Kalbhen (1980). Bleaching agents such as hydrogen peroxide, benzoyl peroxide and oxidizing acids such as perchloric and nitric acids commonly used in certain sample preparation techniques have been identified as culprits causing chemiluminescence, particularly when mixed with alkaline tissue solubilizers as reported by Peng (1976). But chemiluminescence has been reported even under simpler circumstances such as in the mixing of dioxane-based fluor cocktails with alkaline solubilizers (Kalbhen, 1967).

Chemiluminescence can persist in experimental samples, producing count rates at levels appreciably above background for hours or even days. Some decay curves are given by Kalbhen (1980). The decay of chemiluminescence can be accelerated by temperature elevation.

Means of avoiding and suppressing chemiluminescence are reviewed by Peng (1976). Among these may be listed acidification of alkaline tissue digests to pH below 7.0 with non-oxidizing acids, followed by heating to 50 °C prior to the addition of the fluor, as suggested by Neame (1975), as well as the storing of samples at elevated temperatures prior to counting to allow for chemiluminescence decay.

Kalbhen (1980) has made a thorough study of the pulse height spectra of chemiluminescence photons and shows that they are located predominately in the energy range of tritium. Thus, for the radioassay of low levels of tritium, chemiluminescence can be a serious problem. For the assay of ^{14}C or even higher energy beta-particle-emitting radionuclides, the lower level discriminator may be raised to eliminate any pulses which might originate from chemiluminescence, at a cost of reduced counting efficiency. An awareness of chemiluminescence is of ultimate importance for radiotracer samples of low activity, and observations should be made to detect its presence.

Phosphorescence or photoluminescence can occur upon the excitation of the scintillator fluor by ultraviolet light originating from sunlight, or even by fluorescent lamps or other artificial laboratory lighting. As reported by Kalbhen (1980), phosphorescence reactions decay completely in 20–30 min, and simply storing experimental samples in the liquid scintillation counter in the dark for about 30 min is sufficient to alleviate this problem.

Horrocks and Kolb (1981) report that "static", the build-up and subsequent discharge of static electricity on liquid scintillation vials, can be a serious source of erroneous counts. This is reported to occur most commonly in dry, hot climates or in heated rooms with low humidity, and particularly with plastic vials. The recommended method of detecting "static" and chemiluminescence is to prepare non-radioactive controls in an identical manner to the experimental samples containing radiotracer. Horrocks and Kolb (1981) recommend the removal of "static" by wiping the liquid scintillation vial with a cloth or paper towel dampened with water or ethanol.

VI. Cherenkov counting

A. Theory

Photons are produced when a charged particle travels through a transparent medium at a speed greater than the speed of light in that medium. These photons are referred to as Cherenkov radiation in honour of Cherenkov for his basic research on this phenomenon (Cherenkov, 1934). The observation of Cherenkov radiation, however, was reported as early as 1929 by Mallet, whose studies were not as extensive as those of Cherenkov. Cherenkov radiation consists of a continuous spectrum of wavelengths concentrated mainly in the u.v. region and extending into the visible part of the spectrum. Only a negligible amount of photon emissions is found in the i.r. or microwave regions. This photon emission is the result of a coherent disturbance of adjacent molecules of the medium by the travelling charged particle, and has been described by Marshall (1952) as the electromagnetic "shock wave" that is analogous to the sonic boom created by supersonic aircraft.

Cherenkov radiation, when produced at significant levels, can be employed for the efficient radioassay of radiotracers with a conventional liquid scintillation counter. Although the production of Cherenkov radiation does not involve the scintillation phenomenon, a conventional liquid scintillation counter can detect and count Cherenkov photons emitted from a given sample in a standard counting vial. The radiotracers may be counted in water or some other suitable transparent liquid without the use of scintillation fluor or any other chemical reagents. This leaves the sample in an uncontaminated state which is often suitable for any other chemical tests deemed necessary in a given research study. This technique is popular for the radioassay of relatively high-energy beta-particle-emitting tracers in agriculture, biology and chemistry. The ease of sample preparation, low cost (no need for fluors and fluor solvents) and the automatic counting techniques available with a liquid scintillation counter are among the principal reasons for the wide-spread use of this counting method.

The formation of Cherenkov radiation in a given transparent medium rests on equation (5.16) as a threshold condition:

$$\beta n = 1, \qquad (5.16)$$

where β is the relative phase velocity of the particle, i.e. the velocity of the particle divided by the speed of light in a vacuum, and n is the refractive index of the medium (i.e. the ratio of the velocity of light in a vacuum to its velocity in the medium). The value of β of the charged particle is dependent on its kinetic energy, as reported by Jelley (1962):

$$E = m_0c^2\left[\frac{1}{(1-\beta^2)^{1/2}} - 1\right], \tag{5.17}$$

where m_0 is the particle rest mass in grams, c is the velocity of light, equal to 2.99×10^{10} cm s^{-1}, and E is the kinetic energy in ergs (where 1.602×10^{-12} ergs = 1 electron volt). The value of the rest energy for an electron is

$$\begin{aligned} m_0c^2 &= (9.11 \times 10^{-28}\ \text{g})(2.99 \times 10^{10}\ \text{cm s}^{-1})^2 \\ &= 8.14 \times 10^{-7}\ \text{ergs}. \end{aligned}$$

The energy in ergs may be converted to electron volts as follows:

$$\frac{8.14 \times 10^{-7}\ \text{ergs}}{1.602 \times 10^{-12}\ \text{ergs eV}^{-1}} = 511\ \text{keV}.$$

Substituting the value of 511 keV for m_0c^2 in equation (5.17) and solving for β gives

$$\beta = \left[1 - \left(\frac{1}{E/511 + 1}\right)^2\right]^{1/2}. \tag{5.18}$$

Thus, the value of β where electrons or beta particles are concerned is dependent upon the electron energy (E) in electron volts according to equation (5.18). If water is taken as the transparent medium, $n = 1.332$ and β must exceed 0.7508 according to equation (5.16) if Cherenkov radiation occurs. Substituting 0.7508 for β in equation (5.18) and solving for E gives 263 keV (or 0.263 MeV) as the minimum or threshold electron energy necessary for the production of Cherenkov radiation in water.

From equations (5.16) and (5.18) it is evident that the energy threshold for the production of Cherenkov radiation by electrons (beta particles) is a function of the refractive index. Consequently, an increase in the refractive index of the medium should lower the energy threshold and increase the detection efficiency for a beta-particle-emitting radionuclide. This is because the beta-particle energy spectrum of all beta-particle-emitting nuclides ranges from a small value above zero to E_{max}. In practice, the Cherenkov effect and detection efficiencies are significantly increased only for radionuclides which have a low beta-particle E_{max}, as illustrated in Table 5.2. This may be explained by the fact that only a small part of the total beta-particle emission energy spectra of high energy beta-particle-emitting nuclides is below 263 keV. The use of solvents of high refractive index as Cherenkov counting media for low energy beta-particle emitters such as ^{99}Tc is not a practical approach since very high counting efficiencies are obtained by liquid scintillation techniques.

Table 5.2 Cherenkov response as a function of solvent refractive index (Ross, 1969)

Nuclide	E_{max} (MeV)	Detection efficiency (%)			
		$n = 1.3220^a$	$n = 1.4026^b$	$n = 1.4353^c$	$n = 1.4644^d$
^{99}Tc	0.295	0.01	—	—	1.02
^{204}Tl	0.765	16.1	19.6	21.4	23.2
^{32}P	1.71	50.2	50.9	51.3	51.6

[a] Water. [b] 53 wt % glycerol. [c] 75 wt % glycerol. [d] 95 wt % glycerol.

An important property of Cherenkov radiation which should be considered is the directional emission of the Cherenkov photons. Cherenkov photons are emitted as a cone at an angle θ to the direction of the beta particle as depicted in Huygen's construction of the Cherenkov wavefront, illustrated in Fig. 5.24 and described by Marshall (1952). The angle θ is defined by the equation

$$\cos\theta = \frac{(c/n)t}{\beta ct} = \frac{1}{\beta n}, \tag{5.19}$$

where β and n are as defined for equations (5.16) and (5.18).

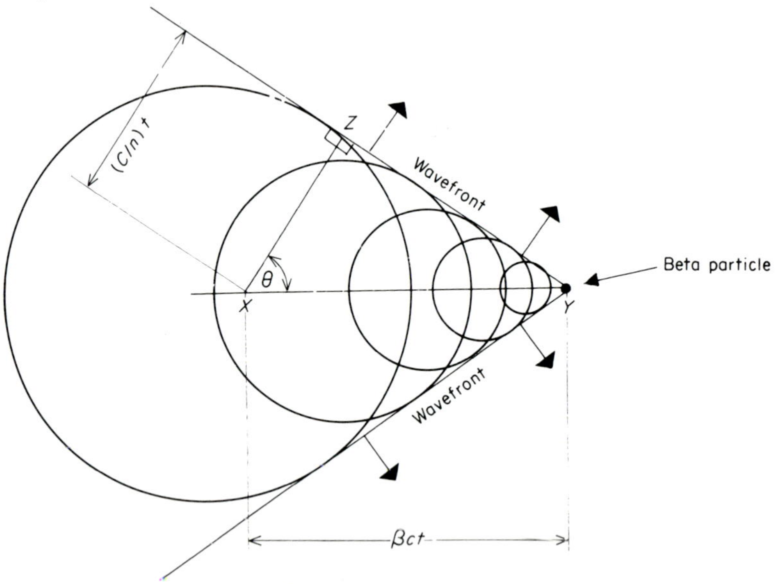

Fig. 5.24 Huygen's construction of a conical Cherenkov wavefront. The position of the charged particle at time $t = 0$ is x; the distance of travel of the particle from x to y in time t is βct; the distance of travel of the Cherenkov photon from x to z is $(c/n)t$. The angle of emission of Cherenkov photons is represented by θ.

As the beta-particle energy (E) of equation (5.18) increases, the value of β approaches unity and cos θ approaches 0.7508. Consequently, the maximum angle of emission of a Cherenkov photon is 41.3°. Cherenkov photons are emitted at a more acute angle for electrons just above the threshold energy and cannot exceed an emission angle of 41.3°, even for the most energetic electrons. This directional emission of Cherenkov photons is a disadvantage when conventional liquid scintillation spectrometers are used for counting. The photocathodes of most liquid scintillation counters consist of two PM tubes positioned at 180° relative to each other (Fig. 5.9). This is not an optimum arrangement for the detection of those photons which may be admitted to these photocathodes at other angles. Thus, when conventional liquid scintillation counters are employed for Cherenkov counting, the counting efficiencies are inferior to the theoretical maximum efficiencies. For optimum counter response Ross (1969) suggests that four photocathodes be operated in coincident pairs, each pair located at 180° to the other.

Extensive treatments of the origin and interpretation of Cherenkov radiation are given by Marshall (1952), Jelley (1962) and Gruhn and Ogle (1980).

B. Quenching and quench correction

Chemical quenching in Cherenkov counting is non-existent because Cherenkov photons arise from a physical disturbance of solvent molecules rather than from chemical fluorescence, as occurs in the liquid scintillation phenomenon. However, colour quenching can be very significant in Cherenkov counting, greater than that which may occur in standard liquid scintillation counting techniques, as reported by Elrick and Parker (1968), since Cherenkov photons are weaker and cover a wider spectrum of wavelengths than those produced in the scintillation phenomenon. The quench corrections and counting efficiencies in Cherenkov radioassay are efficiently determined by one of the following methods (see also Section V.D):

(i) *Internal standardization* (see also Section V.D.1) This method may be highly accurate but is time consuming. An added disadvantage of this correction technique in Cherenkov counting is the possible contamination of the sample by the addition of the internal standard, i.e. addition of the standard could render the sample unsuitable for subsequent tests unless a carrier-free standard is employed.

(ii) *Sample channels ratio* (see also Section V.D.2) The sample channels ratio technique has been successfully applied to the Cherenkov counting of ^{32}P. Stubbs and Jackson (1967) prepared a series of vials each containing ^{32}P

of the same activity (50.0 nCi) and increasing amounts of Scarlet R or Naphthol Yellow S dye as colour quenchers. The quantity of dye per vial ranged from 0.01 to 3.00 mg. Channel A was set successively to give channel widths from 50–100 to 50–600, where a wide-open channel of 50–1000 was chosen for channel B. The detection efficiency was calculated on the basis of the counting rate observed in channel B. For each width selected for channel A, the channels ratio of A:B was plotted against detection efficiency. After testing various widths for channel A, the optimum width was selected as that which gave the most linear plot over the greatest range of detection efficiency. For the Cherenkov counting of ^{32}P or other beta-particle-emitting nuclides, the above procedure should be tested before rejecting the possibility of applying this rapid and simple method to quench correction determinations. The channels ratio technique has also been successfully employed by Moir (1971) for counting efficiency determinations in the Cherenkov counting of ^{42}K in biological samples.

The author has successfully employed the above quench correction technique for the assay of ^{32}P in water using 0.01 % CrO_3 as quenching agent. Increasing amounts of CrO_3 solution, which is yellow in colour, is added to a series of liquid scintillation vials, each containing the same activity of ^{32}P standard (0.1 μCi ml^{-1}) as described in Table 5.3. Water is added to bring the quenched standards to a uniform and optimum volume.

A typical quench correction curve obtained from these quenched standards is illustrated in Fig. 5.25. The yellow colour produced by the CrO_3 quenching agent is similar to the slight yellow colour of ashed plant material

Table 5.3 Constituents of ^{32}P non-quenched (vial no. 1) and quenched standards in water for Cherenkov counting

Vial no.	Vial constituents (ml)		
	H_2O	Quencher (0.01 % CrO_3)	Standard [^{32}P]phosphate
1	14.0	0.0	1.0
2	13.5	0.5	1.0
3	13.0	1.0	1.0
4	12.0	2.0	1.0
5	11.0	3.0	1.0
6	10.0	4.0	1.0
7	9.0	5.0	1.0
8	8.0	6.0	1.0
9	6.0	8.0	1.0
10	4.0	10.0	1.0
11	2.0	12.0	1.0
12	0.0	14.0	1.0

dissolved in dilute acid (0.1 N HCl). Although a yellow colour is used in the above procedure, Stubbs and Jackson (1967) and Bem *et al.* (1980) report identical quench correction curves when either a yellow or a red dye is employed as the quenching agent. The sample channels ratio technique for quench correction has been successfully employed for the assay of ^{32}P and ^{210}Pb as reported by Fric and Palovcikova (1975) and Karamanos *et al.* (1975).

(iii) *External standard channels ratio* (see also Section V.D.3) The suitability of this technique was tested by Elrick and Parker (1968) by assaying coloured samples of ^{204}Tl using an external standard of ^{226}Ra. The method is considered as practical and most sensitive for higher energy beta-particle-emitting radionuclides. The usefulness of this quench correction technique for the radioassay of higher energy beta-particle emitters in plant material was demonstrated by Brownell and Lauchli (1969).

Ross (1980) and Smith (1981) report that the use of external standards is often not applicable to quench correction in Cherenkov counting because the external standards commonly supplied with liquid scintillation counters are of such a low energy (e.g. ^{241}Am, ^{133}Ba and ^{137}Cs) that they produce Compton electrons below the Cherenkov threshold energy. A ^{60}Co external standard was successfully employed by Nascimiento Filho (1977) for the determination of ^{32}P Cherenkov counting efficiencies in soil extracts.

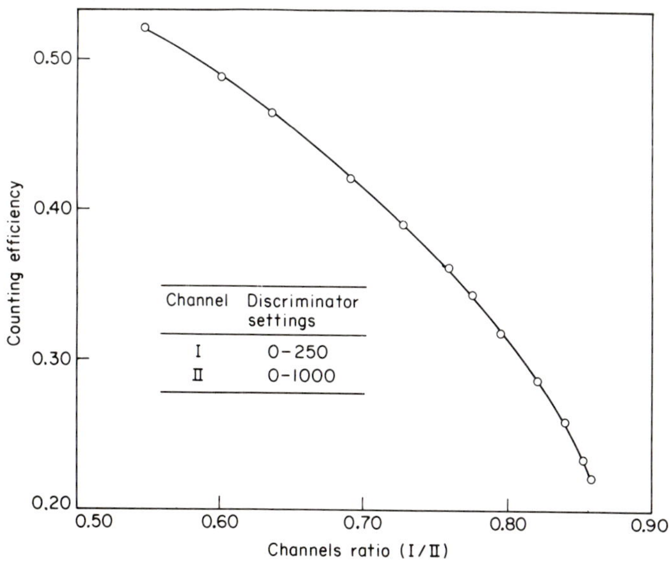

Fig. 5.25 Typical quench correction curve for the Cherenkov counting of ^{32}P in water.

The detection efficiencies in Cherenkov counting may be optimized by controlling several counting parameters in addition to the previously mentioned refractive index of the counting medium and the geometric orientation of the photocathode tubes. Among these other parameters are the following, which are simple and easily controlled if considered appropriate:

(i) *Sample volume* The detection efficiency of Cherenkov photons is highly dependent on the sample volume. Ross (1969) reports a maximum counting efficiency at 8 ml of sample volume for various nuclides tested; however, the optimum volume may differ from one instrument to another. For example, Clausen (1968) reports an optimum volume of 10–11 ml for the Cherenkov counting of ^{32}P with the liquid scintillation counter used in his experiments. Obviously, it is necessary that the optimum counting volume be determined experimentally for the particular instrument employed, and that this volume be used for every sample if reproducible counting efficiencies are to be maintained.

(ii) *Counting vials* Because a large fraction of Cherenkov radiation is emitted in the u.v. region it is normally not transmitted by glass vials. The counting efficiency can therefore be increased by employing quartz or plastic vials. Plastic vials may be more appropriate because of the high cost of the quartz type. Ross (1980) reports that polyethylene vials give somewhat higher counting efficiencies and lower backgrounds than glass vials. Further, polystyrene vials can offer an improved counting efficiency which may be due to a weak scintillation effect in the polystyrene. Nylon and polystyrene vials are useful for Cherenkov counting, with nylon producing higher counting efficiencies for ^{36}Cl, as indicated by Lauchli (1969).

A comparison of Teflon and polyethylene vial materials was made by Calf (1977) for the Cherenkov counting of ^{32}P (Table 5.4) and ^{36}Cl, which shows that Teflon gives higher counting efficiencies (E) and lower backgrounds (B) for these two nuclides. Teflon gives a 10-fold greater figure-of-merit (E^2/B) for the Cherenkov counting of ^{36}Cl when compared to that of glass. The applications of figure-of-merit are discussed further on in this chapter.

Table 5.4 Cherenkov counting of ^{32}P in water (Calf, 1977)

Vial	Volume (ml)	Efficiency (%)	Background (cpm)	Figure-of-merit, E^2/B
Glass	10	36.9	12.4	110
Polyethylene	10	41.3	7.3	234
Teflon	10	44.6	5.2	343

Glass vials can give high counting efficiencies if wavelength shifters are used (see (iii) below).

(iii) *Wavelength shifters* The u.v. component of Cherenkov radiation which may pass the quartz or plastic vials can be partially absorbed by the glass of the photomultiplier tube. The sodium and potassium salts of 7-amino-1,3-naphthalenedisulphonic acid (ANDA) in 100 mg l^{-1} concentration are successfully employed as wavelength shifters by absorbing the u.v. photons and re-emitting them isotropically in the visible region, as demonstrated by Parker and Elrick (1966) and Elrick and Parker (1968). Thus the directional character of Cherenkov photons and the absorption of their ultraviolet wavelengths are reduced by the emission of the radiation in all directions at wavelengths more efficiently detected by the photomultiplier tubes. Some experimentally determined Cherenkov counting efficiencies of aqueous samples with and without a wavelength shifter are given in Table 5.5.

Lauchli (1969, 1971) reports that weak beta-particle emitters such as ^{36}Cl require 5 mM ANDA, and strong beta-particle emitters such as ^{86}Rb and

Table 5.5 Experimentally determined Cherenkov counting efficiencies with and without a wavelength shifter (Parker and Elrick, 1970)

Nuclide	E_{max} (MeV)	Counting efficiency (% of disintegration)[a]		
		S11 photocathode		RCA 4501-V3 photocathode, aqueous sample
		Aqueous sample	Wavelength shifter	
^{204}Tl	0.77 (98 %)	1.3	2.6	—
^{137}Cs	0.51 (92 %), 1.17 (8 %)	2.1	—	—
^{36}Cl	0.71 (98.3 %)	2.3	4.7	5.3
^{198}Au	0.96 (99 %)	5.4	—	—
^{47}Ca	0.66 (83 %), 1.94 (17 %)	7.5	14.8	—
^{40}K	1.31 (89 %)	14	31	34.3
^{24}Na	1.39 (100 %)	18	40	—
^{86}Rb	0.68 (8.5 %), 1.77 (91.5 %)	23	46	—
^{32}P	1.71 (100 %)	25	50	43.3
^{144}Ce–^{144}Pr	2.98 (97.7 %)	54	75	59.2
^{42}K	2.0 (19 %), 3.6 (82 %)	60	85	76.3
^{106}Ru–^{106}Rh	2.0 (3 %), 2.4 (12 %)	62	85	75.3

[a] The S11 photocathode is obsolete; however, useful comparisons can be made with the effect of a 100 mg l^{-1} ANDA wavelength shifter. Data obtained with the more efficient RCA photocathode is for samples without a wavelength shifter; and higher counting efficiencies would be expected if a wavelength shifter were employed.

^{42}K require 2.5 mM ANDA for optimum counting efficiencies. ANDA is useful for the routine counting of ^{36}Cl, ^{40}K and ^{86}Rb in plant materials, with counting efficiencies of 13%, 79% and 61%, respectively, for samples containing 1 g of digested plant tissue, and 23%, 93% and 70%, respectively, for samples not containing plant tissue. Large increases in the counting efficiencies are obtained by this technique, which is to be highly recommended if the sample is not needed for further studies or chemical analysis (the wavelength shifter may act as a contaminant). ANDA may also be employed successfully as a wavelength shifter in the Cherenkov counting of ^{59}Fe in plant materials, as reported by Kannan (1975).

Fric and Finocchiaro (1975) found anthranilic acid and quinine to be effective as wavelength shifters for the Cherenkov counting of ^{32}P.

Table 5.6 Effect of different wavelength shifters, dissolved in distilled water, on the counting of Cherenkov radiation from ^{32}P, using polyethylene counting vials, vial contents of 10 ml, an amplifier gain of 30% and discriminator settings of 20–1000 (van Ginkel, 1980)

Compound	Concentration used ($g\,l^{-1}$)	Counting efficiency (%)
Water in glass vial		51.0
Water in quartz vial		53.0
Water in polyethylene vial		55.5
β-Methylumbelliferone[a]	0.2	67.0
Esculin[a]	0.4	73.0
Quinine sulphate[b]	1.0	68.4
Quinine[b]	1.0	66.2
	10.0	76.0
	25.0	77.0
Thymine	0.5	65.9
Alloxazin	0.1	55.1
Pyrimidine-HCl	5.0	56.4
Quinolinic acid	0.5	53.3
Sodium dihydrobenzoate	1.0	66.3
Pentobarbitol	1.0	57.3
Phenobarbitol	1.0	57.0
Lumichrome	0.001	57.2
Sodium salicylate[c]	0.5	66.8
	1.0[d]	68.0
	100[d]	76.0
	1000[d]	85.0

[a] Stable in the pH range 3–11.
[b] Can only be used at pH 0–3.
[c] Highly sensitive to pH change and salt concentration.
[d] Refractive index is increased.

Concentrations of 10^{-3} M anthranilic acid and 2×10^{-3} M quinine increased the counting efficiency of ^{32}P from 47.2% to 58.4% and from 47.2% to 54%, respectively. Bezaguet *et al.* (1979) report an increase in the light output by a factor of about 3.8 when ANDA is employed as a wavelength shifter at a concentration of 10 mg l^{-1}. Many chemical compounds have been tested for their wavelength shifting properties. Some are sensitive to pH, salt concentration and storage in the sample solution for periods of a few days or weeks. van Ginkel (1980) has tested the stability and wavelength shifting properties of several organic compounds, and some of their properties are listed in Table 5.6.

Although the use of wavelength shifters can significantly increase the counting efficiency, we must keep in mind that the detection process is no longer a purely physical one, and thus chemical quenching is now possible. A further disadvantage is that the counting sample may be rendered useless for subsequent chemical tests since the wavelength shifter may act as a contaminant. For this reason, the use of wavelength shifters is at times referred to as Cherenkov scintillation counting.

To have the advantages of using a wavelength shifter while still counting under pure Cherenkov conditions free of chemical quenching, Ross (1976) has fabricated the counting vial illustrated in Fig. 5.26. The vial consists of an outer chamber used to contain the wavelength shifting solution, and the inner chamber of ~10 ml volume contains the counting sample of radionuclide and solvent. The entire vial is made of quartz so that the u.v. radiation can pass the inner chamber wall before reaching the solution of wavelength shifter. The vials are reusable, and to prevent contamination the counting

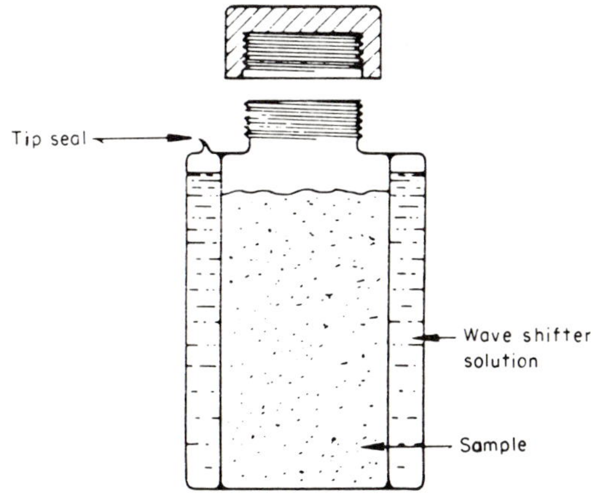

Fig. 5.26 Quartz counting vial (side view) with isolated wavelength shifter compartment for Cherenkov counting (Ross, 1976).

sample is contained in a polyethylene bag placed within the inner chamber. Ross (1976) reports an increase in counting efficiency of ^{89}Sr by a factor of 1.88 when dimethyl-POPOP is used in the wavelength shifting chamber as compared to pure water. Greater improvements in counting efficiency would be expected for lower energy beta-particle emitters such as ^{36}Cl. The only disadvantage of this technique is that the volume of sample that may be conveniently counted is reduced by a factor of two.

Wiebe and Ediss (1976) report very high Cherenkov counting efficiencies for ^{36}Cl using a high refractive index methylsalicylate ($n = 1.522$) as the counting medium. The high counting efficiencies are due not only to the increased refractive index (thus lowered Cherenkov threshold energy) but also to the "wavelength shifting" properties of this solvent. Counting efficiencies of 82.4 and 91.6% are reported depending on the liquid scintillation spectrometer employed.

(iv) *Summation circuitry* Due to the directional characteristics of Cherenkov photon emission, higher counting efficiencies are obtained if the two PM tubes of conventional liquid scintillation spectrometers are coupled through a summation circuit without coincidence circuitry. By this technique Parker and Elrick (1966) increased the Cherenkov counting efficiency of ^{32}P to 70%. The backgrounds may be 100 times higher because of the high noise level of PM tubes, and consequently it is not a practical method for the assay of samples of low activity. Elrick and Parker (1968) recommend that care be exercised in employing this pulse summation technique because anomalous counting efficiencies can occur due to afterpulses (satellite pulses, described by Morton *et al.* (1967), produced in PM tubes) which are counted when the coincidence circuitry is not employed.

Cherenkov counting of ^{32}P in aqueous extracts has become most popular, particularly because of the easy and inexpensive sample preparation techniques involved as well as the relatively high counting efficiencies obtained. As chemical quenching does not exist in Cherenkov counting (without wavelength shifters) sample preparation techniques may be employed with little concern for the type of reagents used, and sample colour may be bleached by chemicals with no quenching effect. The popularity of Cherenkov counting of ^{32}P is illustrated by reports by Wiebe *et al.* (1971), Smith *et al.* (1972), Fric and Palovcikova (1975), Rowinska *et al.* (1975), Bem *et al.* (1980), Chow (1980) and Lefebvre and Glass (1981).

Glass (1978) describes a very simple and rapid sample preparation technique where ^{32}P- or ^{86}Rb-labelled plant material is ashed directly in glass liquid scintillation vials. An aqueous solution of 2.5 mM ANDA is then added before counting. Other sample preparation techniques may be obtained from a previous book (L'Annunziata, 1979).

The radioassay of beta-particle-emitting radiotracers other than ^{32}P by

Cherenkov counting has been found advantageous. Bem *et al.* (1978) describe the Cherenkov counting of ^{115m}Cd in blood and animal tissue with counting efficiencies of $\sim 35\%$. Cherenkov counting has also been applied to the assay of ^{89}Sr without interference from the soft beta particles of ^{90}Sr as reported by Carmon (1979).

Cherenkov counting has been found useful for the analysis of double labels such as ^{32}P–^{45}Ca and ^{32}P–^{86}Rb. Bem *et al.* (1980) describe the technique for the radioassay of ^{32}P and ^{45}Ca in biological samples encountered in studies of phosphorus and calcium metabolism in animals. The activity of ^{32}P of the double label is first determined by Cherenkov counting and the sum of the activities of the ^{32}P and ^{45}Ca nuclides are then determined by liquid scintillation counting after the addition of a scintillation fluor miscible with water. The analysis of ^{32}P and ^{86}Rb double labels in plant material is reported by Lefebvre and Glass (1981). The total activity of ^{32}P and ^{86}Rb is determined by Cherenkov counting. The sample is subsequently submitted to anion exchange column chromatography to separate the ^{86}Rb from the double label and then determine its activity by Cherenkov counting. The activity of ^{32}P is then obtained by difference.

It is obvious that Cherenkov counting cannot compete with liquid scintillation counting with respect to the highest possible counting efficiency. In addition, Cherenkov counting cannot be used for very low-energy beta-particle emitters. However, if very high counting efficiencies are not necessary, Cherenkov counting of relatively high-energy beta-particle emitters has the following advantages over liquid scintillation counting: (i) samples may be counted directly in aqueous solution or in organic solvents; (ii) no fluors or other compounds need be added, and thus the sample may be used for subsequent studies or analysis; (iii) a large volume of sample may be counted since fluor solution is not needed; (iv) chemical quenching is negligible; (v) much time is saved in sample preparation.

VII. Solid scintillation counting

Solid scintillation is a phenomenon whereby the absorption of bombarding alpha, beta or gamma radiation by certain crystalline inorganic or organic materials results in the emission of flashes of visible light from the solid absorbing material. Because alpha and beta particles have low penetration power in matter, solid scintillation is not generally employed for the quantitative measurement of these radiations.

Unfortunately, the detection and analysis of electromagnetic radiation by organic materials is generally not efficient due to the high penetrating power of electromagnetic radiation (except in those cases involving low-energy gamma or X-radiation). The relative low density of organic materials (anthracene, 1.3 g cm^{-3}, and PTP, 1.2 g cm^{-3}) does not permit the absorption of a significant quantity of high-energy gamma radiation. However, the higher density of certain inorganic crystalline materials such as NaI (3.7 g cm^{-3}), LiI (4.1 g cm^{-3}) or CsI (4.5 g cm^{-3}) attribute to these solids a greater "stopping power", i.e. a greater ability to absorb energy of impinging gamma radiations. The NaI, LiI or CsI must be in the solid crystalline state and are usually prepared with a 0.1 % content of thallium iodide or europium iodide. Such crystalline detectors are identified as NaI(Tl), LiI(Eu) and CsI(Tl). The thallium or europium ions serve as activation centres in the inorganic crystals where de-excitation of absorbed gamma-ray energy and subsequent emission of visible light or scintillation occurs.

A. Principles of solid scintillation gamma-ray detection

1. *The scintillation process*

The scintillation process is explained in detail by Dietrich *et al.* (1973), Kabler (1975) and Heath *et al.* (1979). In brief, the principal mechanism of photon emission is an electron–hole pair production in the NaI crystal followed by electron–hole diffusion to Tl sites where recombination occurs. The Tl is normally found in the crystal lattice of Tl^+ ions. An electron (e^-) may diffuse and combine with Tl^+ to form Tl^0, and a hole (h^+) may diffuse and combine with Tl^0 to produce Tl^+ at the excited state $(Tl^+)^*$, which emits luminescent light ($h\nu$) upon de-excitation. This process is described by equations (5.20) and (5.21):

$$e^- + Tl^+ \rightarrow Tl^0 \tag{5.20}$$

$$h^+ + Tl^0 \rightarrow (Tl^+)^* \rightarrow Tl^+ + h\nu. \tag{5.21}$$

The reverse sequence may also occur; that is, a hole may diffuse and be trapped by a Tl^+ ion to produce Tl^{2+}. An electron may then combine with Tl^{2+} to yield an excited Tl^+ ion or $(Tl^+)^*$ which emits light upon falling to the stable state. This sequence may be illustrated by the following equations:

$$h^+ + Tl^+ \rightarrow Tl^{2+}, \tag{5.22}$$

$$e^- + Tl^{2+} \rightarrow (Tl^+)^* \rightarrow Tl^+ + h\nu. \tag{5.23}$$

Although of slightly lower density, a NaI(Tl) crystal is most commonly used for gamma-ray detection, rather than LiI(Eu) or CsI(Tl), due to its shorter decay time, i.e. the time necessary for deactivation. The decay time of a NaI(Tl) crystal is 0.3 μs, whereas that of a LiI(Eu) or CsI(Tl) crystal is ~1.1 μs. This shorter decay time permits a greater resolution of succeeding deactivation events in the detector crystal. Consequently, higher count rates may be obtained with a NaI(Tl) detector without coincidence loss.

2. *Conversion of scintillations to voltage pulses*

The employment of the solid scintillation phenomenon for the analysis of gamma radiation requires the conversion of the visible photon emissions of the inorganic crystal to voltage pulses of sufficient magnitude to be registered by a scaler. As in the case of liquid scintillation, a PM tube with a photosensitive cathode is used to convert a given photon emission to photoelectrons which are amplified by a series of dynodes to an avalanche of electrons equivalent to an amplification of 10^6–10^9. The magnitude of the amplification is proportional to the initial energy of the photon emitted by the NaI(Tl) crystal.

The photon intensities emitted by the NaI(Tl) crystal detector depend on the gamma-ray energies absorbed by the crystal and on the mechanisms of interaction of the bombarding gamma rays with the crystal detector. The predominant interactions of gamma rays with matter are the photoelectric effect, the Compton effect, pair production and annihilation, which are described in a previous book (L'Annunziata, 1979).

These interactions, as mechanisms of gamma-ray energy absorption in a crystal detector, are shown in Fig. 5.27. Gamma rays γ_1, γ_2 and γ_3, as emitted from a given sample illustrated in Fig. 5.27, are monoenergetic in the case of gamma-ray-emitting nuclides such as ^{22}Na, ^{54}Mn, ^{60}Co, ^{65}Zn, ^{85}Sr, ^{113}Sn and ^{137}Cs, among many others. Such gamma radiation may undergo several mechanisms of energy dissipation within the crystal lattice. Gamma-ray γ_1, for example, is shown interacting with the crystal via the photoelectric effect. Consequently, the kinetic energy of the electron in this case is almost equivalent to the gamma-ray energy, i.e. the gamma-ray energy less the binding energy of the ejected electron. The activation and deactivation occurring within the crystal as a result of this interaction produce photons of visible light with an intensity proportional to the gamma-ray energy. If all gamma-ray interactions occurred by the photoelectric effect and multiple interactions in the crystal detector did not occur, the output pulses produced would be of a fixed magnitude as a result of the emission of photons of equal intensity from the crystal. This is so because the number of photons emitted by the crystal detector is proportional to the amount of excitation energy dissipated in the crystal by the gamma rays.

The simple hypothetical case presented above unfortunately does not exist due to other mechanisms of gamma-ray interaction which occur in addition to the photoelectric effect. For example, γ_2 of Fig. 5.27 is illustrated interacting with an electron according to the Compton effect in which it imparts only a fraction of its energy to a crystal electron and is deflected as γ'_2. A

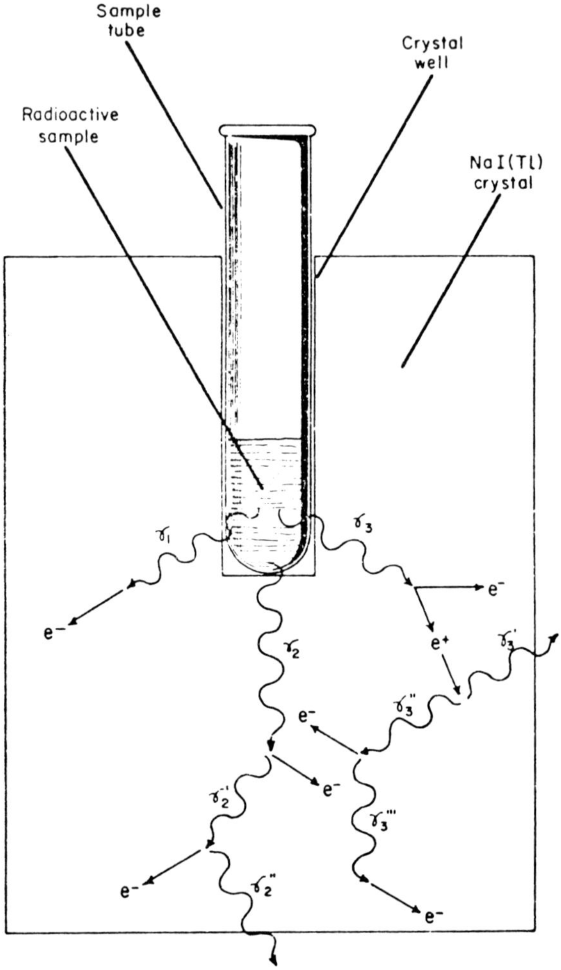

Fig. 5.27 The possible interactions of three gamma rays, γ_1, γ_2 and γ_3, of the same energy with an inorganic crystal.

subsequent interaction of γ_2' via the Compton effect is depicted in Fig. 5.27 with the deflection of γ_2'', which travels out of the crystal without suffering a further interaction. Obviously then, the magnitude of the gamma-ray energies passed on to the crystal electrons by γ_2 and γ_2' is only a fraction of that imparted by the previous example of γ_1. Thus, photons of visible light which are lower in number (intensity) than those produced by photoelectric interactions are also emitted by the crystal. A gamma ray may also dissipate all of its energy in the crystal detector by means of a series of Compton effects (not illustrated in Fig. 5.27). Such an occurrence would produce a pulse height of similar magnitude as that resulting from the photoelectric effect.

Gamma rays of 1.02 MeV energy or greater may also dissipate their energy within the crystal via pair production as illustrated by γ_3. The positron, e^+, of the electron pair combines with an electron and is annihilated with the production of two gamma rays of 0.51 MeV energies, shown as γ'_3 and γ''_3. These gamma rays may pass through the crystal detector without causing any interaction, as depicted by γ'_3, or they may undergo Compton and photoelectric interactions as depicted by γ''_3 and γ'''_3, respectively.

Because a gamma ray, originating either from the sample or from annihilation radiation, may impart any fraction of its energy to crystal electrons in Compton interactions, a wide range of photon intensities of visible light is emitted from the crystal detector. Consequently, a spectrum of pulse magnitudes is produced as a function of time. All pulses produced may be amplified and then registered as counts with an electronic counter or scaler. The instrumental components used to count such pulses as an assay of gamma radiation intensity are discussed in the following section.

B. Components of a solid scintillation counter

The basic components of a solid scintillation counter are the crystal detector, PM tube, high voltage supply, preamplifier, pulse height analyser, scaler and timer. In modern instruments the scaler and timer are integral parts of a microprocessor to which is linked an automatic sample changer and printer for the readout of count rates and related data. A block diagram illustrating the arrangement of these components is given in Fig. 5.28. The crystal detector, PM tube and preamplifier are shown as a single unit since these components are kept as close as possible to each other, firstly, to reduce to a minimum loss of photon intensity between the crystal detector and PM tube and, secondly, to reduce any pulse degradation that would result from excessive cable capacitance from the PM tube to the preamplifier.

The crystal detector is most commonly 7.5 cm × 7.5 cm in size. Smaller crystals are also used but they are less efficient in gamma-ray detection since a greater portion of the gamma radiation may pass the crystal without dissipating all of its energy within the crystal lattice. Larger crystals are obviously more efficient to a certain extent, although more costly, and oversized crystals are characterized by an enhanced background without increased sensitivity.

Because these crystals are hygroscopic, they are protected from atmospheric moisture by an aluminium coating ($\sim$0.05 cm thick). The aluminium also serves as a shield for alpha and low-energy beta radiation and

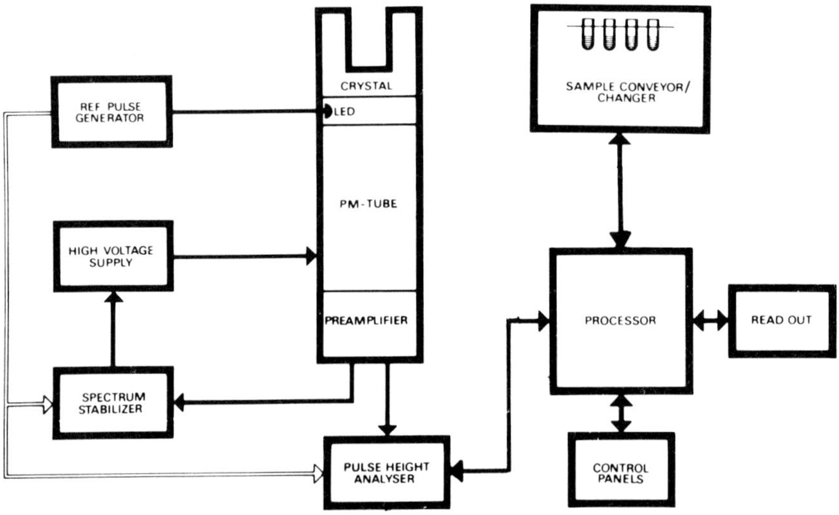

Fig. 5.28 Block diagram of a solid scintillation well counter with gain stabilization and automatic control systems. (Courtesy of Wallac Oy, Finland.)

consequently can, in certain cases, permit the exclusive detection of gamma radiation. The inner part of the aluminium is often silvered or whitened in order to reflect photons of visible light towards the PM tube, which would otherwise be absorbed in part by the aluminium coating of the crystal wall.

Crystal detectors may be employed in the form of a block. The planar surface of a crystal of this type can detect, at best, only 50% of the gamma radiation emitted from a given sample since the crystal can be exposed to a maximum of 180° of a possible 360° of radiation emission. This counting geometry is often referred to as 2π geometry. It is a relatively inefficient means of radiation detection. The optimum counting geometry is obtained by completely surrounding the sample with the radiation detector, which can receive all of the possible radiation emissions (360°) of a given sample. Consequently, in solid scintillation counting it is very common to employ a crystal detector with a cavity drilled at the centre, as illustrated in Figs 5.27 and 5.28. The sample to be counted is enclosed within a plastic or glass tube and placed within this cavity, which is referred to as the sample well. In this manner the crystal detector almost completely surrounds the sample and the detection geometry approaches that of 4π.

Placed in direct contact with the crystal detector is the PM tube. Transparent silicon grease and a plastic light pipe are used to ensure a good contact between the crystal and PM tube which permits a most efficient capture of the photons of visible radiation emitted from the crystal. An illustration and description of the function of a PM tube is given in the section on liquid scintillation counting. The PM tube may someday be replaced by the solid-state light amplifier reported by Orphan *et al.* (1978). As described by Heath *et al.* (1979), the light amplifier consists of a conventional photocathode coupled to a silicon diode. The photoelectrons produced at the photocathode are accelerated at a potential of 15 keV and focused onto the reversed-biased p–i–n silicon junction diode. Because ~3.6 eV are required to produce an electron–hole pair in silicon, the light amplifier produces a net single stage gain of ~4000 (i.e. 15 000 eV/3.6 eV). A subsequent slight gain produces an amplification equivalent to that of a PM tube.

The preamplifier amplifies the signals (pulses) by increasing the signal-to-noise ratios before they enter a cable leading to the amplifier. Preamplifier output signals commonly consist of a random step function proportional to input amplitudes. The rise time for each output pulse may be of the order of magnitude of 100 ns or less with an exponential decay requiring ~200 μs to fall within 2 % of the baseline. This produces a pulse commonly referred to as a tail pulse. Since the decay time is relatively slow in comparison to the rise time, normal count rates (50 000–100 000 cpm) may cause pulses to rise from the tail of a preceding pulse.

The preamplifier output is linearly amplified by a subsequent amplifier which contains a pulse-shaping circuit operating with time constants much shorter than the decay time of the preamplifier signal output and much longer than its rise time. This in effect removes the slow component of the preamplifier output signal and produces uniformly shaped and individual output pulses as illustrated in Fig. 5.29. Thus, the amplitudes of these pulses do not usually overlap, and their magnitudes are used to convey the energies of the gamma-rays dissipated within the crystal detector.

Subsequent to amplification and pulse-shaping, the pulse height discriminator (analyser) permits the passage of pulses which fall within predetermined amplitude levels. Pulse height discrimination is discussed in more detail in the section on liquid scintillation counting in this chapter.

Conventional scintillation counters have gain controls which permit the manual adjustment of the magnitude of amplification of output voltage pulses. The modern scintillation counters such as that illustrated in Fig. 5.28 have automatic gain control. The automatic gain stabilization system described by Soini and Rawlins (1975) controls the gain of the detector by regulating the high voltage to the PM tube according to the amplitude of a reference pulse. The reference pulses are produced in the Wallac Oy solid scintillation spectrometers by two GaAs–P light-emitting diodes (LED) mounted within the crystal assembly. The LED is driven by pulses from a reference pulse generator which switches the LED on for about 1 μs. The

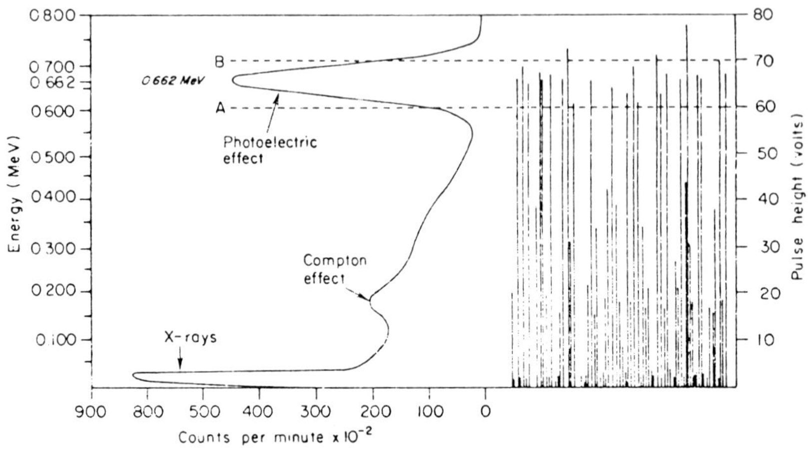

Fig. 5.29 The gamma-ray energy spectrum of ^{137}Cs in relation to the distribution of a fraction of many thousands of pulse heights produced in a single-channel solid scintillation spectrometer. Pulse height discriminators of variable height and width are marked A and B.

pulse generator simultaneously disables the pulse height analyser for about the same period of time to allow the spectrum stabilizer to measure the LED pulses from the photomultiplier. The spectrum stabilizer will either increase or decrease the high voltage until the LED pulse amplitude is equal to the reference level.

Modern spectrometers are equipped with a microprocessor which can control the sample changer and pulse height analyser (subsequently discussed) as well as perform computations to reduce data to count rate and even disintegration rate. The microprocessor also commands the printout or plot of sample data. Modern solid scintillation spectrometers can process automatically as many as 900 samples in tubes. The samples are changed and counted; and the sample activities or count rates are registered automatically.

C. Single-channel gamma-ray spectrometry

The positive linear correlation between the gamma-ray energy dissipation within the crystal detector and pulse height amplitude output from the PM tube, preamplifier and linear amplifier may be utilized to measure the gamma-ray energies of a given radionuclide. For this purpose a single-channel analyser may be employed.

A single-channel analyser contains two electronic discriminators set at two variable pulse amplitude levels (Fig. 5.29), often referred to as discriminator A (lower level) and discriminator B (upper level), as described previously for liquid scintillation. Those pulses which have amplitudes below the level of discriminator A are rejected and thus do not produce output pulses. Only those pulses with amplitudes falling within the lower and upper discriminator levels, referred to as the discriminator window, produce output pulses which pass on to the scaler to be registered as counts. The difference in levels between discriminators B and A (in volts) is referred to as the window width, and the level of discriminator A is known as the base level or window height.

The spectrum of pulse heights in volts, being directly proportional to the gamma-ray energy dissipation within the crystal detector, gives the gamma-ray energy spectrum of the radiation source. Such an energy spectrum is illustrated in Fig. 5.29 in relation to the spectrum of pulse heights produced. The gamma-ray spectrum is simply a plot of the relative abundances or populations of pulses of a given magnitude (voltage). The portion of the peak enclosed within the 10 V window of Fig. 5.29 describes the distribution of pulse amplitudes found between the A and B discriminators. This peak is the result of the interaction of 0.662 MeV gamma-rays of ^{137}Cs with the crystal detector via the photoelectric effect, or a combination of Compton effects resulting in the total dissipation of the gamma-ray energy in the crystal. The entire gamma-ray energy spectrum illustrated on the left-hand part of Fig. 5.29 is obtained by varying the window height at sufficiently small intervals over the entire pulse height range, from 0 to 80 V in this example.

Gamma ray spectra are very useful in the identification of unknown gamma-ray-emitting nuclides. For example, the peak due to the photoelectric effect, usually the highest energy peak, gives a measure of the gamma-ray energy of an unknown nuclide. The value of the gamma-ray energy, 0.662 MeV in the example given in Fig. 5.29, may be used to identify the nuclide.

Gamma-ray spectrometry may also be applied to assay more than one gamma-ray-emitting nuclide in a mixture. It is at times advantageous to employ more than one radiotracer in a given experiment. If these nuclides have distinct gamma-ray energies whose photoelectric peaks are well separated, they may be assayed separately by alternately focussing the discriminator window directly over the peak of each nuclide. However, the analysis of many gamma-ray-emitting nuclides in a mixture is most effectively carried out with the high resolution obtained with semiconductor detectors (see Section IV.B) employed with multichannel gamma-ray spectrometers.

D. Multichannel gamma-ray spectrometry

With the single-channel analyser previously discussed pulses are counted in numerous channel positions corresponding to definite pulse height ranges in order to construct energy spectra. Multichannel analysers permit the simultaneous accumulation of pulses in many thousands of energy levels (channels) in a semiconductor core memory.

The voltage pulses, which result from gamma-ray interactions with the radiation detector, pass on to an analogue-to-digital converter (ADC) where each pulse amplitude is converted into an equivalent time interval by the production of an equivalent number of pulses from a periodic oscillator. The number of pulses produced by the periodic oscillator is proportional to the amplitude of the converted voltage pulse. An address register then counts the number of pulses produced by the periodic oscillator, which gives a quantified measure of the amplitude of each input signal. This quantification identifies the channel among many thousands within which the pulse is to be stored in the core memory. In some analysers as many as 10^{20} counts may be stored in each channel. The contents of the core memory are viewed on an oscilloscope and, if desired, they may be registered by a printer or recorder. Such a simultaneous accumulation and readout of pulses in many thousands of channels corresponding to pulses of defined amplitudes is very time saving, an advantage which is often necessary for isotopes of very short half-life. This advantage offered by multichannel analysers can be particularly appreciated when high resolution semiconductor detectors are utilized and several thousands of channels with narrow window widths are required. The ^{60}Co gamma-ray spectrum obtained with a 4000-channel analyser is illustrated in Fig. 5.6.

A review of the developments and state of the art in multichannel spectrometers is given by Kandiah (1979).

E. Counting efficiencies

Efficiencies of solid scintillation counting are often defined in two fashions, either as the photoelectric efficiency or as the total efficiency. The photoelectric efficiency is defined as the fraction of the total number of photons emitted by the radionuclide which produce pulses under the photopeak. The photopeak, as described earlier, encompasses pulses which arise from the photoelectric effect or a series of Compton interactions equivalent to the complete absorption of photon energy in the crystal. The total efficiency is the fraction of the total number of gamma photons, which give rise to pulse events in the crystal detector. The detection efficiency used by an investigator will, of course, depend on whether he sets the discriminator window to encompass only the photopeak or the entire photon spectrum. Other counting efficiency factors may also be used; these are governed by the discriminator settings (counting window) which may be set to encompass any portion of the photon spectrum. The magnitude of the counting efficiencies obtained with a given solid scintillation detector are dependent upon several factors, among which are the energy of the photons emitted from the nuclide, the sample density and volume, and the crystal detector size.

The relationship between photon energy and counting efficiency is nearly linear when plotted on logarithmic series (Brinkman and Veenboer, 1979). As illustrated in Fig. 5.30, the counting efficiency decreases as the photon energy increases. Thus, nuclides which emit high-energy gamma radiation (e.g. ^{24}Na, 1.38 MeV and 2.76 MeV) are detected with lower counting efficiencies than nuclides which emit lower energy gamma radiation (e.g. ^{54}Mn, 0.83 MeV). Such comparisons can be made, of course, when there is negligible self-absorption among the nuclide samples. The energy values graphed in Fig. 5.30 cover the range from 0.40 to 4.0 MeV. Even higher

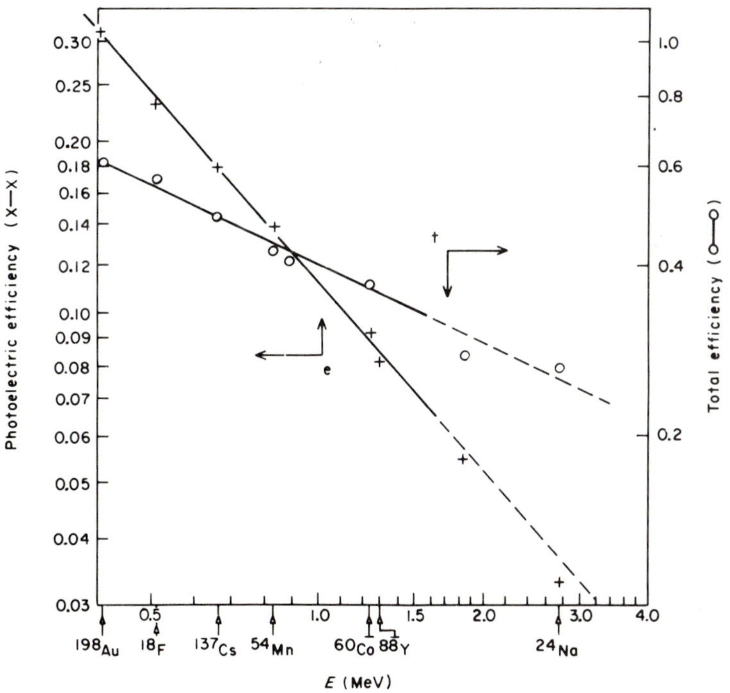

Fig. 5.30 Photoelectric (e) and total efficiency (t) curves for samples placed on the bottom of the well of a 60 mm × 60 mm NaI(Tl) crystal: diameter of well, 17 mm; depth of well, 38 mm (Brinkman and Veenboer, 1979).

counting efficiencies are expected for the detection of nuclides which emit lower energy photons. For example, ^{125}I emits X-radiation of 0.035 MeV and is detected with an optimum total counting efficiency of ~ 0.74 or 74%. The relationship between counting efficiency and photon energy is explained by the fact that most of the photons of low energy ($\leqslant 0.100$ MeV) hitting the detector are absorbed due to high absorption coefficients (Holmberg *et al.*, 1972). As the photon energy increases, the absorption coefficients decrease or, in other words, the "stopping-power" of the detector crystal decreases.

The sample volume can have a marked effect on the counting efficiency of the well-type solid scintillation detectors. The highest counting efficiencies are obtained from point sources situated at the bottom of the detector well. As the nuclide source is dispersed, either in a liquid or solid material, further away from the well floor of the detector, the near 4π geometry is reduced and counting efficiency decreases. Calibration curves of counting efficiency as a function of sample volume should be prepared to provide a measure of this effect of sample geometry on counting efficiency. The experimental procedure used for the preparation of such a calibration curve is given by L'Annunziata (1979).

The counting efficiencies of experimental samples can be affected significantly by self-absorption, particularly when the nuclide tracer emits low-energy X- or gamma radiation. Self-absorption may be defined as the absorption of radiation by the sample emitting the radiation. This absorbed radiation does not reach the detector and, if appreciable, counting efficiencies can be markedly reduced. Self-absorption is a function of sample density, thickness and photon energy. Photons of low energy such as the 0.035 MeV X-radiation emitted from ^{125}I are much more susceptible to self-absorption than the higher energy gamma rays from ^{57}Co (0.122 MeV) or ^{60}Co (1.17 and 1.33 MeV), and can be attenuated significantly before escaping from a typical 1 cm diameter sample tube (Malcolme-Lawes and Massey, 1980). The effect of salt concentration and ions of different atomic mass absorption coefficients on the solid scintillation counting efficiency of ^{125}I is illustrated in Fig. 5.31. Malcolme-Lawes and Massey (1980) report a channels ratio counting technique as a monitor for such effects on the counting efficiency of ^{125}I. Rieppo *et al.* (1975) and Rieppo (1976a,b) have studied extensively the relationship of counting efficiency and sample density (e.g. water, bone and aluminium) for photons in the energy range 0.010–0.200 MeV.

It was stated earlier in this chapter that crystal detector sizes are commonly 75 mm × 75 mm. Sizes inferior to this give lower gamma-ray counting efficiencies because the probabilities become greater that a gamma-ray may pass the crystal detector without dissipating its energy within. Belle *et al.* (1974) demonstrated that oversized crystals are characterized by an enhanced background without increased sensitivity. They conclude that a 70 mm ×

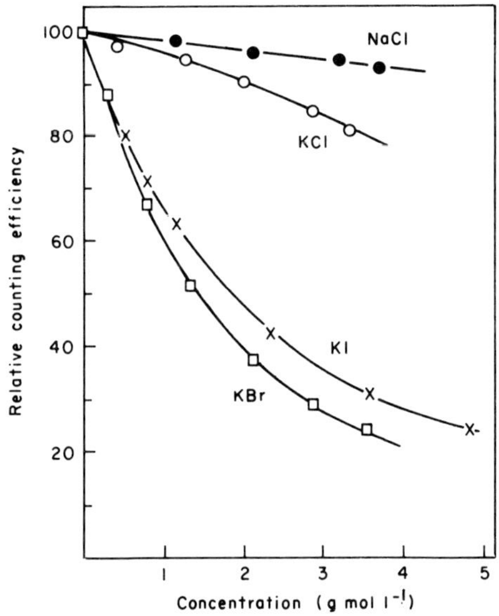

Fig. 5.31 Variation of counting efficiency for ^{125}I (relative to efficiency in water = 100) with concentration of materials in aqueous samples. Sample volumes are 5 ml; bottle diameter 1.3 cm (Malcolme-Lawes and Massey, 1980).

70 mm NaI(Tl) crystal is best with gamma-ray energies up to 0.4 MeV, whereas a 150 mm × 150 mm crystal is recommended for gamma-ray energies between 0.4 and 3.0 MeV.

VIII. Figure-of-merit

The comparative performance of counters or the relative effectiveness of sample preparation techniques for counting are at times evaluated in terms of the figure-of-merit. The former is called the instrument figure-of-merit and the latter is referred to as the sample figure-of-merit.

The instrument figure-of-merit is E^2/B, where E is the percentage counting efficiency of a given nuclide and B is the background count rate. This figure is used as a measure of the optimization of instrument response to sample radiation while taking into account the background level. It can be used as a measure of the optimization of counting parameters (see Table 5.4) or instrument settings such as optimal window widths for highest counting efficiencies at minimal backgrounds. Some examples of optimal instrument figures-of-merit reported are from 175 to >200 for ^{3}H and >350 for ^{14}C

with conventional liquid scintillation counters and >200 for ^{125}I counting with conventional solid scintillation counters. These figures can be considerably lower, for example, when experimental samples are counted at less than optimal conditions.

The sample figure-of-merit is $E \times$ sample size, where the sample size is expressed in units of weight or volume. It is a relative measure of the sample conditions which affect the detector (counter) response. For example, the optimum amount of sample dissolved in a fluor cocktail can be determined by comparing the figures-of-merit of samples varying in size (Herberg, 1960; Painter, 1976). Likewise, the quality of different fluor cocktails can be measured by comparing their sample figures-of-merit for a given sample size.

As explained by Kobayashi and Maudsley (1974), the expression of the figure-of-merit has arisen from the need to reduce the counting time required to achieve an acceptable statistical number of counts, which is a function of the counting efficiency and background. As the figure-of-merit is increased, less counting time is required to reach any level of statistical accuracy desired on the sample counts. A treatment of statistical computations and criteria in radionuclide counting is given in the Appendix in Volume 2.

IX. Autoradiography

Autoradiography may be defined as the production of an image on a photographic emulsion (a suspension of silver bromide crystals in gelatin) by nuclear radiation emitted from a particular body or material. The production of this image by nuclear radiation is explained by the theory of Gurney and Mott (1938) as refined by Rogers (1967) and others. The proposed mechanism is that the nuclear radiation causes ionization of bromide ions within the silver bromide crystals of the photographic emulsion by the ejection of an orbital electron from a bromide ion. Ejected electrons may travel from one ion to another until they are trapped at sites which consist of defects within the silver bromide crystal structure. At this structural defect a silver ion absorbs the electron to become a silver atom. This affected crystal becomes visible only after the application of a chemical "developer" which reduces all of the silver ions of the affected crystal to metallic silver (observed as black grains). The crystals which do not contain reduced silver are dissolved out of the emulsion by the chemical "fixer", leaving a black image produced by the nuclear radiation.

The limitations of space for this chapter will not allow any detailed treatment of autoradiography; however, a review of the theory and practice of autoradiography is given in a previous book by L'Annunziata (1979).

A. Autoradiographic resolution

Of utmost importance in autoradiography is the resolution of the image produced by the nuclear radiation. Resolution is often defined by those working with autoradiography as a measure of the ability to distinguish between two closely situated radioactive sources. The ability to distinguish between radiotracer-labelled pubescent hairs of a plant leaf, its transport components and the remaining leaf material may be considered as a figurative measure of the resolution with a given macroautoradiograph of a plant leaf. In the simplest case, the distance by which two radioactive sources must be separated to distinguish their autoradiographs on a given emulsion is considered as the resolution. This aspect of resolution has been studied by Gomberg (1951), who defined resolution as the distance which separates two sources of equal strength where the grain density between the two sources is 50% of the density encountered above each of the sources.

The resolution in autoradiography is governed by several factors among which are, in brief, (i) the range of the ionizing particles, (ii) the type and thickness of the emulsion, (iii) the distance between the specimen and the emulsion, (iv) the time of exposure and (v) the time of development (Levi *et al.*, 1963). In general, X-ray emulsions used for low-resolution autoradiography are 10–30 μm thick and contain approximately six grains of 1–4 μm diameter per cubic micrometre, yielding an optimum resolution of 25–30 μm (Levi *et al.*, 1963). Using the best emulsions and development procedures presently available, optimum resolution of an autoradiograph may be obtained by controlling variables (i), (iii) and (iv), which are discussed in more detail below.

1. *Range of the ionizing particles*

The range of an ionizing particle emitted by a given radionuclide obviously cannot be controlled; however, if more than one radioisotope of a given element is available, the isotope emitting particulate radiation of lower energy and consequently of shorter range should be chosen for autoradiography when high resolution is required (e.g. microautoradiography). For example, ^{33}P (E_{max} = 0.25 MeV) is preferable to ^{32}P (E_{max} = 1.71 MeV) in high-resolution autoradiograph studies involving phosphorus. The greater the range of the ^{32}P beta particles in an emulsion would cause a broadening of the grain density distributions, resulting in poorer resolution. High-resolution autoradiography with ^{14}C (E_{max} = 0.15 MeV) obviously produces more highly resolved autoradiographs than those obtained with ^{33}P, other experimental conditions being equal. Even more highly resolved autoradiographs are obtained with ^{3}H (E_{max} = 0.018 MeV) due to the low energy and consequent short range of its beta particles. However, in macroautoradiography ^{3}H has the limitation of very high self-absorption within the sample to be autoradiographed due to the very low energy and short range of its beta particles. Consequently, ^{3}H is not an effective tracer for the autoradiography of organisms which are not sectioned, e.g. whole-plant material.

2. *Distance between the specimen and the emulsion*
Optimum resolution is obtained when the emulsion is placed in direct contact with the radioactive source. Optimization of contact between the source and the emulsion in macroautoradiography is obtained by keeping the source as flat as possible and by applying pressure to the emulsion above the source. Slight pressure is usually applied to the film and source material during the period of exposure in order to optimize the contact. Excessive pressure can produce on the film emulsion an impression of the object with which it is in contact. Consequently, it is usually necessary to expose film simultaneously to a similar object not containing a radiotracer as a control. In microautoradiography intimate contact between the film emulsion and a thin section of sample on a microscope slide is achieved by use of liquid emulsion, as discussed later in this chapter.

3. *Exposure time*
The exposure of film to radioactive point sources for periods of time greater than the optimum exposure time would produce excessive grain densities (blackening) from the distance closest to the point sources to those distances equivalent to the maximum range of the beta particles. The probability of the production of silver grains close to the point sources is greater than that at further distances. However, if sufficient time is allowed, a complete saturation (blackening) of all of the silver grains would occur within the range of the beta particles. Because the number of crystals within an emulsion is limited, the number of saturated silver grains produced will be such that the distinguishing of two neighbouring point sources, even when these are separated by a distance greater than the resolution obtainable, will be impossible.

The amount of time required to expose emulsion to a given sample is difficult to determine with precision. However, the time may be estimated with the aid of a calibration curve consisting of a plot of the radioactivity of a given sample type against the time required to produce a clearly discernible autoradiograph. Tsuk *et al.* (1964) have shown that the logarithm of the ^{14}C activity applied to a paper chromatogram is linear between 0.2 and 0.01 μCi, if plotted against the logarithm of the required time of exposure, found to range between 1 and 20 h for the above-mentioned range of activity, and for the particular X-ray film used. In preparing such a curve it is recommended that the necessary exposure times for definite levels of activity be determined within a definite area, such as the exposure time necessary to produce a discernible image from a 0.01 μCi cm^{-2} spot.

B. Macroautoradiography

In macroautoradiography one determines the distribution of a radiotracer over a relatively large area such as an entire plant or animal section. In brief, a dry and flat (pressed) plant or a dried thin (5–100 μm) animal section is placed in contact with the emulsion of X-ray film in a dark room. After a given period of exposure time, the film is developed and fixed in the dark room.

An image produced by the nuclear radiation is observed on the developed film and illustrates the position or distribution of the radiotracer in the material placed in contact with the photographic emulsion. The time of exposure required to produce a discernible image is a function of the level of radiation absorbed by the emulsion. Examples of the autoradiography of plant and animal tissue are given in Figs 5.32 and 5.33, respectively. The

Fig. 5.32 Autoradiograph of part of a soybean plant which had absorbed a ^{14}C-labelled herbicide by root uptake. The autoradiograph was taken with Kodak SB-54 single-coated blue-sensitive X-ray film. (From unpublished research of M. F. L'Annunziata and M. G. Parkins, courtesy of Amchen Products, Inc., Ambler, PA.)

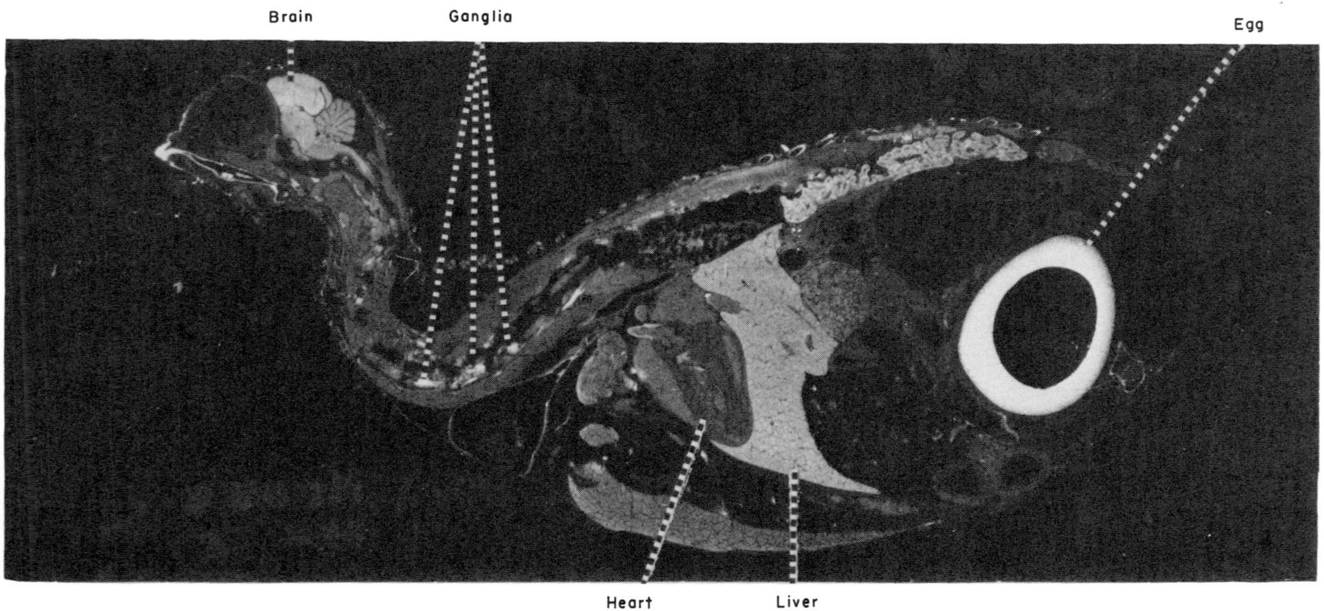

Fig. 5.33 Contact print of an autoradiograph of a hen quail 10 days after injection of ^{203}Hg-labelled methyl mercury. There is a strong accumulation in the white of the egg in the oviduct. The use of methyl mercury as a fungicide in the treatment of seed has led to the death of many bird embryos, which get a large dose of methyl mercury when they use the eggwhite for their nutrition. A strong accumulation is also seen in the ganglia and proximal parts of the nerves in the lower neck and thoracical region, which may be related to the symptoms of lameness due to methyl mercury poisoning. A slight accumulation is also seen in brain and liver. Light regions on contact prints correspond to dark regions on autoradiographs (Ullberg and Dencker, 1977).

degree of blackening in the autoradiograph is a measure of the relative amounts of radiotracer in the plant or animal section.

More detailed information on the macroautoradiography of plant matter is given by Levi (1962, 1966), Walter *et al.* (1970) and Fiveland *et al.* (1972), and detailed treatments of macroautoradiography of animal sections may be obtained from (Ullberg (1954, 1962), Gahan (1972), Cross *et al.* (1974), Wegner and Patzschke (1976) and Ullberg and Dencker (1977). A review of both techniques is given by L'Annunziata (1979).

C. Microautoradiography

Microautoradiography involves the use of photographic emulsion to determine the distribution of a radiotracer over a very small area observable only with the light or electron microscope.

1. *Stripping film technique*

When the light microscope is employed, stripping film may be used as a source of emulsion in microautoradiography. Stripping film, such as Kodak AR 10, consists of a thin (~5 μm) layer of emulsion and an underlying thicker (~10 μm) layer of gelatin set on a glass support. In the darkroom under safelight illumination the emulsion is stripped off its glass support and floated (emulsion downward) on the surface of a 0.05% aqueous solution of KBr. The floating emulsion is subsequently picked up with a glass microscope slide containing a histological section of a plant or animal so that the emulsion comes into contact with the specimen section. The slide and its clinging emulsion are left to dry prior to storage in the darkroom for the required period of exposure. The histological sections with their attached and developed autoradiograph emulsions may be stained as an aid in locating the plant or animal cells, their nuclei and other cell organelles. The darkened silver grains are observed simultaneously and counted. A grain count over a cell or organelle compared to the background count gives the relative amount of radiotracer. The scientific techniques involved in the preparation of the histological material, the stripping film and the autoradiographs, and in the staining, are discussed in detail by Appleton (1972), Gey (1974) and Maraldi (1976). With all of the factors governing resolution, at its optimum the resolution obtainable by this technique is dependent upon the resolving power of the light microscope.

2. *Liquid emulsion technique and high-resolution autoradiography*

Optimum resolution is obtained by maintaining an intimate contact between the emulsion and specimen. Because solid emulsions and specimens are not perfectly smooth, only close rather than intimate contact can be achieved by the contact methods previously discussed. An intimate contact is reached, however, by the application of an emulsion in liquid form. The liquid emulsion can enter into and fill any microscopic crevice or uneven character of the specimen. The following steps describe the general procedure involved in liquid emulsion autoradiography:

(i) A histological specimen embedded in paraffin or plastic resin is cut on a microtome to a thickness of ~0.5–5 μm and mounted on a microscope slide.
(ii) The specimen may be stained at this point or after the emulsion has been developed and fixed.
(iii) The mounted slide is dipped into a batch of melted bulk emulsion and dried to give an ~5 μm thick emulsion layer over the specimen.
(iv) The glass slide containing the specimen and emulsion is left in the darkroom for the required exposure time.
(v) The emulsion is then developed and fixed according to the procedure described by the emulsion manufacturer.
(vi) The developed emulsion and stained specimen are finally mounted under a coverslip prior to the microscopic localization and counting of individual blackened silver grains and the determination of their cellular position relative to the stained specimen.

Liquid emulsion autoradiography is employed with both light microscopy and high-resolution electron microscopy. Specific techniques and detailed procedures for the preparation of samples and emulsion layers for the light and electron microscope are given by Bogoroch (1972) and Budd (1972), respectively. The resolution obtainable by the light microscope, reported by Budd (1972) to be 75–184 μm, is limited by the optical resolution of the microscope and the minimum observable grain size. However, with the electron microscope, resolutions ranging from 0.1 to 0.2 μm have been reported by Bachmann *et al.* (1967) and Budd (1972); and this has been demonstrated by Prabhu Gaunkar *et al.* (1979) to be the limit of resolution with currently available emulsions and developers. An example of an electron micrograph is illustrated in Fig. 5.34. The scanning electron microscope has also been applied to the analysis of biological autoradiographs. This technique is discussed by Hodges and Muir (1976).

Relative levels of radiotracer in cells or cell organelles may be determined by grain counting, which is also referred to as quantitative microscopy autoradiography. The individual silver grains in the liquid emulsion located above a particular cellular body (e.g. mitochondria) may be counted in the microscope field and correlated with grain counts obtained from other cell

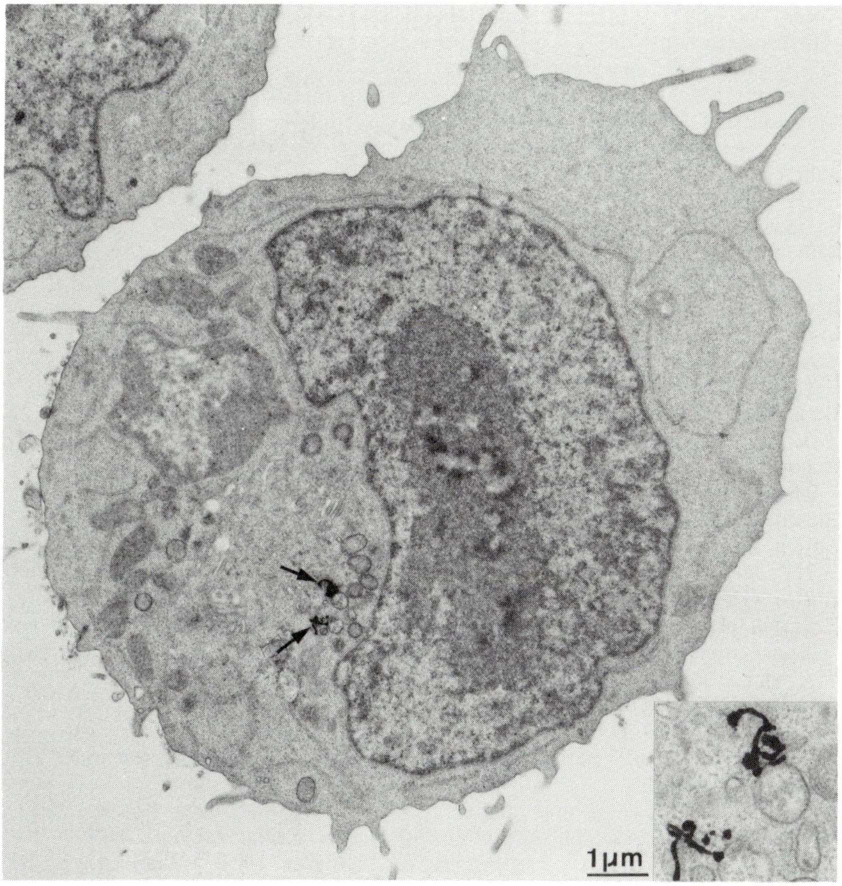

Fig. 5.34 Electron microscope autoradiograph general view at low magnification (× 7000) of a lymphocyte which was incubated with [^{125}I]anti-insulin receptor antibody at 37 °C for 120 min. Two autoradiographic grains (arrows) are seen in the Golgi region of the cell. This region is typically localized in the indentation of the nucleus. The grains are seen close to membrane-bounded structures with morphological characteristics of lysomes. *Inset,* × 21 700 (Carpentier *et al.*, 1981).

parts. The technique and statistical considerations involved in grain counting in the light and electron microscope fields are described by Caro *et al.* (1962), Dupuy-Coin and Bouteille (1975), Dadoune (1976), Doermer (1976) and Hardt and Panijel (1976).

References

Abson, W., Allsworth, F. L. and Salmon, P. C. (1968). In "I.E.E.E. Nucleonic Instrumentation Conference", IEEE Publication no. 47, p. 206, IEEE, London.

Appelgren, K. L., Lewis, D. H. and Haggendal, E. (1965). *Scand. J. Clin. Lab. Invest.* **17**, 511–512.

Appleton, T. C. (1972). *In* "Autoradiography for Biologists" (P. B. Gahan, ed.), pp. 33–49, Academic Press, New York and London.

Ashcroft, J. (1970). *Anal. Biochem.* **37**, 268–275.

Bachmann, L., Salpeter, M. M. and McHenry, F. (1967). *J. Cell Biol.* **33**, 299–305.

Barrows, G. H., Samols, E. and Becker, B. (1976). *J. Nucl. Med.* **17**, 1017–1018.

Belle, Y. S., Zlobin, L. I., Moroz, G. L. and Yushkevich, G. F. (1974). *Spectrom. Methody Anal. Radioakt. Zagryaz. Pochv. Aerozolei, Dokl. Vses. Soveshch.* 11–14.

Bem. E. M., Bem, H. and Reimschuessel, W. (1978). *Bull. Environm. Contam. Toxicol.* **19**, 677–683.

Bem, E. M., Bem, H. and Reimschuessel, W. (1980). *Int. J. Appl. Radiat. Isotopes* **31**, 371–374.

Bezaguet, A., Geles, C., Reucroft, S., Marini, G. and Martellotti, G. (1979). *Nucl. Instrum. Meth.* **158**, 303–306.

Bogoroch, R. (1972). *In* "Autoradiography for Biologists" (P. B. Gahan, ed.), pp. 65–94, Academic Press, New York and London.

Bojsen, J. and Rossing, N. (1978). *Nucl. Instrum. Methods* **150**, 49–53.

Bojsen, J., Groth, S. and Rossing, N. (1981). *Int. J. Appl. Radiat. Isotopes* **32**, 719–727.

Borosi, G., Cazzola, M. and Perugini, S. (1980). *In* "Liquid Scintillation Counting, Recent Applications and Development", Vol. II, "Sample Preparation and Applications" (C.-T. Peng, D. L. Horrocks and E. L. Alpen, eds), pp. 517–523, Academic Press, New York and London.

Boynton, G. R. (1975). *Nucl. Instrum. Meth.* **123**, 599–603.

Bransome Jr., E. D. (1973). *Semin. Nucl. Med.* **3**, 389–399.

Brinkman, G. A. and Veenboer, J. Th. (1979). *Int. J. Appl. Radiat. Isotopes* **30**, 171–176.

Brock, M., Ingvar, D. H. and Jacobsen, C. W. (1967). *Exp. Brain Res.* **4**, 126–137.

Brownell, J. R. and Lauchli, A. (1969). *Int. J. Appl. Radiat. Isotopes* **20**, 797–798.

Budd, G. C. (1972). *In* "Autoradiography for Biologists" (P. B. Gahan, ed.), pp. 95–118, Academic Press, New York and London.

Caine, S., Holzer, A., Beinglass, I., Dishon, G., Lowenthal, E. and Schieber, M. (1978a). *Nucl. Instrum. Meth.* **150**, 83–89.

Caine, S., Holzer, A., Beinglass, I., Schieber, M. and Lowenthal, E. (1978b). *I.E.E.E. Trans. Nucl. Sci.* **25**(1), 649–652.

Calf, G. E. (1977). *Int. J. Appl. Radiat. Isotopes* **28**, 737–738.

Carmon, B. (1979). *Int. J. Appl. Radiat. Isotopes* **30**, 97–100.

Caro, L. G., van Tubergen, R. P. and Kolb, J. A. (1962). *J. Cell Biol.* **15**, 173–188.

Carpentier, J.-L., Van Obberghen, E., Gorden, P. and Orci, L. (1981). *Exp. Cell. Res.* **134**, 81–92.

Cerretelli, P., Blau, M., Pendergast, D., Eisenhardt, C., Rennie, D. W., Steinbach, J. and Entine, G. (1978). *I.E.E.E. Trans. Nucl. Sci.* **25**(1), 620–623.
Chase, G. D. (1980). *In* "Liquid Scintillation Counting, Recent Applications and Development", Vol. II, "Sample Preparation and Applications" (C.-T. Peng, D. L. Horrocks and E. L. Alpen, eds), pp. 489–502, Academic Press, New York and London.
Cherenkov, P. A. (1934). *Compt. Rend. Acad. Sci. U.R.S.S.* **2**, 451–454.
Chow, P. N. P. (1980). *In* "Liquid Scintillation Counting, Recent Applications and Development", Vol. I, "Physical Aspects" (C.-T. Peng, D. L. Horrocks and E. L. Alpen, eds), pp. 387–395, Academic Press, New York and London.
Clausen, T. (1968). *Anal. Biochem.* **22**, 70–73.
Cosolito, F. J., Cohen, N. and Petrow, H. G. (1968). *Anal. Chem.* **40**, 213–215.
Cramer, C. F., Nicholson, M., Moore, C. and Teng, K. (1971). *Int. J. Appl. Radiat. Isotopes* **22**, 17–20.
Cross, P. and McBeth, G. W. (1976). *Health Phys.* **30**, 303–306.
Cross, S. A. M., Groves, A. D. and Hesselbo, T. (1974). *Int. J. Appl. Radiat. Isotopes* **25**, 381–386.
Cummins, E. G. and Horne, C. G. (1977). *In* "Liquid Scintillation Counting", Vol. IV (M. A. Crook and P. Johnson, eds), pp. 252–259, Heyden, London.
Dadoune, J. P. (1976). *Exp. Cell Res.* **97**, 31–41.
Deme, S. (1967). "Semiconductor Detectors for Nuclear Radiation Measurement", Wiley–Interscience, New York.
Dern, R. J. and Hart, W. L. (1961a). *J. Lab. Clin. Med.* **57**, 322–330.
Dern, R. J. and Hart, W. L. (1961b). *J. Lab. Clin. Med.* **57**, 460–467.
Dietrich, H. B., Purdy, A. E., Murray, R. B. and Williams, R. T. (1973). *Phys. Rev. B* **8**, 5894–5901.
Dobbs, H. E. (1965). "Dispensing Solutions for Liquid Scintillation Counting", Memorandum no. AERE-M1574, UK Atomic Energy Research Establishment, Harwell.
Doermer, P. (1976). *Acta Histochem. Suppl.* **16**, 39–50.
Dupuy-Colin, A. M. and Bouteille, M. (1975). *Exp. Cell Res.* **90**, 111–118.
Eakins, J. D. and Brown, D. A. (1966). *Int. J. Appl. Radiat. Isotopes* **17**, 391–397.
Elrick, R. H. and Parker, R. P. (1968). *Int. J. Appl. Radiat. Isotopes* **19**, 263–271.
Everett, L. J., Ring, J. G. and Nguyen, D. C. (1980). "Liquid Scintillation Counting, Recent Applications and Development", Vol. I, "Physical Aspects" (C.-T. Peng, D. L. Horrocks and E. L. Alpen, eds), pp. 119–128, Academic Press, New York and London.
Ewan, G. T. (1979). *Nucl. Instrum. Meth.* **162**, 75–92.
Fiveland, F. J., Erickson, L. C. and Seely, C. I. (1972). *Weed Res.* **12**, 155–163.
Flynn, K. F., Glendenin, E. P. and Wright, P. M. (1964). *Nucl. Instrum. Meth.* **27**, 13–17.
Fric, F. and Finocchiaro, V. (1975). *Radiochem. Radioanal. Lett.* **21**, 205–210.
Fric, F. and Palovcikova, V. (1975). *Int. J. Appl. Radiat. Isotopes* **26**, 305–311.
Gahan, P. B. (ed.) (1972). "Autoradiography for Biologists", Academic Press, New York and London.
Gey, W. (1974). *In* "Methods in Human Cytogenetics" (H. G. Schwarzacher, U. Volf and E. Passarge, eds), pp. 119–134, Springer Verlag, New York.
Gibson, J. A. B. (1976). *In* "Liquid Scintillation, Science and Technology" (A. A. Noujaim, C. Ediss and L. I. Wiebe, eds), pp. 153–166, Academic Press, New York and London.

Gibson, J. A. B. (1980). *In* "Liquid Scintillation Counting, Recent Applications and Development", Vol. I, "Physical Aspects" (C.-T. Peng, D. L. Horricks and E. L. Alpen, eds), pp. 153–172, Academic Press, New York and London.
Glass, A. D. M. (1978). *Int. J. Appl. Radiat. Isotopes* **29**, 75–76.
Gomberg, H. J. (1951). *Nucleonics* **9**, 28–43.
Grangsjo, G., Ulfendahl, H. R. and Wolgast, M. (1966). *Nature (Lond.)* **211**, 1411–1412.
Grau-Malonda, A. (1982). *Int. J. Appl. Radiat. Isotopes* **33**, 371–375.
Gringorten, J. L. (1983). *Int. J. Appl. Radiat. Isotopes* **34**, 1367–1376.
Gruhn, C. R. and Ogle, W. (1980). *In* "Liquid Scintillation Counting, Recent Applications and Development", Vol. I, "Physical Aspects" (C.-T. Peng, D. L. Horrocks and E. L. Alpen, eds), pp. 357–374, Academic Press, New York and London.
Guilmette, R. A. and Bay, A. S. (1981). *Anal. Chem.* **53**, 2354–2356.
Gurney, R. W. and Mott, N. F. (1938). *Proc. Roy. Soc. Lond. A* **164**, 151–167.
Gutcho, S., Johnson, J. and McCarter, H. (1973). *Clin. Chem.* **19**, 998–1001.
Hardt, N. and Panijel, J. (1976). *Exp. Cell Res.* **101**, 78–86.
Heath, R. L., Hofstadter, R. and Highes, E. B. (1979). *Nucl. Instrum. Meth.* **162**, 431–456.
Helman, E. Z. and Spiehler, V. (1974). *Clin. Chem.* **20**, 516.
Hendee, W. R., Ibbott, G. S. and Crusha, K. L. (1972). *Int. J. Appl. Radiat. Isotopes* **23**, 90–95.
Herberg, R. J. (1960). *Anal. Chem.* **32**, 42–46.
Hodges, G. M. and Muir, M. D. (1976). *In* "Principles and Techniques of Scanning Electron Microscopy", Vol. V (M. A. Hayat, ed.), pp. 78–93, Van Nostrand Reinhold, New York.
Holmberg, P., Rieppo, R. and Rassi, P. (1972). *Int. J. Appl. Radiat. Isotopes* **23**, 115–120.
Horrocks, D. L. (1966). *Int. J. Appl. Radiat. Isotopes* **17**, 441–446.
Horrocks, D. L. (1971). *Int. J. Appl. Radiat. Isotopes* **22**, 258–260.
Horrocks, D. L. (1976a). *In* "Liquid Scintillation Science and Technology" (A. A. Noujaim, C. Ediss and L. I. Wiebe, eds), pp. 1–16, Academic Press, New York and London.
Horrocks, D. L. (1976b). *In* "Liquid Scintillation Science and Technology" (A. A. Noujaim, C. Ediss and L. I. Wiebe, eds), pp. 185–198, Academic Press, New York and London.
Horrocks, D. L. (1978a). US Patent No. 4,075,480.
Horrocks, D. L. (1978b). *In* "Liquid Scintillation Counting" (M. A. Crook and P. Johnson, eds), pp. 145–168, Heyden, London.
Horrocks, D. L. (1978c). *J. Radioanal. Chem.* **43**, 489–521.
Horrocks, D. L. and Kolb, A. J. (1981). *Laboratory Practice* **30** (5), Beckman Instruments, Irvine, Calif.
Ihle, H. R., Karayannis, M. and Murrenhoff, A. P. (1965). *In* "IAEA Symposium on Radioisotope Sample Measurement Techniques in Medicine and Biology (Vienna)", pp. 485–503, IAEA, Vienna.
Jelley, J. V. (1962). *Contemp. Phys.* **3**, 45–57.
Jordan, W. C., Spiegler, V., Haendiges, R. and Helman, E. (1974). *Clin. Chem.* **20**, 733–737.
Kabler, M. N. (1975). *In* "Proceedings of the NATO Advanced Study Institute on Radiation Damage in Material, Corsica 1973", Hordhof International Publ.

Kalbhen, D. A. (1967). *Int. J. Appl. Radiat. Isotopes* **18**, 655–656.
Kalbhen, D. A. (1980). *In* "Liquid Scintillation Counting, Recent Applications and Development", Vol. II, "Sample Preparation and Applications" (C.-T. Peng, D. L. Horrocks and E. L. Alpen, eds), pp. 273–280, Academic Press, New York and London.
Kallman, H. (1950). *Phys. Rev.* **78**, 621–622.
Kandiah, K. (1979). *Nucl. Instrum. Meth.* **162**, 699–718.
Kannan, S. (1975). *Int. J. Appl. Radiat. Isotopes* **26**, 557–558.
Karamanos, R. E., Bettany, J. R. and Rennie, D. A. (1975). *Can. J. Soil Sci.* **55**, 407–413.
Kato, T. and Hatagami, T. (1980). *Anal. Chem.* **52**, 586–587.
Keough, R. F. and Powers, G. J. (1970). *Anal. Chem.* **42**, 419–421.
Kerr, V. N., Hayes, F. N. and Ott, D. G. (1957). *Int. J. Appl. Radiat. Isotopes* **1**, 284–288.
Khouri, E. M., Olsson, R. A., Bedynek, J. L. and Bass, B. G. (1977). *Am. J. Physiol.* **232**, H95–H98.
Kobayashi, T., Sugita, T., Takayanagi, S., Iio, M. and Yanada, H. (1972). *Phys. Med. Biol.* **47**, 656–660.
Kobayashi, T., Sugita, T., Takayanagi, S., Iio, M. and Sasaki, Y. (1973). *I.E.E.E. Trans. Nucl. Sci.* **20**(1), 310–317.
Kobayashi, T., Sugita, T., Sasaki, T., Iwase, T., Sassa, R. and Iio, M. (1974). *I.E.E.E. Trans. Nucl. Sci.* **21**(1), 735–740.
Kobayashi, T., Sugita, T. and Matsuo, N. (1980). Brit. Pat. No. 1,559,664 (Cl. HO1L31/02).
Kobayashi, Y. and Maudsley, D. V. (1974). "Biological Applications of Liquid Scintillation Counting", Academic Press, New York and London.
Kolb, A. J. (1978). "Quench Monitoring with *H* Number, Part One: The Advantage of *H* Number", Report no. T1283-NUC-78-18 Beckman Instruments, Irvine, Calif.
Kolb, A. J. and Fujimoto, C. S. (1978). "Quench Monitoring with *H* Number, Part II: *H* Number Reproducibility and the Application of a Standard Quench Set", Report no. T1315-NUC-78-20, Beckman Instruments, Irvine, Calif.
Kolb, A. J. and Fujimoto, C. S. (1979). "Quench Monitoring with *H* Number, Part III: The Accuracy and Precision of *H* Number in the Presence of ^{32}P", Report no. Ti341-NUC-79-21, Beckman Instruments, Irvine, Calif.
Konishi, T., Hammricks, P. E. and Walst, P. J. (1978). *Acta Otolaryngol.* **86**, 22–34.
Kumpulainen, L. H. and Saukkonen, H. A. (1979). *Int. J. Appl. Radiat. Isotopes* **30**, 407–410.
L'Annunziata, M. F. (1979). "Radiotracers in Agricultural Chemistry", Academic Press, New York and London.
Laney, B. H. (1976). *In* "Liquid Scintillation Science and Technology" (A. A. Noujaim, C. Ediss and L. I. Wiebe, eds), pp. 135–152, Academic Press, New York and London.
Laney, B. H. (1977). *In* "Liquid Scintillation Counting", Vol. IV (M. A. Crook and P. Johnson, eds), pp. 74–84, Heyden, London and New York.
Lauber, A. (1970). *In* "Proceedings of the Symposium on Semiconductor Detectors for Nuclear Radiation, Munich, 1970" (P. Eichinger and G. Keil, eds), Technical University, Munich.
Lauber, A. (1972). *Nucl. Instrum. Meth.* **101**, 545–550.
Lauber, A. and Malmsten, B. (1970). *Nucl. Instrum. Meth.* **81**, 77–84.

Lauber, A. and Rosencrantz, B. (1964). Report no. AE-162, Atomenergi, Studsvik, Sweden.
Lauber, A. and Wolgast, M. (1972). *Phys. Med. Biol.* **17**, 525–537.
Lauchli, A. (1969), *Int. J. Appl. Radiat. Isotopes* **20**, 265–270.
Lauchli, A. (1971). *In* "Organic Scintillators and Liquid Scintillation Counting" (D. L. Horrocks and C.-T. Peng, eds), pp. 771–782, Academic Press, New York and London.
Lefebvre, D. D. and Glass, A. D. M. (1981). *Int. J. Appl. Radiat. Isotopes* **32**, 116–117.
Levi, E. (1962). *Science* **137**, 343–344.
Levi, E. (1966). *In* "Proceedings of the Symposium on Isotopes in Weed Research", pp. 189–194, Vienna.
Levi, H., Rogers, A. W. M., Bentzon, M. W. and Nelson, A. (1963). *Mat.-Fys. Meddr.* no. 33.
Lindenbaum, A. and Lund, C. J. (1969). *Radiat. Res.* **37**, 131–140.
Long, E. C. (1977). "Application of Quench Monitoring by Compton Edge: the *H#*", Report no. 1096-NUC-77-2T, Beckman Instruments, Irvine, Calif.
Lundqvist, H., Johanson, K. J. and Jonsson, G. (1976). *Int. J. Appl. Radiat. Isotopes* **27**, 233–237.
McDowell, W. J. (1980). *In* "Liquid Scintillation Counting, Recent Applications and Development", Vol. I, "Physical Aspects" (C.-T. Peng, D. L. Horrocks and E. L. Alpen, eds), pp. 315–332, Academic Press, New York and London.
McDowell, W. J. and Weiss, J. F. (1976a). "Liquid Scintillation Alpha Counting and Spectrometry and its Application to Bone and Tissue Samples", Report no. CONF-760640-1 (available from INIS).
McDowell, W. J. and Weiss, J. F. (1976b). *In* "Liquid Scintillation Science and Technology" (A. A. Noujaim, C. Ediss and L. I. Wiebe, eds), pp. 17–31, Academic Press, New York and London.
McKenzie, J. M. (1979). *Nucl. Instrum. Meth.* **162**, 49–73.
McKlveen, J. W. (1976). *Radiat. Res.* **66**, 199–214.
McQuarrie, S. A., Wiebe, L. I. and Ediss, C. (1980). *In* "Liquid Scintillation Counting, Recent Applications and Development", Vol. I, "Physical Aspects" (C.-T. Peng, D. L. Horrocks and E. L. Alpen, eds), pp. 291–299, Academic Press, New York and London.
Malcolme-Lawes, D. J. and Massey, S. (1980). *Int. J. Appl. Radiat. Isotopes* **31**, 155–158.
Mallet, L. (1929). *Compt. Rend. Acad. Sci. Paris*, **188**, 445–447.
Maraldi, N. M. (1976). *In* "Principles and Techniques of Electron Microscopy", Vol. VI (M. A. Hayat, ed.), pp. 271–289, Van Nostrand Reinhold, New York.
Marshall, J. (1952). *Phys. Rev.* **86**, 685–693.
Martini, M. (1973). *I.E.E.E. Trans. Nucl. Sci.* **20**(1), 294–309.
Meyer, E., Martin, M. and Sternberg, J. (1972). *I.E.E.E. Trans. Nucl. Sci.* **19**(3), 237–242.
Meyer, E., Desaulniess, G. and Sternberg, J. (1975). *Int. J. Appl. Radiat. Isotopes* **26**, 697–702.
Miller, G. L. (1977). *I.E.E.E. Trans. Nucl. Instrum. Meth.* **162**, 113–123.
Miller, M., Kereiakes, J. G. and Friedman, B. I. (1969). *Int. J. Appl. Radiat. Isotopes* **20**, 133–135.
Moir, A. T. B. (1971). *Int. J. Appl. Radiat. Isotopes* **22**, 213–216.
Morley, B. J. and Parker, R. P. (1967). *Phys. Med. Biol.* **12**, 511–512.

Morton, G. A., Smith, H. M. and Wasserman, R. (1967). *I.E.E.E. Trans. Nucl. Sci.* **14**, 443–448.
Nascimiento Filho, V. F. (1977). *Int. J. Appl. Radiat. Isotopes* **29**, 789–790.
Neame, R. D. (1975). *Anal. Biochem.* **64**, 521–529.
Nibeck, J. I., Bares, S. L. and Williams, E. S. (1980). *In* "Liquid Scintillation Counting, Recent Applications and Development", Vol. I, "Physical Aspects" (C.-T. Peng, D. L. Horrocks and E. L. Alpen, eds), pp. 59–72, Academic Press, New York and London.
Noakes, J. E. (1976). US Patent No. 3,960,756.
Noakes, J. E. and Menefee, J. M. (1975). U.S. Patent No, 3,898,460.
Orphan, V., Polichar, R. and Ginaven, R. (1978). *Trans. Am. Nucl. Soc.* **28**, 119–121.
Painter, K. (1976). *In* "Liquid Scintillation Science and Technology" (A. A. Noujaim, C. Ediss and L. I. Wiebe, eds), pp. 283–289, Academic Press, New York and London.
Parker, R. P. and Elrick, R. H. (1966). *Int. J. Appl. Radiat. Isotopes* **17**, 361–362.
Parker, R. P. and Elrick, R. H. (1970). *In* "The Current Status of Liquid Scintillation Counting" (E. D. Bransome, Jr, ed.), pp. 110–122, Grune and Stratton, New York and London.
Parmentier, J. H. and ten Haaf, F. E. L. (1969). *Int. J. Appl. Radiat. Isotopes* **20**, 305–334.
Pehl, R. H. (1978). *I.E.E.E. Trans. Nucl. Sci.* **25**(2), 902–905.
Peng, C.-T. (1976). *In* "Liquid Scintillation Science and Technology" (A. A. Noujaim, C. Ediss and L. I. Wiebe, eds), pp. 313–329, Academic Press, New York and London.
Perry, S. W. and Warner, G. T. (1963). *Int. J. Appl. Radiat. Isotopes* **14**, 397–400.
Petit, C. M. and van de Geijn, S. C. (1978). *Planta* **138**, 137–143.
Pinter, G. G. (1969). *Experientia* **25**, 113–224.
Prabhu Gaunkar, G. V., Huntz, A. M. and Lacombe, P. (1979). *Int. J. Appl. Radiat. Isotopes* **30**, 761–780.
Prokop, E. K., Kearns, D. S., Noakes, J. E. and Menefee, J. M. (1980). *In* "Liquid Scintillation Counting, Recent Applications and Development", Vol. II, "Sample Preparation and Applications" (C.-T. Peng, D. L. Horrocks and E. L. Alpen, eds), pp. 503–514, Academic Press, New York and London.
Reynolds, G. T., Harrison, F. B. and Salvine, G. (1950). *Phys. Rev.* **78**, 488.
Rhodes, B. A. (1965). *Anal. Chem.* **37**, 995–997.
Rieppo, R. (1976a). *Int. J. Appl. Radiat. Isotopes* **27**, 457–459.
Rieppo, R. (1976b). *Int. J. Appl. Radiat. Isotopes* **27**, 491–493.
Rieppo, R., Blomster, K. and Holmberg, P. (1975). *Int. J. Appl. Radiat. Isotopes* **26**, 558–561.
Ring, J. G., Nguyen, D. C. and Everett, L. J. (1980). *In* "Liquid Scintillation Counting, Recent Applications and Development", Vol. I, "Physical Aspects" (C.-T. Peng, D. L. Horrocks and E. L. Alpen, eds), pp. 89–104, Academic Press, New York and London.
Rogers, A. W. (1967). "Techniques of Autoradiography", Elsevier, New York.
Ross, H. H. (1969). *Anal. Chem.* **41**, 1260–1265.
Ross, H. H. (1976). *In* "Liquid Scintillation Science and Technology" (A. A. Noujaim, C. Ediss and L. I. Wiebe, eds), pp. 79–91, Academic Press, New York and London.

Ross, H. H. (1980). *In* "Liquid Scintillation Counting, Recent Applications and Development", Vol. I, "Physical Aspects" (C.-T. Peng, D. L. Horrocks and E. L. Alpen, eds), pp. 375–385, Academic Press, New York and London.
Rowinska, L., Jaskolska, H. and Radwan, M. (1975). *Isotopenpraxis* **11**, 223–225.
Sakai, E., Terada, H. and Katagiri, M. (1976). *I.E.E.E. Trans. Nucl. Sci.* **23**(1), 726–733.
Sasaki, J. and Pitt, B. (1976). *Cardiovasc. Res.* **10**, 499–502.
Schnerr, G. J., Goldman, L. H. and Ryge, P. (1980). *I.E.E.E. Trans. Nucl. Sci.* **27**(11), 240–248.
Seidel, A. and Volf, V. (1972). *Int. J. Appl. Radiat. Isotopes* **23**, 1–4.
Siffert, P. (1978). *Nucl. Instrum. Meth.* **150**, 1–12.
Smith, B. S. W. (1981). *Int. J. Appl. Radiat. Isotopes* **32**, 760–762.
Smith, G. H., Kidwell, W. R. and Schneider, W. C. (1975). *Exp. Cell Res.* **96**, 321–334.
Smith, R. D., Anderson, J. J. B., Ristic, M. M. and Huxsoll, D. L. (1972). *Int. J. Appl. Radiat. Isotopes* **23**, 513–517.
Soini, E. and Rawlins, T. (1975). *In* "Radioimmunoassay and Related Topics in Clinical Biochemistry" (C. A. Pasternak, ed.), pp. 61–69, Heyden, London.
Stanley, P. E. (1976). *In* "Liquid Scintillation Science and Technology" (A. A. Noujaim, C. Ediss and L. I. Wiebe, eds), pp. 103–116, Academic Press, New York and London.
Stanley, P. E. (1980). *In* "Liquid Scintillation Counting, Recent Applications and Development", Vol. I, "Physical Aspects" (C.-T. Peng, D. L. Horrocks and E. L. Alpen, eds), pp. 257–271, Academic Press, New York and London.
Stubbs, R. D. and Jackson, A. (1967). *Int. J. Appl. Radiat. Isotopes* **18**, 857–858.
Taguchi, T., Shirafuji, J., Kobayashi, T. and Inuishi, J. (1975). *Proc. Conf. Solid State Devices* **7**, 267–272.
Takayanagi, S., Kobashi, T., Ueda, H., Kameda, H., Sassa, R., Iio, M. and Sasaki, Y. (1968). *J. Nucl. Med.* **9**, 354.
ten Haaf, F. E. L. and Kuipers, J. G. H. (1977). *In* "Liquid Scintillation Counting", Vol. IV (M. A. Crook and P. Johnson, eds), pp. 260–266, Heyden, London and New York.
ten Haaf, F. E. L. and Verheijke, M. L. (1977). *In* "Liquid Scintillation Counting", Vol. IV (M. A. Crook and P. Johnson, eds), pp. 63–73, Heyden, London and New York.
Thomas, R. C., Judy, R. W. and Harpvotlian, H. (1965). *Anal. Biochem.* **13**, 358–360.
Ting, P. (1973). Clinical Brief no. 6, Beckman Scientific Instruments, Fullerton, Calif.
Tothill, P. (1977). *In* "Liquid Scintillation Counting", Vol. IV (M. A. Crook and P. Johnson, eds.), pp. 155–164, Heyden, London and New York.
Tsuk, G. R., Castro, Th., Laufer, L. and Schwarz, D. R. (1964). *In* "Proceedings of the Conference on Methods of Preparing and Storing Marked Molecules" (J. Sirchis, ed.), Euratom (EUR 1625e), Brussels.
Ueda, H., Sasaki, Y., Iio, M., Kaihara, S., Machida, K., Ito, I., Takayanagi, S., Kobayashi, T. and Sugita, T. (1969). *J. Nucl. Med.* **10**, 713–721.
Ullberg, S. (1954). *Acta Radiol, Suppl.* **188**, 21–31.
Ullberg, S. (1962). *Biochem. Pharmacol.* **9**, 29–38.
Ullberg, S. and Dencker, L. (1977). *Proc. Europ. Soc. Nucl. Meth. Agric.* **1977**, 26–33.
van de Geijn, S. C. and Petit, C. M. (1978). *Planta* **138**, 145–151.
van Ginkel, G. (1980). *Int. J. Appl. Radiat. Isotopes* **31**, 307–312.
Walter, J. P., Eastin, E. F. and Merkle, M. G. (1970). *Weed Res.* **10**, 165–171.

Wegner, L. A. and Patzschke, K. (1976). *Int. J. Appl. Radiat. Isotopes* **27**, 573–583.
Whited, R. C. and Schieber, M. M. (1979). *Nucl. Instrum. Meth.* **162**, 113–123.
Wiebe, L. I. and Ediss, C. (1976). *In* "Liquid Scintillation Science and Technology" (A. A. Noujaim, C. Ediss and L. I. Wiebe, eds), pp. 93–102, Academic Press, New York and London.
Wiebe, L. I., Noujaim, A. A. and Ediss, C. (1971). *Int. J. Appl. Radiat. Isotopes* **22**, 463–467.
Wolgast, M. (1968). *Acta Physiol. Scand. Suppl.* 313, 97–101.
Yang, W. and Lee, E. K. C. (1969). *J. Chem. Educ.* **46**, 277–283.

6

The Measurement of ^{15}N†

Rudolf Fiedler

This chapter is dedicated to the memory of Günter Proksch. In Günter I lost a very dear friend and colleague. He generously shared with me his broad knowledge and experience in the field of stable isotope measurement. Our collaboration was very fruitful, and part of this chapter is a condensed version of some of that work.

† Taken in part from Fiedler and Proksch (1975), with the permission of the Elsevier Scientific Publishing Co., Amsterdam.

I. Introduction

The use of an isotope as a tracer was introduced in 1913 by Hevesy and Paneth. They used the isotope ^{212}Pb to study the solubility of lead salts. Since then the use of tracer methods has become a standard technique in many research fields, such as agriculture, biology, chemistry, medicine and physics. Based on the isotope dilution law, a stable or radioactive isotope tracer may be used to evaluate quantitatively the pathways of a non-labelled element in a system. The isotope dilution law states that in a system which is in equilibrium the ratio of the tracer isotope to the carrier isotope stays constant. Exceptions to this law occur with "isotope effects" which, in the case of ^{15}N, are considered to be negligible.

The longest-lived radioactive isotope of nitrogen is ^{13}N, which has a half-life of 10 min. This fact limits tracer studies with radioactive nitrogen. To carry out tracer experiments of reasonable duration the stable isotopes of nitrogen must be used. Nitrogen consists of two stable isotopes, ^{14}N and ^{15}N. Nitrogen-15 is present in nature only with an abundance of 0.3663 ± 0.0004 atom %. Tracer experiments carried out with nitrogen require the use of depleted or enriched ^{15}N. The advantage of using stable isotopes for tracer techniques is that there is no limitation in time for performing an experiment. The disadvantage is the more difficult, complicated, and also more expensive method of detection.

Physicochemical methods or chemical exchange processes can be used for ^{15}N enrichment. The latter method is used commercially and enables, by recycling, enrichments up to 99 atom %. Depleted material down to 0.01 atom % is a byproduct of this method.

The amounts of ^{14}N and ^{15}N in a nitrogen sample can be determined by means of mass or emission spectrometric techniques. The number of ^{15}N atoms in the total number of nitrogen atoms in a sample expressed as a percentage is called the atom % or percentage abundance. When the ^{15}N percentage abundance of natural nitrogen, i.e. 0.366 %, is subtracted from the ^{15}N percentage abundance of the sample, the ^{15}N atom % excess is obtained. This is analogous to the specific activity of a radioactive sample. In contrast to radiation detection equipment, which can quantify only the radioactive isotope and not the carrier element, mass or emission spectrometers can determine the ^{15}N and ^{14}N abundance in one operation.

II. Principle of detection

Nitrogen atoms in a gas are paired to form the nitrogen molecules $^{14}N^{14}N$, $^{14}N^{15}N$ and $^{15}N^{15}N$, which can also be expressed as $^{28}N_2$, $^{29}N_2$ and $^{30}N_2$.

Mass spectrometry and emission spectrometry provide output signals proportional to the number of these three molecular types present in a sample. In any nitrogen gas mixture, regardless of whether the gas mixture is in equilibrium or not, the ^{15}N percentage abundance can be derived from the formula

$$\frac{^{29}N_2 + 2(^{30}N_2)}{2(^{28}N_2 + {^{29}N_2} + {^{30}N_2})} \times 100. \tag{6.1}$$

This represents the ratio of the ^{15}N atoms to the total number of nitrogen atoms in this gas mixture expressed in per cent. When the intensities of $^{28}N_2$, $^{29}N_2$ and $^{30}N_2$ derived from mass or emission spectrometric methods are substituted in the above equation, the ^{15}N percentage abundance can be calculated.

Equilibrium exists between $^{28}N_2$, $^{29}N_2$ and $^{30}N_2$ molecules according to the equation

$$^{28}N_2 + {^{30}N_2} \rightleftharpoons 2\,{^{29}N_2}. \tag{6.2}$$

It is not necessary to measure the output signals of the three nitrogen molecules in the sample; only the measurement of the $^{28}N_2$ and $^{29}N_2$ signals is required. The measurement of these signals is done by taking the peak heights from a recorder or values obtained from a peak integrator connected to the spectrometer. The ^{15}N percentage abundance can then be calculated by the equation:

$$^{15}\text{N percentage abundance} = \frac{100}{2R + 1}. \tag{6.3}$$

where R is given by (intensity $^{28}N_2$/intensity $^{29}N_2$) and is the ratio of peak heights found for $^{28}N_2$ and $^{29}N_2$.

Precise determinations of ^{15}N variations in natural abundance experiments are determined by the following equation:

$$\delta\,^{15}N = \left(\frac{(^{15}N/^{14}N)_{\text{sample}}}{(^{15}N/^{14}N)_{\text{standard}}} - 1\right) \times 1000. \tag{6.4}$$

In the case of depleted ^{15}N material the standard used is air at its natural nitrogen composition. The theoretical background and derivation of equation (6.4) is given in the Appendix in Volume 2.

III. Sample preparation techniques

A. Basics of the conversion to nitrogen gas

For the determination of $^{15}N:^{14}N$ ratios, it is necessary to convert all of the nitrogen atoms in the sample to a single compound. For both mass and emission spectrometry, nitrogen gas is preferred for the following reasons:

(i) It is much easier to generate N_2 from different inorganic and organic compounds than NO, NO_2 or NH_3. When nitrogen is produced as the gas, it is easily separated from the remainder of the sample.

(ii) Nitrogen gas is inert with respect to all materials normally involved in mass and emission spectrometry.

(iii) The interpretation of the isotopic data is simple as there are no interferences from other elements. In the case of NO, for example, the mass spectrometric peak of $^{14}N^{17}O^+$ would be identical with that of $^{15}N^{16}O^+$.

Organically and inorganically bound nitrogen can be converted to nitrogen gas by three methods. One method is based on a modification of the Kjeldahl digestion (Kjeldahl, 1883). Digestion of the sample is accomplished in sulphuric acid with additives such as hydrogen peroxide or suitable catalysts. Organic and inorganic nitrogen is converted to ammonia. The ammonia is recovered by steam-distillation of the acid mixture to which an excess of sodium hydroxide is added. The ammonia gas is trapped in standard hydrochloric acid solution. For the determination of the total nitrogen content, the excess of acid is back-titrated with sodium hydroxide. The titrated solution is then acidified with hydrochloric acid and evaporated to a small volume. Nitrogen in the sample is converted to N_2 by the Rittenberg method (Rittenberg, 1948), which involves the oxidation of the ammonia in the extract by sodium hypobromite according to the following reaction:

$$2\,NH_3 + 3\,BrO^- \rightarrow N_2 + 3\,H_2O + 3\,Br^-. \tag{6.5}$$

The nitrogen gas produced can be used either for mass or emission spectrometric analysis.

The second method is based on a modification of the Dumas reaction (Dumas, 1834). The method involves dry combustion of the organic and/or inorganic nitrogen-containing compounds with copper oxide, copper, or a mixture of these. The following possible reactions occur which serve as examples:

$$CO(NH_2)_2 + 3\,CuO \rightarrow N_2 + CO_2 + 2\,H_2O + 3\,Cu, \tag{6.6}$$

$$2\,KNO_3 + 5\,Cu \rightarrow N_2 + K_2O + 5\,CuO. \tag{6.7}$$

Water and carbon dioxide formed by the combustion are absorbed on calcium oxide. The Dumas reaction can be carried out in many ways. The modified Dumas procedure allows one to convert all bound nitrogen into gas. The measurement of the pressure enables the determination of the total nitrogen content of the same sample to be analysed for ^{15}N abundance. It is also possible to use a combination of both the Kjeldahl and Dumas methods. By this combined method, ammonia originating from the Kjeldahl digestion is converted to nitrogen by Dumas combustion.

The method of converting nitrogen compounds and materials into nitrogen gas should be simple and involve a minimum of manpower and time. The dilution of ^{15}N by cross-contamination, which would alter the ^{15}N abundance of the sample, must be kept to a minimum. The use of well mixed, finely ground material minimizes sub-sampling errors. The potential inhomogeneity of agricultural samples requires that sub-samples as large as possible be used. Thus, taking more sample than is required for the analysis reduces the possibility of dilution or cross-contamination from micro-leaks or memory effects and decreases the sampling errors. The stepwise procedure to be followed in carrying out the sample preparation techniques can be obtained from Fiedler and Proksch (1975). A brief treatment of these techniques is given below.

B. The Kjeldahl–Rittenberg method

This method was introduced by Rittenberg (1948). The Kjeldahl method was used to convert the sample nitrogen to ammonia which was distilled into hydrochloric or sulphuric acid. After concentration of the acid distillate, the ammonia was oxidized to nitrogen with an alkaline solution of sodium hypobromite in a vacuum. The following steps are necessary for the analysis: (i) digestion of the sample with concentrated sulphuric acid; (ii) separation of the ammonia by distillation after addition of alkali; (iii) determination of the ammonia by back-titration, colorimetry, etc.; (iv) acidification; (v) evaporation to a small volume; and (vi) oxidation of the ammonia to nitrogen.

A detail account of each of these steps will be given in the succeeding sections. Modifications of the method, e.g. micro or sub-micro analyses of nitrogen by diffusion techniques (Conway, 1947), or separate determinations of ^{15}N in different chemical forms, will be included. Sources of contamination and their elimination will be mentioned at the end of the corresponding analytical step.

In general, for the digestion, quantities sufficient for triplicate isotopic analysis are normally used. For the oxidation process, an aliquot amount of extract can be used since in this step the ammonium salt or nitrogen gas is taken from a uniformly distributed sample and no sub-sampling errors arise.

1. *Digestion*

The digestion method includes the conversion of organically and inorganically bound nitrogen in the sample to ammonia by boiling with sulphuric acid. Although the Kjeldahl method has been used for about 90 years and several hundreds of papers on the technique have been published, there is still no universally accepted digestion method. The best compromise seems to be to adapt the digestion technique to the nature of the sample material. Several attempts have been made to decrease the digestion time required for quantitative reaction. Pretreatment of the sample, addition of salts to increase the boiling point of the digest, catalysts, and oxidizing agents (other than the permanganate suggested by Kjeldahl) have been examined.

Pretreatment is necessary for nitrates and nitrites, for compounds containing the pyridine heterocyclic ring and for compounds such as lysine which, when treated with concentrated sulphuric acid, are converted to pyridine-like compounds. Nitrates must be reduced to ammonia with, for example, iron powder and acid; nitrites are first oxidized with, say, permanganate to nitrate and then treated as above (Bremner, 1965). Pyridine (which does not char with sulphuric acid), and compounds forming it, must be first reduced; the addition of charring organic compounds such as dextrose also helps (Middleton and Stuckey, 1951).

Potassium sulphate is used to raise the digestion temperature, which lies between 390 and 410 °C. With 2 g of K_2SO_4 per millilitre of sulphuric acid, complete recovery of nitrogen at a temperature of 390 °C can be obtained. This temperature is lower than the boiling point of the mixture. Ogg and Willits (1950), McKenzie and Wallace (1954) and Baker (1961) describe the effect of potassium sulphate on the temperature of digestion.

Catalysts are selenium, mercury(II) oxide and copper sulphate. The least effective catalyst is copper sulphate, but it is still in use. Mercury(II) oxide has been found to be the best catalyst for protein and nicotinic acid digestion. Complete recovery from tryptophan can be obtained by using a higher boiling point mixture by adding potassium sulphate and mercury(II) to the sulphuric acid (McKenzie and Wallace, 1954). Higher than the normal concentration of HgO can affect the complete recovery of the nitrogen after a digestion time of about 20 min from resistant compounds such as nicotinic acid, tryptophan and other nitrogen-containing substances (Perrin, 1953).

Selenium alone does not always result in complete recovery of sample nitrogen. It probably causes some decomposition of ammonia. Middleton and Stuckey (1951) and Baker (1961) reported losses of nitrogen when selenium was used at temperatures above 387 °C. The use of selenium alone seems justified only when the heating is not continued after decolorization of the sulphuric acid, and when it is known that this is sufficient to give quantitative recovery of ammonia. A mixture of selenium and mercury(II) oxide at 365 °C results in maximum recovery of nitrogen (Baker, 1961).

Of all the oxidizing agents tested, hydrogen peroxide has been found to be the best (Kirk, 1950). McKenzie and Wallace (1954) developed a semimicro and micro Kjeldahl method which was used for the complete digestion of tryptophan (98–99 % recovery). The method involved 20 additions of 0.05 ml of 30 % hydrogen peroxide at 120–140 °C (decomposition temperature of H_2O_2) to the digest.

Depending on the amount of sample, the Kjeldahl digestion can be divided into three types: macro, semimicro and micro. In principle, the digestion proceeds more rapidly when small sub-samples are used. The macro method is used when only relatively large samples can be guaranteed to be homogeneous. Kirk (1934) used 1–2 μg of nitrogen for ultramicro Kjeldahl work with the same precision as the macro method. In this method complete recovery of nitrogen was easier to obtain than with larger samples. McKenzie and Wallace (1954) recommended the use of 0.2–2 mg of nitrogen per sample, and Baker (1961) recommended 5 mg. The use of microgram samples or less naturally only makes sense when it is possible to distill quantitatively such small amounts of nitrogen as ammonia. This was first demonstrated in 1933 by Conway (1947), who found that diffusion techniques are a more satisfactory means of removing ammonia than distillation from a micro Kjeldahl digest. Kirk (1934) removed the ammonia (4–20 μg) from the digest by a combination of boiling and aeration. Later, amounts of nitrogen of the order of 0.1 μg (Brüel *et al.*, 1946) and 0.5 μg (Tompkins and Kirk, 1942) were handled with 1 % accuracy by applying a vacuum diffusion technique.

It is important to control the consumption of sulphuric acid; otherwise, the boiling point will increase when the digest becomes clear if potassium sulphate is used, and losses of nitrogen may occur. The amount of acid needed for a particular compound can be calculated by the method of Middleton and Stuckey (1951):

> "The amount of acid required for any compound is readily calculated from its molecular formula by deducting from the latter the elements of ammonia and carbon dioxide (or water) corresponding to the number of atoms of N and O, respectively. Any other elements present may be allowed for in a similar manner; for example, sulphur, when combined with O as in sulphonic acids, may be taken as converted to sulphur dioxide, while sulphur in a sulphide group is assumed to be eliminated unoxidized. In the cases of salts of alkali metals, allowance should be made both for the additional acid used in forming the sulphate and for the extra alkali sulphate present. By the subtraction of the elements

Table 6.1 Factors for calculating acid consumption

1 g sample	Amount of sulphuric acid allowed (ml)
CH_2	11.4
CH	10.2
$C_{1.5}H$	9.9
C_2H	9.6
$C_{2.5}H$	9.4
C_3H	9.3
C_4H	9.1
C	8.9
Carbohydrate	4.0
Fat	9.7
Protein	4.9
Zn	0.8
Na	1.2
K	0.7
Salicylic acid	5.5

> of ammonia, water, carbon dioxide, sulphur dioxide, HCl, sulphur, sodium, etc., from the formula there is left a residual formula containing C and H".

The factor used in the calculation depends on the C:H ratio of this residual formula, and is taken from Table 6.1, which also gives the general values for carbohydrates, fats and proteins, and the factor for zinc for use when it is added in the determination of nitrogen present as nitrate. The acid consumption per gram is then equal to

$$\frac{\text{Residual formula weight}}{\text{Molecular weight}} \times \text{Factor},$$

and the amount of acid (in millilitres) to be added for a particular assay is then equal to 6 + (acid consumption per gram × weight of substance taken). Actually, great accuracy is not generally required in the calculation of the factor, and the total volume of acid need not be measured with an accuracy greater than, say, ± 0.25 ml. For ill-defined samples, Table 6.2 gives some indication of sulphuric acid requirements.

2. *Distillation*

After digestion the ammonia is liberated by addition of an alkali. Distillation can be carried out conventionally or by steam-distillation. Steam-distillation has the advantage that "bumping" is prevented in the presence of

Table 6.2 Amounts of sulphuric acid needed for various materials during Kjeldahl digestion (Bremner, 1965)

Material	Acid consumption (ml 18 M H_2SO_4 per gram of material)
Soil organic carbon	10.0
Soil organic matter	5.8[a]
Al_2O_3	1.6
Fe_2O_3	1.04[b]
Clay	0.6
$CaCO_3$	0.6
Silt	0.3
Sand	0
Salicylic acid	6.8
$Na_2S_2O_3 . 5H_2O$	0.6
Reduced iron	1.5[b]

[a] Calculated from organic carbon value on assumption that soil organic matter contains 58% carbon.

[b] Calculated on the assumption that digestion product is iron(III) sulphate.

Example: Sulphathiazole ($C_9H_9O_2N_3S_2$) has an atomic mass of 255.3. Subtract three NH_3, SO_2 and S (one S atom is not linked to O), leaving a C_9 atomic mass of 108. Then the acid consumption is $(108/255) \times 8.9 = 3.8$ mg g^{-1}.

precipitates in the extract. The type of distillation apparatus chosen depends on personal choice. In some cases, the Kjeldahl flask itself can be incorporated into the system; others prefer to transfer the extract to a separate distillation flask. The size of the apparatus, of course, depends on the sample size used.

Hoskins (1944) developed an interchangeable macro and semimicro apparatus. Kirk (1950), McKenzie and Wallace (1954), Bremner (1965) and many other workers have developed different types of steam-distillation apparatus.

The receiving liquid of the distillation is normally boric acid, but for mass and emission spectrometry isotope determination hydrochloric or sulphuric acid are preferable. The smaller solubility product of the borates causes them to precipitate when the solution is evaporated to a small volume for the determination of the ^{15}N:^{14}N ratio, and losses of ammonia can also occur (Wingo *et al.*, 1950).

The actual determination of ammonia is normally done by titration or colorimetry (Bremner, 1965). In case of separation by diffusion techniques, however, capillary burettes are necessary for the titration.

For the determination of ^{15}N:^{14}N ratios in distillates, interferences are introduced in the normal Kjeldahl analysis. These may originate from foreign sources or from the sample itself. The foreign sources normally dilute the percentage of ^{15}N, whereas the ^{15}N content of other samples may cause cross-contamination. The causes of interference with the $^{15}N:^{14}N$ ratio determination associated with digestion and distillation are as follows: (i) dilution or contamination from outside sources, such as absorption of ammonia or amines from air, nitrogen impurities in the chemicals, and the nitrogen content of the indicator; (ii) cross-contamination from poorly cleaned laboratory equipment, especially the distillation apparatus; (iii) losses of nitrogen by incomplete conversion to ammonia of certain resistant nitrogen-containing compounds or groups – this gives rise to false ^{15}N percentages.

In order to prevent interferences from contamination the following rules should be observed:

(a) The laboratory atmosphere must be free of ammonia and amine vapours; no bottle or container of such liquids should be stored in the working area where ^{15}N preparation work is done. Only pure chemicals (e.g. reagent grade) should be used. The acids (sulphuric and standard) always must be kept in tightly stoppered bottles to prevent absorption of ammonia and amines. The dilution of ^{15}N from a sample by nitrogen from chemicals in a normal preparation procedure has been found to be of the order of 2% (Hüser, 1963). Proksch (1969) obtained a percentage ^{15}N abundance value by the Kjeldahl–Rittenberg procedure which was 2% lower than the value obtained by direct combustion with a modified Dumas method. This was attributed to nitrogen impurities in the chemicals which were not detectable by normal chemical methods. The proper choice of an indicator must also be considered from this point of view. Dilution of the ^{15}N percentage can occur because of the nitrogen content of the indicator. This is especially important in the case of emission spectrometric determination of ^{15}N where only very small samples are necessary and the amount of sample nitrogen may be in the same order of magnitude as the indicator nitrogen. The often-used Tashiro indicator, 0.2% methylene blue solution in 95% ethanol and 0.2% methyl red solution in 95% ethanol (McKenzie and Wallace, 1954), contains so much nitrogen that it may dilute the ^{15}N percentage of the sample to half of its value, as has been reported by Faust (1969). Faust, however, found no release of nitrogen from the Tashiro indicator during distillation or Conway diffusion and no decomposition with alkaline hypobromite solution when the extract was oxidized after titration and evaporation. It is also of importance to have the end-point of the titration in the acid region to prevent losses of ammonia at the end of the titration.

(b) All equipment must be cleaned with detergents or acids. The cleaning of the distillation apparatus is not easy and even steaming for several minutes may not be sufficient to remove all nitrogen which is adsorbed on the glass walls and tubes. The degree of cross-contamination which can arise has been investigated by several authors and is illustrated in Table 6.3. This table shows that cross-contamination becomes appreciable when samples with large differences in ^{15}N percentage abundances or total nitrogen content are distilled one after the other. This cross-contamination may be prevented by distilling 10–20 ml of ethanol between samples (Bremner, 1965).

(c) Incomplete digestion or losses of ammonia during the digestion can be prevented by choosing a suitable digestion method. When selenium is used as a catalyst or potassium sulphate for raising the temperature, an excessive digestion time or overheating should be avoided. The combination of adequate digesting agents, catalysts, temperature and time of digestion has to be established for each particular kind of material, e.g. seeds, leaves, roots, etc.

3. *The oxidation of ammonia to nitrogen gas*

The oxidation of ammonia from the Kjeldahl digestion with hypobromite takes place in alkaline medium in an evacuated container. The oxidation proceeds according to reaction (6.5) previously described.

Hypobromite is a rather unstable compound and the reagent should be kept in strongly alkaline medium, stored in a refrigerator. Hüser *et al.* (1960) found that the full oxidizing power lasted only 1 week, but it is said to be possible to keep the solution for 6 months without too much loss of oxidizing power (Bremner, 1965). The freshly prepared solution develops some oxygen but this stops after a few days. To inhibit the formation of oxygen during the oxidation of ammonia, potassium iodide can be added (Sims and Cocking, 1958), but this does not prevent it completely (Balestieri, 1968). In addition to oxygen, N_2O^+ is also formed (Clusius and Rechnitz, 1953; Capindale and Tomlin, 1957), especially with freshly prepared hypobromite solution. NO^+, which is formed from N_2O during ionization, interferes with ^{15}N measurements in the mass spectrometric analysis (Holt and Hughes, 1955), and a liquid nitrogen trap should be built into the inlet system of the mass spectrometer or the transfer line for the preparation of the emission spectrometric discharge tubes, where N_2O is readily frozen out.

Bromine may also be formed in the solution. After the titrimetric determination of ammonia, the solution should be acidified to prevent ammonia losses. If too great an excess of hydrochloric or sulphuric acid is used, bromine may be formed according to the reaction

$$Br^- + OBr^- + 2H^+ = H_2O + Br_2. \tag{6.8}$$

The excess of acid should not be higher than 0.35 ml of 0.1 N HCl or H_2SO_4 (Bremner, 1965).

Table 6.3 Cross-contamination in the distilling apparatus

Investigator	First distillation		Second distillation[a]			Nitrogen retention	
	Amount of N (mg)	Percentage ^{15}N in sample (%)	Amount of N (mg)	Percentage ^{15}N in sample (%)	Percentage ^{15}N in distillate (%)	Amount (μg)	Percentage (%)
Bremner (1965)	1	30	1	0.366	0.396	1	0.1
Faust (1969)	0.5	55	0.05	0.366	2.05	1.6	0.3
Hüser (1963)	3	48.5	3	0.362	0.704	22	0.7
Hüser[b] (1963)	5	96.5	5	0.362	0.374	2.7	0.05

[a] Second distillation follows immediately after the first one.
[b] Steaming for 1 h after the first distillation.

The preparation of the alkaline hypobromite solution is as follows (Bremner, 1965): Dissolve 200 g of sodium hydroxide in 300 ml of water and cool the solution in ice. Transfer half of the cooled solution to a 500 ml wide-mouth Erlenmeyer flask, immerse the flask in crushed ice and add 60 ml of bromine over a period of 30 min. Stir the solution vigorously during the addition of bromine and regulate the rate of addition in such a way that the temperature of the solution does not exceed 5 °C. When the addition of bromine has been completed, add the remainder of the alkaline solution and after stirring the mixture for a few minutes, stopper the flask and place it in a refrigerator for 4–6 days. Remove the precipitate of sodium bromide which forms during this period by filtration under suction through a glass-fibre filter, and dilute the filtrate with an equal volume of a solution prepared by dissolving 2 g of potassium iodide in 1 litre of water. Store the hypobromite–iodide solution in a tightly stoppered bottle in a refrigerator. Care should be taken during preparation and storage of the reagent to protect it from atmospheric CO_2. One millilitre of this solution will oxidize 5–6 mg of ammonia to nitrogen, and the reagent will retain its activity for at least 6 months if stored in a refrigerator.

The preparation of nitrogen gas samples is described subsequently under specified techniques for mass and emission spectrometry. The procedures vary somewhat depending on the type of spectrometer employed.

C. The modified Dumas method

The original Dumas method (Dumas, 1834) involved the oxidation of organic material with copper oxide in an atmosphere of carbon dioxide which is then passed through a strong potassium hydroxide solution. All elements present (e.g. C, H, N, Cl, S, O, P, etc.) are converted to forms which either react with copper oxide or are very soluble in strong alkali solution, except nitrogen, the volume of which is measured at the end of the reaction. This method has been used in many modifications for determinations of organically bound nitrogen.

In the modified Dumas method, organic matter is oxidized with copper oxide and the generated gases, except nitrogen, are absorbed by calcium oxide. For nitrates and nitrites, it is necessary to replace a part of the copper oxide by copper to reduce these salts to nitrogen. The reaction is carried out in vacuum in a closed tube (Cook, *et al.*, 1967, 1968; Faust, 1967; Middelboe, 1970; Perschke *et al.*, 1971). The Dumas method offers certain advantages. It involves less preparation time and manpower than other methods since only two working steps are involved. As each sample is contained in a separate tube and the amount of chemicals involved is smaller than for the Kjeldahl–Rittenberg process, dilution and cross-contamination risks are smaller.

Samples may be in the form of organic or inorganic compounds, and all kinds of biological material, including Kjeldahl concentrates, can be treated.

Improvements to the method have been developed for total nitrogen determination. The combustion temperature for samples treated in Pyrex glass ampoules is 570 °C and is therefore relatively low. Without mixing during the time of combustion, only 70–80 % of the sample is converted to nitrogen. Mixing of the sample increases the conversion up to 90 %. By replacing 50 % of the copper oxide with Cuprox (1:1) and adding small amounts of potassium perchlorate, the nitrogen conversion reaches 99 %. Cuprox consists of copper oxide containing platinum catalyst on alumina (Coleman reagent). A detailed description of this method is given under mass spectrometry.

Due to different needs for the two ^{15}N determination methods, sample preparation differs in several steps.

D. Sample preparation for mass spectrometry

1. *Kjeldahl–Rittenberg method*

Figure 6.1 illustrates a vacuum system for the preparation of samples as Kjeldahl extracts. The vacuum system consists of valves with conical joints for the connection of the sample vessels, a cold trap (liquid air or liquid nitrogen cooled), which has a valve on its bottom, and a rotary pump. The valve on the cold trap allows drainage of water when the ice collected in the trap has melted after sample preparation (Reid *et al.*, 1967). The vacuum in the system is determined by the vapour pressure of the solution in the sample

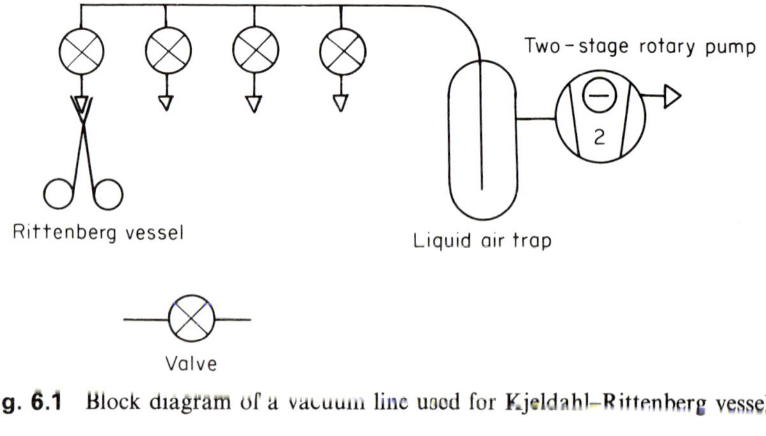

Fig. 6.1 Block diagram of a vacuum line used for Kjeldahl–Rittenberg vessels.

vessel (Rittenberg vessel); therefore, only a rotary pump is necessary to obtain the required vacuum. The Rittenberg vessel is illustrated in Fig. 6.2.

An adequate amount of Kjeldahl extract (up to 2 ml) is pipetted into one arm of the Rittenberg vessel. Normally, 0.5–2 mg of nitrogen is sufficient for routine mass spectrometric determinations of ^{15}N. In principle, twice the minimum amount of nitrogen required for the analysis of the isotopic ratio should be used in order to reduce the influence of cross-contamination and/or dilution from micro-leaks. The optimal amount of nitrogen is determined by the volume of the inlet system of the mass spectrometer and the Rittenberg vessel. In practice, an amount of solution necessary to obtain the optimal nitrogen gas pressure is added to one arm of the Rittenberg vessel; into the other one the required amount (up to 2 ml) of alkaline hypobromite solution is pipetted. The Rittenberg flasks are closed with the upper cap, using high-vacuum grease. For this purpose the conical joint is lubricated with four lines of grease every 90°, and the cap is heated with a hair dryer to about 40 °C and then pressed on the joint so that all enclosed air bubbles are removed. The side arm must correspond with the outlet in the ground joint so that the inside of the vessel is open to the air. When the grease has cooled the vessel can be attached to the vacuum line. The stopcocks of

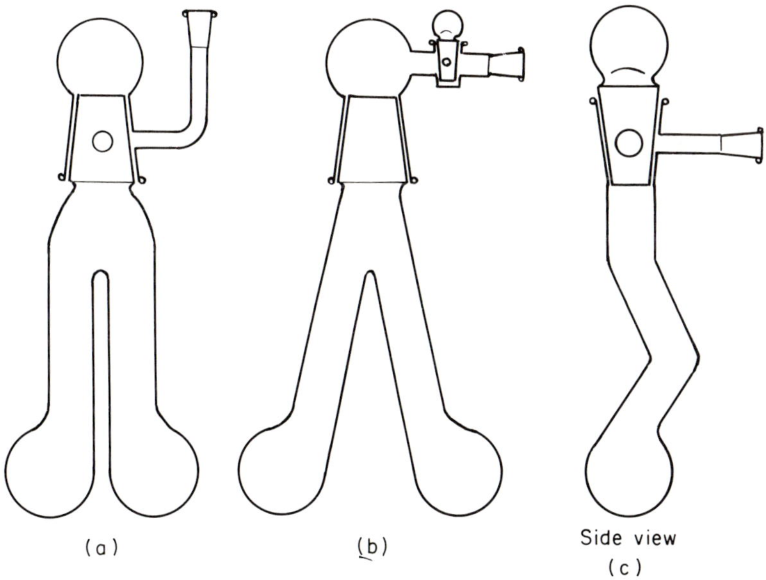

Fig. 6.2 Different types of Rittenberg vessel used for mass spectrometry, which can be rotated on a vacuum line without removal, as employed by (a) Bremner (1965), (b) Balestieri (1968) and (c) Smith *et al.* (1963).

the vacuum line, preferably made of Teflon (greaseless), are open partially to the vacuum which removes the dissolved air from the solutions. Soon after this is complete, the liquid starts to boil. When boiling stops and the liquid freezes, the stopcocks are opened fully for 8–10 min. Then the container is closed by turning it through 180°. Once the ice has melted, the reaction is started by transferring the sample into the alkaline hypobromite solution by tilting the vessel. Care should be taken to transfer the sample and not the hypobromite solution because the reaction must take place in alkaline medium. When the reaction is complete (30 s), the mixture is frozen with liquid air, the Rittenberg vessel is connected to the inlet system of the mass spectrometer and the gas allowed to flow in. The freezing should be done carefully in order to avoid cracking the glass. Freezing should start at the bottom to enable the ice and the upper liquid to expand to the upper part of the arm. Experience has shown that about 2% of the vessels crack during routine operation.

Medina *et al.* (1978a,b, 1979) have modified the Kjeldahl–Rittenberg method for the preparation of very small (0.1 μmol N_2 minimum) biological samples including heterocyclic nitrogen compounds for mass spectrometry. The Kjeldahl digestion of the sample (1–10 mg) is performed in a small test tube containing 0.3–0.5 ml of concentrated H_2SO_4 and 10 mg of selenium dioxide, as illustrated in Fig. 6.3. Many samples can be digested simultaneously on a heating block. After 4–12 h a NaOH solution is layered under the H_2SO_4, and the tube is stoppered with a rubber plug to which is attached a small plastic spoon containing 0.4 N HCl. After 12 h of shaking at 50 °C the NH_3 liberated is collected by microdiffusion into the HCl with a recovery of >98% as NH_4Cl. The isotope effect resulting from the diffusion is negligible ($k_{14}/k_{15} = 1.02 \pm 0.005$) whereby the isotope content of the NH_4Cl is 99.9% of that of the digested sample. The NH_3 is then estimated by microtitration in a plastic Eppendorf vial and the sample is subsequently evaporated to dryness.

The vial, fitted with a rubber O-ring, is connected to a vacuum system (Medina *et al.*, 1979) similar to that described by Ross and Martin (1970). The sample and vacuum apparatus are connected to the inlet system of a double collector isotope-ratio mass spectrometer. After evacuation, the dry sample is oxidized with 10% $Li^+ BrO^-$ solution (stored under He atmosphere) to convert the ammonium nitrogen to N_2. Depending on the amount of nitrogen gas produced, it is expanded or flushed by He into the reservoir of the inlet system of the mass spectrometer and the sample inlet is closed. While the $^{15}N:^{14}N$ ratio of the sample is being determined, the next sample can be connected to the vacuum line and treated with oxidizing solution.

Medina *et al.* (1979) report that there is only a 0.01% memory effect at 25 °C when oxidizing natural and 10-fold natural abundance samples by this

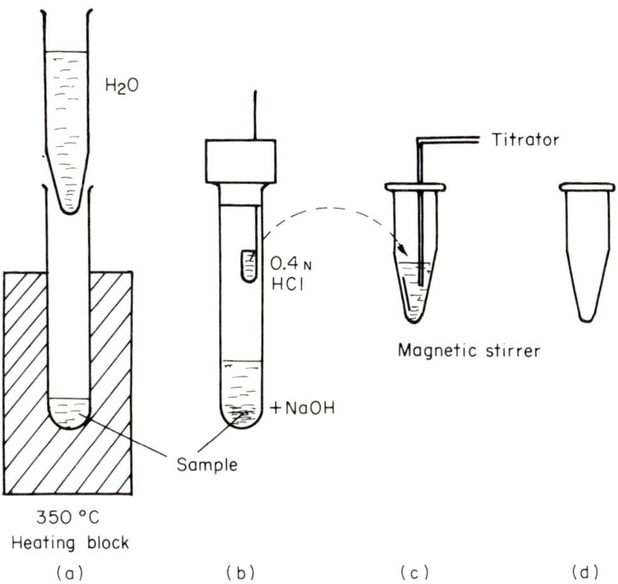

Fig. 6.3 Sample preparation for the ^{15}N determination: (a) Kjeldahl digestion, (b) NH_3 microdiffusion, (c) microtitration in Eppendorf vials and (d) evaporation of the sample and preparation of the vials for the connection to the hypobromite apparatus (Medina *et al.*, 1979).

technique; and at −40 °C there is no memory effect. The time required for hypobromite oxidation and isotope determination is only 12 min. The reproducibility is reported to be ±0.0004 atom % in the range of natural ^{15}N abundance for sample sizes of 10 µmol N_2; however, smaller samples of 1 µmol are analysed with a precision of 1.5 %. When He is used as a carrier of the N_2, it is feasible to determine isotope ratios of samples containing only 0.1 µmol N_2.

2. *Modified Dumas method*

This method is discussed in detail in Rolle (1962), Proksch (1969) and Fiedler and Proksch (1972).

Finely powdered dry material (200 µm or less) or an extract evaporated to dryness, containing 0.5–5 mg of nitrogen, is placed into a Pyrex glass tube. The dimensions of the Pyrex glass tube are 10 mm outer diameter, 1 mm wall thickness, and approximately 150 mm length. The tubes are precleaned by preliminary heating at 500 °C for 3 h together with copper oxide (and copper when nitrate is present) and freshly prepared calcium oxide (activated by heating to 1000 °C). The sample should contain 0.5–5 mg of nitrogen. The amounts of copper oxide and calcium oxide should correspond to about twice the quantities theoretically needed for oxidation and absorption, respectively, in order to ensure complete conversion of sample nitrogen gas. For example, 0.15 g of plant material ground to 200–300 mesh, composed mainly of $(CH_2O)_x$, requires 0.8 g of copper oxide and 0.75 g of calcium oxide for absorption of water and carbon dioxide. Actually, 1.6 g of oxidant and 1.5 g of lime should be placed in the tube, which is then connected to the vacuum line. This consists of no more than a mechanical pump, a liquid air trap and a vacuum manometer.

The end vacuum of the line itself should be about 5×10^{-3} mbar. The tubes are evacuated for 5–15 min until a pressure of 5×10^{-2} or 10^{-2} mbar is attained. The time required and the magnitude of the end vacuum depend on the vapour pressure of the particular kind of sample. Inorganic and organic compounds with low vapour pressure are ready within 5 min at a vacuum of 10^{-2} mbar, whereas plant material needs 15 min for a final vacuum of 5×10^{-2} mbar. The tube is then sealed off with a hand torch, and the top of the tube is annealed for about 2 min to prevent cracking from tension in the relatively thick glass. The contents of the tube are carefully mixed before combustion in a muffle furnace, in which the tubes are placed in a horizontal position for at least 3–4 h. When the samples have cooled to room temperature, they can be introduced into the inlet system of the mass spectrometer, which should contain a liquid air or nitrogen trap.

The temperature of combustion is very important, as the rate of reaction increases with increasing temperature. The maximum temperature, is, however, limited by the properties of the glass of the sample tube. Above 570 °C Pyrex becomes plastic and may explode because of the gas pressure inside the tube. The sample can be safely heated up to 570 °C provided that the temperature control of the oven is satisfactory, and no temperature gradients greater than 5 °C occur. Samples should be placed on an asbestos plate in the centre of the oven. When the temperature cannot be controlled to at least ±5 °C, which is normally the case for ovens working at such temperatures, 550–560 °C is safer. It is also advisable to check the oven temperature with an additional thermocouple placed in the same region as the samples. Although only 70–80 % of the sample nitrogen is converted to nitrogen gas for plant material (though 85–90 % can be achieved by careful mixing), the $^{15}N:^{14}N$ ratios obtained after combustion at 550–560 °C for 3–4 h agree satisfactorily with those obtained by the Kjeldahl–Rittenberg method. In work on grain analysis, a check was made in which the $^{15}N:^{14}N$ ratios (above 2 % excess ^{15}N) from grain burned to yield 70–80 % of the total nitrogen agreed completely with ratios obtained from a combustion in which there was 97 % recovery of nitrogen.

The use of special glasses, such as Supremax and Pyrex 1720 with softening points as high as 650 °C, has more disadvantages than advantages: Supremax glass is about three times as expensive as Pyrex and the necessary glass-blowing technique is more complicated, since the glass has to be carefully and slowly heated to the working temperature to prevent breaking. The glass tends to react with lime, which results in cracks when the tubes are cooled afterwards. Pyrex 1720 is about 16 times more expensive to the author than normal Pyrex, and tubes of the required dimensions are not all readily available commercially. The use of quartz is even more costly; it reacts with lime above 600 °C and cracks, because of the formation of calcium silicate.

E. Total nitrogen determination before mass spectrometry

In the modified Dumas method, if it is assumed that the total nitrogen in the sample is converted to nitrogen gas and all other gases formed are absorbed by the lime, the pressure of the nitrogen can then be taken as a measure of the total nitrogen content of the sample. The pressure of the generated nitrogen gas is measured after the sample tube has been crushed in an evacuated chamber of known volume or a volume previously calibrated by nitrogen samples of known nitrogen content. The inlet system of the mass spectrometer can be used for this pressure analysis. The pressure can be read from a mercury manometer normally connected to the inlet system, by a membrane manometer, or by a capacity pressure meter.

The pressure measured is proportional to the nitrogen present in the sample. The total nitrogen can be calculated as follows:

$$\text{Percentage N} = \frac{\text{Amount of nitrogen}}{\text{Amount of sample material}} \times 100 \tag{6.9}$$

or

$$\text{Percentage N} = \frac{K \times \text{Pressure of nitrogen}}{\text{Amount of sample material}} \times 100, \tag{6.10}$$

where K is a constant depending on the volume (about 60 ml) and the temperature of the system. The temperature and the volume can easily be kept constant, so that equation (6.10) can be used for the measurements when K is known. The manometer used has a working range of 0–50 mmHg, and the recorder is best calibrated to give a 40 mm reading for 1 mg of nitrogen. Normally, 150 mg of sample material is used for the combustion. The maximum indication on the recorder allows the measurement of samples up to 4.6 mg of nitrogen or approximately 3 % total nitrogen. For samples with higher total nitrogen, 120–100 mg of sample material is used so that 3.8–4.6 % total nitrogen can be measured. Naturally, for higher nitrogen contents, smaller samples can be used if homogeneity can be guaranteed.

Details on the sample preparation and pressure analysis procedures for total nitrogen determinations are given by Fiedler and Proksch (1972, 1975) and Fiedler *et al.* (1973).

F. Sample preparation for emission spectrometry

1. *Preparation of electrodeless discharge tubes for sample analysis*

The discharge tube required for emission spectrometry should have the following properties:

(i) It should be transparent for ultraviolet radiation as the spectral bands used for the analysis are in this region.
(ii) The softening point should be high because any adsorbed gases must be removed by heating. If the Dumas method is used to convert the sample nitrogen to nitrogen gas, the temperature of the mixture during combustion may be as high as 700–800 °C.
(iii) As the tube is used only once, the price should be as low as possible.

Materials used for discharge tubes are Uviol glass, hard glass (Pyrex, Rasotherm) and quartz glass. These are described below.

Uviol has a transmittance of 90 % for a 1 mm thickness for radiation of 300 nm. The sample tube has a volume of about 2 ml. Because of a restriction in the middle of the tube with a wall thickness of 0.25–0.3 mm, an intense discharge is obtained concentrated around the restriction. The tube is connected to the vacuum line, as suggested by the supplier, by means of black vacuum wax. The softening point of Uviol is 470 °C.

Hard glass tubes are the same shape and size as those of Uviol. Hard glass has a transmittance of 8 % at 300 nm for a thickness of 10 mm (Faust, 1965). The thickness of the glass at the restriction is also 0.25–0.3 mm which is sufficient to allow enough ultraviolet radiation to pass through. The connection to the vacuum line is the same as in the case of Uviol. The softening point is 570 °C. By using a grating monochromator which has a higher luminosity (spectral luminous efficiency) than prism instruments, the loss of light intensity caused by absorption in comparison to Uviol can be compensated.

The quartz tubes have 100 % transmission at 290–300 nm for a 1 mm thickness; the tubes have an internal diameter of 2 mm and a length of 100 mm, and the wall thickness is 2 mm. The volume is about 0.2 ml. Care must be taken in handling the quartz because, for example, each fingerprint which is not washed off with alcohol will carbonize on heating the tube; this makes the quartz milky and less transparent to the ultraviolet radiation. The connection of the tube to the vacuum line can be made by means of a Teflon O-ring joint, as can the Uviol and hard glass tubes, if suitable sized joints are available. Before the tube is connected to the vacuum line, it must be cleaned of all inorganic and organic compounds, especially those which contain nitrogen. This is done by washing the tubes with acid and baking them in a furnace at a temperature of 50 °C lower than the softening point of the glass. To avoid subsequent contamination, the tubes should be kept dust-free in a desiccator. As the amount of sample nitrogen in the discharge tube is of the order of 1–10 μg, nitrogen still adsorbed on the walls of the tube can dilute the nitrogen of the sample significantly. Faust (1967) demonstrated that sample tubes which have not been heated and chemicals which are not degassed can dilute the ^{15}N content up to 17%.

In addition to nitrogen, CO_2, water, oxygen and other gases which are present in the laboratory atmosphere must be removed from the tube as they can disturb the discharge. If CO_2 is present, it decomposes to CO under the influence of the high-frequency discharge. CO has a spectral band near that of $^{15}N/^{14}N$ so that the ^{15}N percentage will apparently be too high. When the level of ^{15}N abundance of the sample is below 1% this effect can be significant; Perschke *et al.* (1971) found ^{15}N values which were 5–10% too high because of this effect. Traces of water cause an unstable discharge, and if more than a half-volume of water vapour is present in the discharge tube, the discharge will extinguish immediately after ignition. Oxygen does not affect the discharge below 15% by volume of the nitrogen, as it is removed together with the other gases during the procedure (Leichnam *et al.*, 1968).

A block diagram of the high-vacuum line for the preparation of the tubes is shown in Fig. 6.4. After the tubes have been cleaned in the muffle furnace, the residual adsorbed gases are removed by connecting the sample tubes to the high-vacuum line (better than 10^{-4} mbar) and heating them with a hand torch at a temperature at which the glass starts to soften (a small deformation can be observed). For Uviol and hard glass, treatment with a Tesla coil (leak detector) after heating releases the remainder of the adsorbed gases. This is not necessary for quartz because of the higher baking temperature which can be reached without danger of collapse.

The pressure of nitrogen in the sample tubes is normally adjusted to 3–5 mbar. Although the light intensity of the discharge is highest at a pressure of 1.5 mbar (Sommer and Kick, 1965), the discharge lasts longer when more nitrogen is present, and a more stable discharge is obtained at 3–5 mbar. This happens because the high-frequency discharge must provide enough energy for the ionized gas molecules which become implanted in the glass on collision with the walls; at very low filling pressures, the amount of nitrogen in the tube decreases very rapidly, which causes instability and rapid dying out of the discharge.

Since much less material is required for emission spectrometry measurement than for mass spectrometry, the direct conversion of nitrogen from a

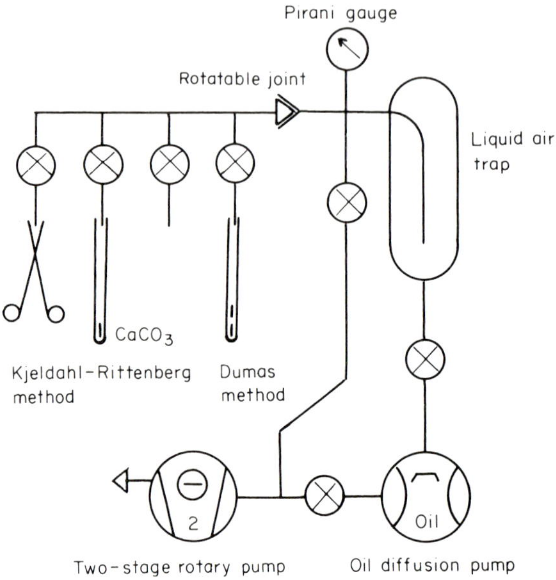

Fig. 6.4 Block diagram of a high-vacuum line for the preparation of electrodeless discharge tubes.

sample should be done only when no subsampling error will be introduced. As the nitrogen content of the sample should yield a pressure of 3–5 mbar in the discharge tube, it is necessary to know the approximate content of the sample to at least $\pm 20\%$. If homogeneity cannot be guaranteed at the sample size required for this method, it is necessary to take a much larger amount, convert it to nitrogen and take an aliquot.

2. *Preparation of electrodeless discharge tubes for standard comparison*
Emission spectrometric curves of the optical signal against ^{15}N content must be prepared and regularly checked against standard discharge tubes filled with gas whose ^{15}N abundance has been established by mass spectrometry. Clearly, the discharge of such tubes should be stable over a long period and their percentage ^{15}N content should correspond to the mass spectrometric value, i.e. there should be no dilution from residual traces of nitrogen from the walls, and carbon dioxide should be absent. They can be prepared as follows: Nitrogen of appropriate ^{15}N enrichment is prepared from a suitable source. A sample is taken for mass spectrometric measurement, and the remainder is used to fill a discharge tube containing a calcium briquette (see Section III.F.4 below) on the normal tube preparation line. The tube walls are heated with the hand torch and treated for 2–3 min with a spark from a

Tesla coil to equilibrate the last traces of nitrogen and impurities on the walls with the sample gas. On cooling, the surface will be coated with a layer of sample gas which will not dilute the standard material. Then the gas is pumped out of the tube, which is immediately thereafter filled again with 3–5 mbar of the nitrogen gas. The tube is then sealed off and is heated for 2 h at 430 °C for Uviol, or 500 °C for hard glass, to remove any carbon dioxide or water from the surface before the actual analysis is made. When this technique is used, the standard tubes obtained will provide at least 5 h of discharge (Fiedler and Proksch, 1972; Ferraris and Proksch, 1972).

3. *Kjeldahl–Rittenberg method*
For emission spectrometric analysis (where about 1/100th of the amount of nitrogen needed in mass spectrometry suffices) 0.1–0.3 ml of solution containing 0.03–0.1 mg NH_4^+–N is placed in a small Rittenberg vessel. Here again, the optimal and minimal amounts of nitrogen are determined by the volume of the transfer line and the discharge tube.

The same basic procedure is followed in the case of the sample preparation for optical ^{15}N determination as for mass spectrometry. The smaller Rittenberg vessel is, however, not removed from the vacuum line, and the mixing is performed by rotating the vessel.

After cooling, the nitrogen gas from the Rittenberg vessel is allowed to flow into a precleaned discharge tube via a liquid air trap which should remove all water. However, micro ice crystals can be carried along with the gas stream and enter the discharge tube. A similar effect has been observed in mass spectrometry. In order to remove this water, an activated calcium oxide briquette, identical to the briquette used for the Dumas method, is added to the tube before it is mounted on the high-vacuum line (Fiedler and Proksch, 1972). After being filled, the tubes are sealed off by a hand torch and heated in a furnace at 500 °C for 2 h (430 °C for Uviol) to complete the lime–water reaction. On cooling, the gas can be analysed for its ^{15}N content. The possible sources of contamination which may occur during oxidation are: (i) air leakage into the Rittenberg vessel; (ii) bad vacuum from a leak in the line; and (iii) cross-contamination from the walls of the Rittenberg vessel. Contamination can be easily prevented by carefully checking the greased ground joint, control of the vacuum during the analysis, and steaming or acid washing of the Rittenberg vessel.

Burridge and Hewitt (1980) report a procedure for filling discharge tubes with ammonia in the pressure range 1–4 mbar. Samples containing 1–30 mg of nitrogen can be analysed. A specially designed vacuum line for filling the discharge tube is described. This apparatus is connected directly to the discharge tube, which is placed in a cavity next to the monochromator, and where the lower end of the tube can be cooled. The sample, as an ammonium solution, is converted to ammonia by (i) freezing the sample, (ii) introduction of sodium hydroxide pellets into the sample container, (iii) evacuation of the sample container, (iv) melting the ammonium solution to start the conversion to ammonia, and (v) introduction of the ammonia into the discharge tube. The measurement of ^{15}N content is carried out on NH_3. The accuracy and precision of this method is suitable for most soil–plant studies.

4. *Modified Dumas method*

The appropriate amount of sample in solution or suspension is placed in a quartz capillary (1.5 mm o.d. and 0.8–0.9 mm i.d.) and dried at 40–50 °C. The dried sample is inserted into the discharge tube together with a 7–10 mm long copper oxide wire (1–2 mg) and a calcium oxide briquette (3 mg CaO). To prevent mechanical disintegration of the briquette when it absorbs water, CO_2, HCl, etc., it should be briquetted with alumina in a 1:1 mixture by applying a pressure of about 10 kg cm^{-2} (Perschke and Proksch, 1971). The discharge tube is cleaned by heating with a hand torch under high vacuum conditions as described before. The copper oxide is heated simultaneously but only to a maximum temperature of 650 °C; otherwise, oxygen will be released. Calcium oxide has to be activated by preheating to 1000 °C to decompose $CaCO_3$. If it is not heated adequately, samples with a ^{15}N abundance below 1% may yield 5–10% (relative) higher spectral values owing to interference by carbon dioxide. During the baking-out process of one end of the tube, the sample-holding capillary is moved to the cool end of the tube to prevent destruction of the sample while heating. When the other end of the tube is heated, the sample is moved to the end already baked out. After the tube has cooled and has been sealed off with a hand torch, the sample is burned at 850 °C for 30 min.

In the case of hard glass tubes, activation of the CuO and CaO can only be made at about 500 °C because of the softening point of the glass, and in the case of Uviol at about 400 °C. The final combustion temperature for Pyrex is 560 °C and for Uviol 440 °C, with heating being continued for 4 h in both cases.

During the preparation, combustion and storage of samples, contamination may occur by incomplete oxidation to nitrogen, air leaks during sealing, and release of nitrogen from CuO and CaO when these chemicals and the tube walls have been insufficiently degassed.

With this method, the isotope ratio can be determined in 1–2 μg of nitrogen. When smaller quantities of nitrogen are used, however, the discharge either does not occur or it is of such short duration that sensitive measurement is impossible. This is due to the relatively fast removal of the nitrogen molecules from the gas phase and implantation in the walls of the tube under the influence of the electrodeless discharge. However, these difficulties can be overcome to some extent by adding certain gases to the tube; isotopic ratios can then be determined on samples down to 0.2 μg of nitrogen (Cook *et al.*, 1967, 1968, 1970).

All of the noble gases sustain a discharge in the plasma of small amounts of nitrogen, but to different extents. The two extremes, helium and xenon, however, are the most interesting. In the presence of xenon, the nitrogen discharge is relatively weak (the ionization potential of xenon is lower than that of nitrogen) but lasts a relatively long time. In the presence of helium, the nitrogen discharge is very intense, but lasts no longer than with nitrogen alone. These results can be interpreted as meaning that xenon is adsorbed on or implanted in the quartz walls more readily than nitrogen and hinders the uptake of the nitrogen. Helium which is not readily taken up by the quartz does not delay nitrogen adsorption, but as its ionizing potential is higher than that of nitrogen, it intensifies the nitrogen by a process of energy transfer. It is therefore advantageous to use both properties to give better nitrogen discharge conditions. Merely coating the walls of the tube with adsorbed xenon and using a helium–nitrogen gas mixture is insufficient, as the discharge lasts only about 3 min; but adding 2% by volume of xenon to the helium (in addition to the xenon coating the tube) enables the discharge to be measured for 10–15 min, which is adequate. The following procedure was therefore developed.

The high-vacuum line used is shown in Fig. 6.5. After removal of adsorbed gases from the walls of the quartz discharge tube, and degassing of copper

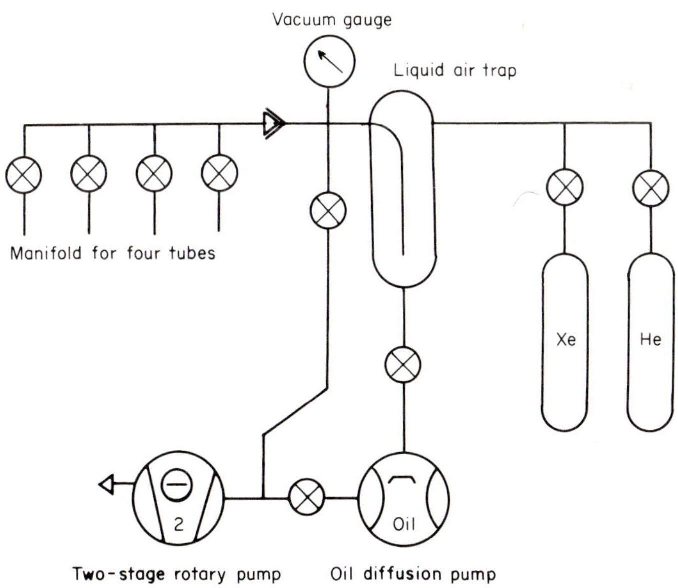

Fig. 6.5 Block diagram of a high-vacuum line for the preparation of electrodeless discharge tubes with noble gases.

oxide and activated calcium oxide by heating with a hand torch has been completed, about 5 mbar of xenon is introduced into the tube. The tube is flamed for about 5 min and cooled to room temperature. During this treatment, remaining traces of impurities on the walls exchange with the xenon which, itself, becomes adsorbed. At the same time, the chemicals are treated in the same way as when using the glass-wall tubes. The temperature for the copper oxide should not exceed 650 °C. The capillary holding the sample is always moved to the cool part of the tube to prevent disintegration of the sample. The xenon still remaining in the gas phase is pumped out and 10 mbar of helium mixed with 2% xenon is introduced. The tube is sealed off and the sample combusted. When this method is used, the $^{15}N:^{14}N$ ratio can be determined in as little as 0.2 μg of nitrogen.

Lloyd-Jones *et al.* (1977) suggest a preparation procedure for samples containing 20–500 μg of nitrogen. This method also allows the determination of the total nitrogen in the sample by use of an ammonium ion-selective electrode. The conversion of sample nitrogen to nitrogen gas is done by the Dumas combustion.

5. *Other methods*

Yamamuro (1981) describes the preparation of combustion tubes for inorganic and organic sample material. The glass tubes (Pyrex, 12 mm o.d. and 10 mm i.d.) and the capillaries used to take the chemical reagents and the samples (6 mm o.d., 4 mm i.d. and 10 mm long) are cleaned in a mixed solution of $K_2Cr_2O_7$ and H_2SO_4, washed with distilled water and then dried in an oven at 120 °C for 2 h. The combustion tubes are prepared by heating with an oxygen-burner flame, producing a restriction in the middle of the tube of 4–5 mm o.d., leaving both ends open. Care should be taken to avoid deformation of the tube. For inorganic samples the following procedure is suggested. The sample is sucked into the capillary and dried under an infrared lamp. Into one part of the combustion tube, wire-shaped $CuO–Cu_2O$ reagent is placed, while the sample capillary and a freshly activated CaO briquette are introduced into the other part. The end containing sample and CaO is closed, and the tube evacuated, carefully heated, and sealed off. The combustion tube is heated in a furnace at 500 °C for 1 min, to change the sample to NH_3 gas. The tube is set between the electrodes of a 27 MHz high-frequency generator. The NH_3 gas is decomposed to N_2 gas, taking care not to heat the reagent tubes inside. The decomposition to N_2 takes 20–40 h.

Organic sample preparation is similar to the procedure for inorganic samples with the exception of the following:

(i) The sample is placed into a capillary closed at one end and with a small hole on the other end. Together with the sample, Pb_3O_4 (50 mg per milligram of sample) is placed into this tube.

(ii) Instead of CuO–Cu_2O wire and CaO, a briquette containing a mixture of CuO and CaO (CuO to be three to five times that of CaO) is used. The CuO–CaO briquette is sealed imperfectly and heated well to about 600 °C immediately before being introduced into the combustion tube.

(iii) Reduced iron metal (50 mg per milligram of sample) is used. Organic samples such as amino acids, proteins, dried plant material and dried animal tissues containing 0.002–0.3 mg of nitrogen can be treated.

All reagents and the sample are introduced into one end of the combustion tube and the tube is evacuated and sealed off. The tube is heated in a furnace at 500 °C for 5 min, and part of the organic matter is melted and gasified. Only the part where the sample and Pb_3O_4 is placed is heated with a flame to 600 °C for 6 min, avoiding the burning of the CuO–CaO briquette. This activates the Pb_3O_4 and decomposes the sample to NO_2, CO_2, SO_2, etc., while the excess O_2 gas is absorbed on Fe. The whole combustion tube is heated again in a furnace at 500 °C for 2 min. If the colour of the discharge, when placed on the electrodes of a high frequency generator, is white, the tube has to be heated for an additional 1–2 min. Continuous treatment for 4 h with high frequency decomposes the NH_3 to N_2.

The described treatment avoids contamination and dilution by nitrogen contained in the reagents. The relationships between the tube size, pressure, colour of the first emission of NH_3, lifetime, and relative intensity of N_2 have been investigated.

IV. Mass spectrometry

A. Principle

A mass spectrometer designed for gas analyses essentially is comprised of five units (Fig. 6.6): (i) an inlet system for the introduction of nitrogen gas; (ii) an ion source where the nitrogen molecules, bombarded with electrons, become charged and accelerated; (iii) a magnetic field in which the charged molecules are separated into different paths according to their momentum – the deflection is proportional to the mass of the charged particle as illustrated in Fig. 6.6, where $m_1 < m_2 < m_3$; (iv) a collector placed at the end of the flight tube, where the ions discharge, and the discharge currents are amplified; (v) a recorder which registers the amplified discharge currents as peaks as the mass range is scanned. The heights of the peaks are proportional to the amounts of the three molecules $^{28}N_2$, $^{29}N_2$ and $^{30}N_2$ in the gas mixture. By simply

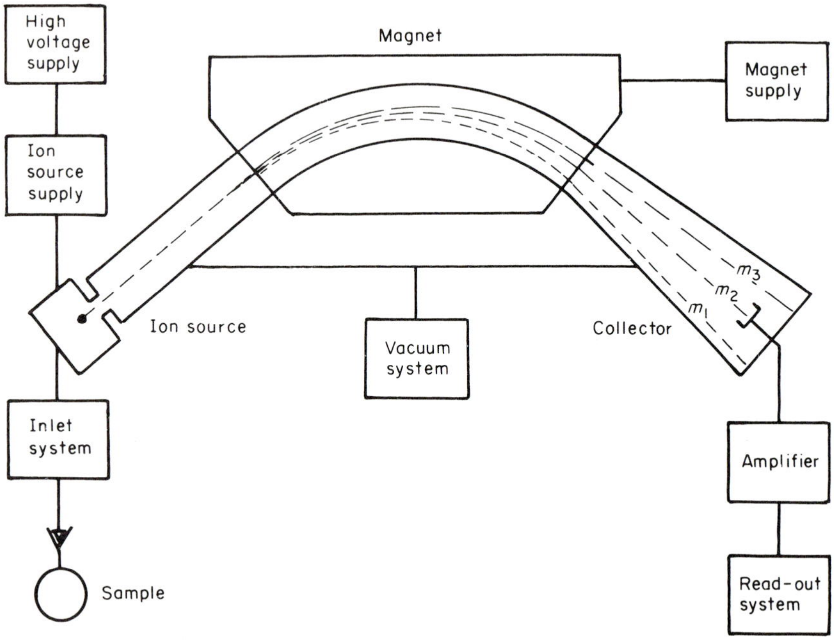

Fig. 6.6 Block diagram of a mass spectrometer for gas analyses.

measuring the peak heights, the percentage ^{15}N abundance can be determined.

Although the principle is simple, the practice of mass spectrometry is complicated. The whole system must be maintained under high vacuum, and the electronic components must be highly stabilized. Artefacts can be produced easily when such instruments are not supervised by highly qualified personnel.

Various types of mass spectrometer are available. The single-focusing mass spectrometer depicted in Fig. 6.6 is most commonly used for nitrogen analysis. Special applications require two-stage and double-focusing mass spectrometers as well as time-of-flight, cycloidal focusing, and radio-frequency instruments. The latter are mass filters better known as quadrupole spectrometers.

The gas sample is introduced into the gas inlet system, where the pressure is reduced; the sample gas passes into the ion source; and the ions formed pass through accelerating slits and become an unresolved ion beam. In the magnetic field the ion travel in deflected circular paths, whose radii are functions of the ion mass:charge ratio. The resolved beam is focused on the detector and the discharge current is amplified and subsequently recorded.

During acceleration through a potential difference V, a kinetic energy eV is given to the ion. All ions having the same charge are accelerated to the same energy level. In a magnetic field with a direction of induction normal to the direction of motion, the ion beam will be deflected in a circular path.

The commonly used equation to describe the interaction is

$$r = \frac{144}{B}\sqrt{\frac{MV}{n}}, \tag{6.11}$$

where r is the radius of deflection in centimetres; M is the mass of an ion in daltons; V is the voltage applied to the electrode of an ion source; B is the magnetic induction in gauss; and n is the electric charge of the ion (generally having the value 1).

Further information on mass spectrometry can be obtained in textbooks by McDowell (1963), Kienitz (1968), Roboz (1968) and Beynon and Brenton (1982).

B. Sample size

The sample size always depends on the volume of the inlet system and of the sample container. Examples for viscous and molecular inlet systems are given below.

1. *Viscous inlet system*

Let the volume of the sample bottle (V_b) be 30 ml, the volume of the inlet system (V_{is}) be 15 ml and the pressure in the inlet system after introduction of the gas sample (P_{is}) be 40 mbar; pV is constant (ideal gas law).

$$V_b P_b = (V_b + V_{is}) P_{is}. \tag{6.12}$$

Thus

$$P_b = \frac{(30 + 15) \times 40}{30} = 60 \text{ mbar}.$$

Recalculated to atmospheric pressure and at a temperature of 0 °C, 60 mbar in a volume of 30 ml gives

$$V_{atm} = \frac{V_b P_b}{P_{atm}} = \frac{30 \times 60}{1000} = 1.8 \text{ ml}. \tag{6.13}$$

Avogadro's law says that the volume of 28 mg of nitrogen is 22.4 ml at a temperature of 0 °C and atmospheric pressure (1000 mbar) or 22.4 ml/28 mg = 0.8 ml per milligram of N_2; moreover, 1.8 ml/0.8 ml mg^{-1} = 2.25 mg. Thus, a minimum of 3 mg of nitrogen would be necessary for this type of inlet system.

The passage of a gas through a duct under conditions such that the mean free path (average distance which a molecule travels between two successive collisions with other molecules of the gas) is very small in comparison with the smallest internal dimension of a transverse section of the duct is known as *viscous flow*, the flow being dependent on the viscosity of the gas.

The viscous inlet system is preferable for mass spectrometric inlet systems.

2. *Molecular inlet system*

Let the volume of the sample bottle (V_b) be 30 ml, the volume of the inlet system up to the standard volume (V_{is}) be 90 ml, the volume of the reservoir (V_r) be 3000 ml, the volume of the standard volume (V_{std}) be 3 ml, and the pressure in the reservoir (P_r) be 10^{-2} mbar. Then

$$P_{std} = \frac{V_r P_r}{V_{std}} = \frac{3000 \times 10^{-2}}{3} = 10 \text{ mbar}; \qquad (6.14)$$

$$P_b = \frac{(V_b + V_{is}) P_{std}}{V_b} = \frac{(30 + 90) \times 10}{30} = 40 \text{ mbar}. \qquad (6.15)$$

Recalculated to atmospheric pressure and a temperature of 0 °C, 40 mbar in a volume of 30 ml gives

$$V_{atm} = \frac{V_b P_b}{P_{atm}} = \frac{40 \times 30}{1000} = 1.2 \text{ ml}.$$

Using the value of 0.8 ml mg^{-1} N_2 calculated in Section IV.B.1, the sample weight is calculated to give 1.2 ml/0.8 ml mg^{-1} = 1.5 mg N_2. The ability to expand the standard volume more than once into the reservoir makes it possible to use only this 1.5 mg of nitrogen to obtain good conditions in the inlet system.

The passage of a gas through a duct under conditions such that the mean free path is very large in comparison with the largest internal dimension of a traverse section of the duct is known as *molecular flow*.

With as much as half of the sample size calculated above, modern mass spectrometers can give sufficient ion intensities for acceptable results. Partial or full computer control of the instruments also allows more precise measurements.

C. Inlet systems

1. *The single inlet system*

The function of any inlet system is to reduce the relative high sample pressure to a working pressure in the range of about 10^{-2} mbar. This pressure is

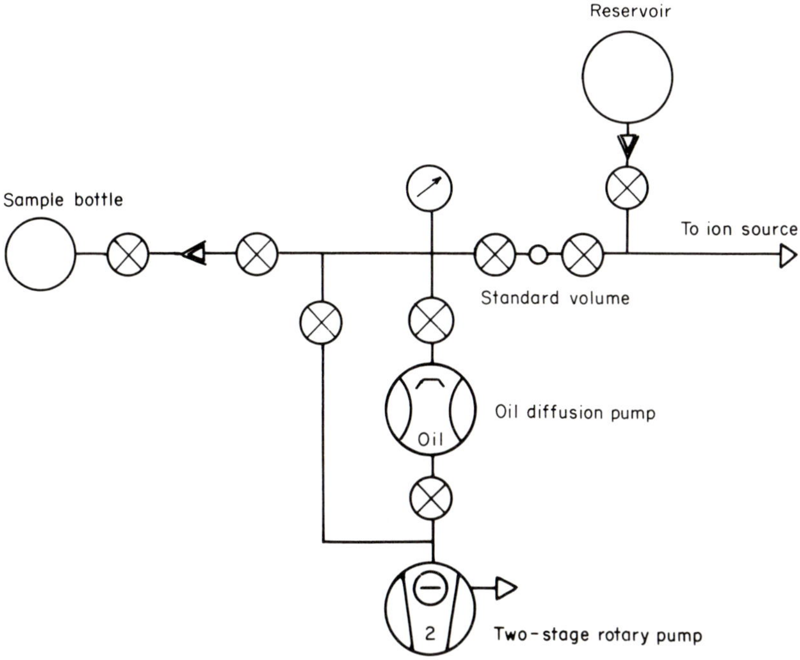

Fig. 6.7 Block diagram of a single inlet system of a mass spectrometer utilizing molecular flow. Standard volume is 1/100th of the reservoir.

necessary to allow a steady stream of gas into the ion source, where the vacuum is in the order of 10^{-5} mbar. The inlet system consists of a vacuum system to which gas samples can be attached. Figure 6.7 shows a block diagram of a simple gas inlet system (molecular inlet system).

The sample is attached to the inlet system and the piping system is evacuated. When the vacuum is below 10^{-3} mbar, the pump valves are closed and the gas is released into the system. According to the ion intensity the instrument needs, the standard volume is expanded one or more times to obtain the reservoir with the necessary pressure. Usually an orifice is placed next to the ion source. This allows a steady and continuous flow from the reservoir to the ion source.

2. *The dual inlet system*

The main difference between the two inlet systems is that the double inlet system is equipped with a reservoir with variable volume (Fig. 6.8). To avoid fractionation of the isotopic sample, viscous flow to the mass spectrometer has to be used. The sample container is attached to the inlet system and the gas is released to the reservoir at its maximum volume. A capillary leak which

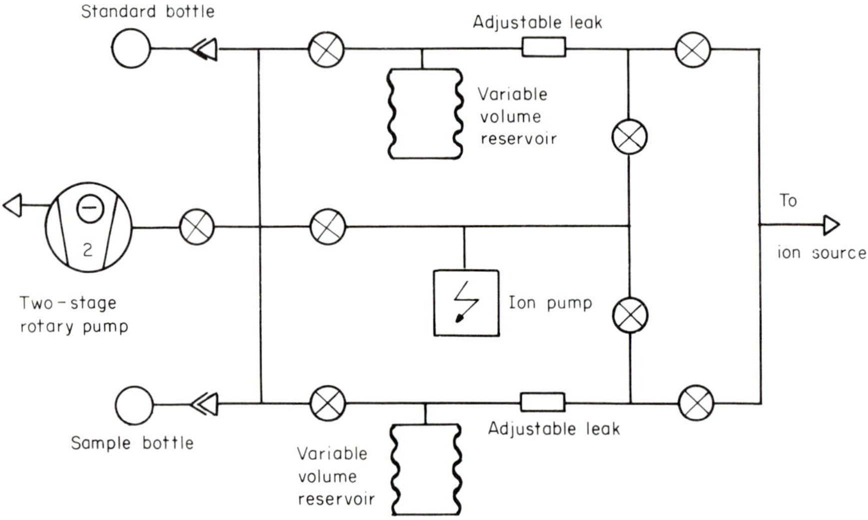

Fig. 6.8 Block diagram of a dual inlet system of a mass spectrometer utilizing viscous flow.

can be adjusted, or a fixed leak, controls the flow of sample gas to the ion source. A valve system allows the sample compartment to be exchanged with an equivalent volume of a standard. The reason for the variable volume is to enable compensation for the decreasing sample pressure during the measurement (reducing the volume to maintain the starting sample pressure). A double inlet system together with a mass spectrometer equipped with a double collector allows a direct comparison of the sample with a standard gas. The sample and the standard gas can be entered into the mass spectrometer alternately for a number of times to obtain close comparative values. This allows very precise measurements of a sample and is particularly useful in the natural abundance range and at low enrichment levels.

3. *Sample introduction into the inlet system*

When the digestion method is used for sample preparation the Rittenberg flask may be utilized as the sample bottle (Fig. 6.2). The ammonium salt solution is placed in one of the side arms, with the hypobromite in the other, and the liquids are mixed shortly before measurement. The ammonium salt solution should always be added to the hypobromite. During the reaction, nitrogen is released. Then the liquid is frozen by immersion in liquid air, which reduces its vapour pressure and prevents unnecessary contamination of the sample. The gas is introduced into the inlet system. A cold trap in the inlet system is required to remove the water vapour, which is always introduced in small quantities with the nitrogen. The storage of Rittenberg vessels which cannot be measured immediately causes problems, because the vessels must be stored only for short periods of time in a refrigerator. Sealing problems caused by shrinking of the vacuum grease may cause the vacuum to be lost. Medina *et al.* (1979) avoid this problem by storing the hypobromite in a big container next to the connection port of the sample bottle. The hypobromite is degassed and kept under helium atmosphere. The sample vial containing the dried sample is evacuated and the dry sample is oxidized by nitrogen-free LiOBr solution. The nitrogen gas is expanded or flushed by a defined amount of helium into the reservoir of the inlet system of a mass spectrometer.

The dry combustion method (i.e. modified Dumas) has the advantage that samples in glass ampoules can be stored easily, and the chance that leaks will develop during storage is small. The main problem is how to open glass ampoules under vacuum conditions. How this is done depends on the number of samples which have to be analysed.

Figure 6.9 shows a crushing apparatus suitable for use when only 10–15 samples per day have to be measured. On the bottom of the device a metal

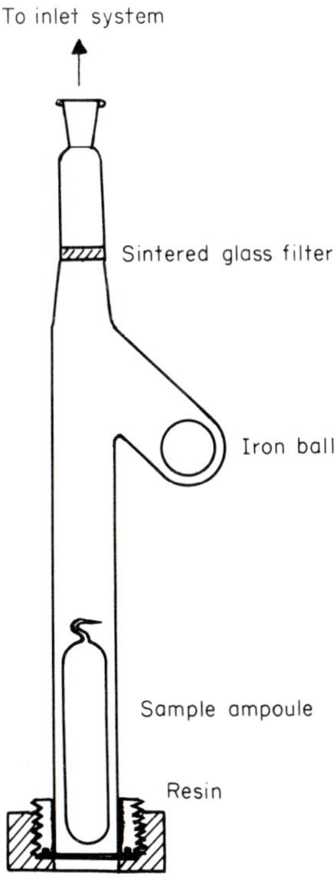

Fig. 6.9 Breaking device for sample ampoules.

thread is connected by a two-component resin to the glass body, sealing being achieved by the resin and an O-ring, and the system is closed by a metal plate at the bottom; the iron ball is then moved by a magnet to break the ampoule. To prevent dust particles from entering the inlet system, a sintered glass filter is placed between the sampler and the inlet system.

The need to analyse a large number of samples led to the development of a special device for breaking sample ampoules. The ampoule is placed in a vacuum chamber in a magazine and destroyed by shooting a piston, which is driven by an electromagnet, into the glass ampoule. A detailed description of this device is given by Fiedler and Proksch (1972).

D. Collector systems

The part of the mass spectrometer at which the ion intensities are measured is called the collector. Single and multiple collectors are used in mass spectrometry.

1. *The single collector*

Figure 6.10 illustrates the principle of operation of a single collector, also known as the "Faraday cup" collector. The ion beam passes through the aperture of the earthed resolving or collector slit, usually adjustable in width to ensure good resolution for higher *m/e* values. The ion beam passes to the suppressor slit which is kept at a potential between −30 and −90 V. Its function is to ensure that secondary electrons ejected from the collector are

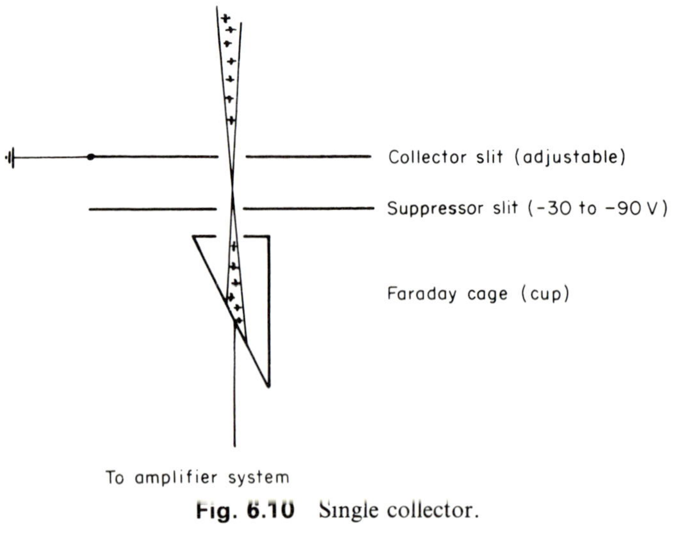

Fig. 6.10 Single collector.

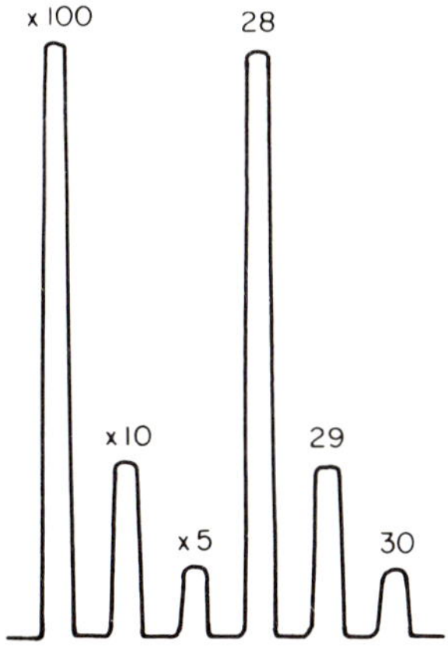

Fig. 6.11 Typical mass spectrum recording of a nitrogen sample after two scans of the ion beam intensities at $m/e = 28$, 29 and 30 ($\times$ 100, $\times$ 10 and $\times$ 5 are attenuation factors).

not detected. The signal is amplified and recorded as a measure of the ion intensity of the beam at a specific m/e value. At $m/e = 28$, the beam can contain CO and C_2H_4 as well as nitrogen, hence the background must be measured before the sample is introduced; this allows the unwanted components of $m/e = 28$ to be corrected for. Figure 6.11 illustrates a typical recording of an enriched nitrogen sample.

The calculation of the ^{15}N atom % from this information is described later in this chapter.

2. *The double collector*

The double-collector method normally requires a dual inlet system to introduce alternately a standard and a sample to the ion source; the principle is shown in Fig. 6.12. The ion beam of the isotope with the lower intensity is focused on collector plate A, while the ion beam of the other isotope is focused on plate B. The discharge of each beam is transferred to the corresponding preamplifier and d.c. amplifier. For measurement, the output signal of amplifier B is reduced by means of a decade potentiometer until the full signal of amplifier A has the same value; a galvanometer or recorder may be used to check the balance of the two signals, the deflection of a decade potentiometer being proportional to the percentage abundance. If the difference between the ratio of a standard and the sample is very small, it can be read directly on the recorder, but if the deflection is more than full scale a change of one or more decades may be required to obtain a reading on scale.

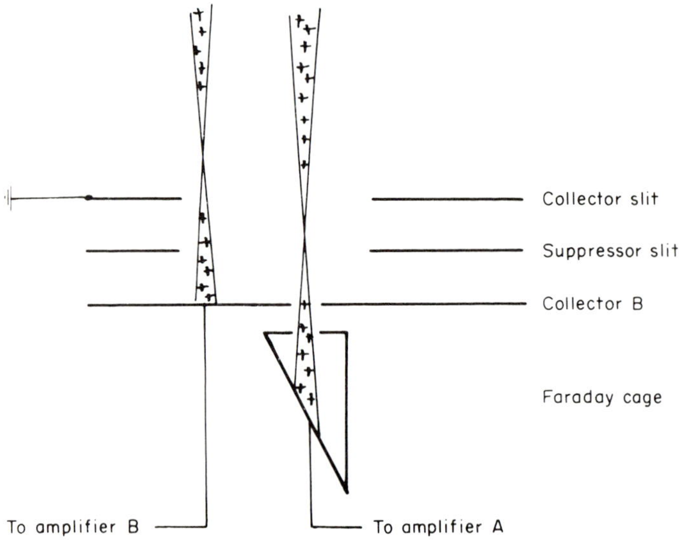

Fig. 6.12 Double collector.

In any case, calibration for this deflection is necessary. Figure 6.13 shows a recording where the unknown sample is compared to a well-known standard.

A multiple collector offers the advantage that ion beams of the isotopes of interest can be collected simultaneously. Variations in ion beam stability do not change the ratio of the beams collected. This permits one to measure small differences in sample isotope ratios more precisely than with a single collector. Here, differences of one part per thousand can be measured very accurately.

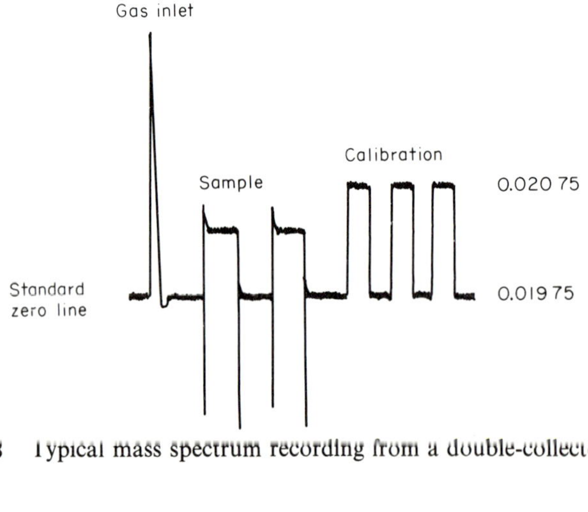

Fig. 6.13 Typical mass spectrum recording from a double-collector system.

Computer-controlled instruments show on the printout only the ratio of the isotopes. In this case it is not necessary to take measurements from a recording of ion currents. Mariotti *et al.* (1978) describe measurements at natural abundances of nitrogen using the double collector. Atmospheric nitrogen was found to be 0.3663 ± 0.0004 atom % ^{15}N. Detection of variations of 7×10^{-5} ^{15}N atom % have been successfully carried out. Faust *et al.* (1980) report a method for the determination of ^{15}N abundances in small amounts of biological material. The sample preparation is carried out in a two-chamber ampoule. The part where the reagents are placed is melted off to avoid interferences of the chemicals used for the combustion. The mass spectrometer used for the measurements was equipped with a double collector.

E. Data evaluation

1. *Single collector measurement*

During successive measurements of one sample, the intensity of the peaks in general decreases steadily. This can be explained by the flow of sample gas from the reservoir to the ion source, thus reducing the gas pressure in the reservoir and resulting in a decreasing number of gas molecules which enter the ion source. In order to allow for this decrease, the ^{15}N percentage abundance can be calculated as follows: As illustrated in Fig. 6.14, successive

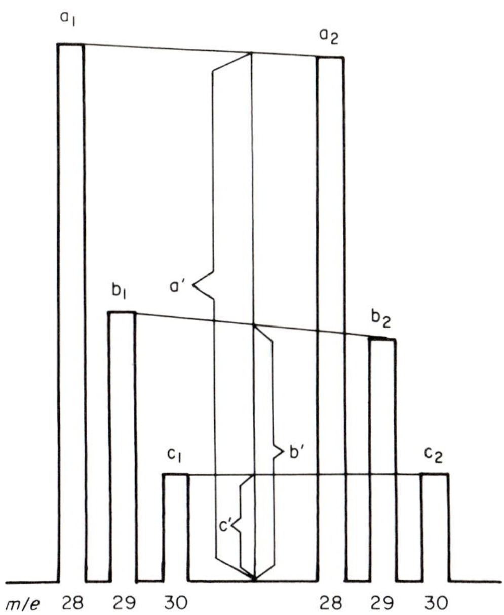

Fig. 6.14 Evaluation of mass spectrometric results.

peaks of $m/e = 28$, 29 and 30 are connected by means of a straight edge, between points a_1–a_2, b_1–b_2 and c_1–c_2, respectively. A vertical line is then drawn at a point equidistant from the peaks of $m/e = 30$ and $m/e = 28$. From the vertical line the peak heights are measured as indicated by a′, b′ and c′ of Fig. 6.14. In this way, the measurements correspond to peak heights, where each isotope was measured at the same time. The calculation of the ^{15}N percentage abundance can then be found according to equation (6.3) or equations (A.10) and (A.11) in the Appendix in Volume 2.

2. *Double collector measurement*
The distance between the zero line and the deflection of the sample is measured first (Fig. 6.13). This is followed by the measurement of the deflection caused by the calibration. The final result is calculated using the following data: calibration value (resistor) at zero line, 0.01975; calibration value changed 0.02075; distance zero line to sample 12 mm; distance for calibration change 30 mm. The value for the sample is then computed as follows:

$$R_{sa} = 0.01975 + \frac{(0.02075 - 0.01975)}{30} \times 12 = 0.02015.$$

The final result according to equation (6.4) is

$$\delta^{15}N = \left(\frac{0.02015}{0.01975} - 1\right) \times 1000 = 20.25‰.$$

V. Emission spectrometry

A. Principle

In emission spectrometry an external energy source is used to bring the nitrogen molecules to an excited state. The excited molecules return to the ground state, and the energy difference is emitted as electromagnetic radiation of specific energy. The small differences in wavelengths of the light emitted by the excited $^{28}N_2$, $^{29}N_2$ and $^{30}N_2$ gas molecules are used for the measurement. The emitted light is resolved by a monochromator and the light intensities of the wavelength can be measured and related to molecules present in the system. The recorded peak heights corresponding to the intensities of the emitted radiations allow the ^{15}N percentage abundance to be calculated in a way similar to mass spectrometry.

Figure 6.15 illustrates a block diagram of an emission spectrometer and consists of the following units: (i) a microwave (high-frequency) power generator for the excitation of electrodeless discharge sample tubes; (ii) an

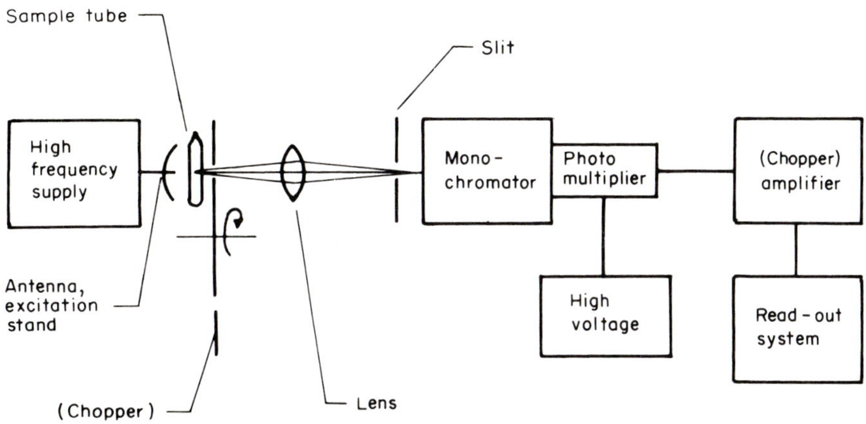

Fig. 6.15 Block diagram of an emission spectrometer for ^{15}N analysis.

excitation stand which consists either of an antenna or a cavity in which the sample tube can be fixed for the transfer of energy from the generator to the nitrogen gas; (iii) a chopper to avoid the influence of other light sources; (iv) a lens to focus the light on the entrance slit of the monochromator; (v) a monochromator to resolve the light consisting of different wavelengths; (vi) a photomultiplier and amplifier for the conversion of the light quanta to electrical signals; and (vii) a recorder or digital integrator with printout, to convert the electrical signals related to light quanta into a readable record. The measurement of peak heights from the recorder or the integrated value to calculate ^{15}N percentage abundance is done in a similar way as for mass spectrometry.

Compared to mass spectrometry, emission spectrometry is much simpler. No high vacuum is involved; the instrument can be installed, maintained and operated by any experienced analyst.

B. Theory of molecular spectra

Isotopes and molecules differ slightly in their respective energy levels; therefore, isotope analysis is possible by emission spectrometry.

Atoms can incorporate energy by raising one or more electrons to a higher energy state. This absorbed energy differs from one isotope to another, and when the electron(s) returns to the ground state(s), the energy is emitted in the form of electromagnetic radiation of distinct wavelengths. Atomic or line spectra are formed where each line arises from a discrete energy difference. The energy difference for identical transitions of isotopes amounts to some hundredths of a nanometre. With molecules, and only di-atomic ones will be considered here, there are five ways in which energy can be incorporated, but only three are of interest for the present spectrometric purpose, namely electronic energy, vibrational energy and rotational energy.

The incorporation of energy may be by (i) raising one or more electrons to a higher orbital similar to the case of atoms; (ii) increased vibration (oscillation) of the two atoms along the internuclear axis, and (iii) increasing the rotation of the molecule around an axis passing through the centre of gravity perpendicular to the internuclear axis.

Each energy change involved in the present process can be explained by a combination of these three kinds of energy storage.

A transition between two electronic states, including the vibrational transitions, would show a line spectrum as in the case of atoms, but superimposed in each vibrational transition are allowed rotational levels which have much smaller energy differences. The whole constitutes an electronic band spectrum, though not every band is equally probable, of which some bands are more intense than others. Each band therefore represents a transition from a vibration state in the excited electronic state to one in the ground state with its associated allowed rotational transitions. In the case of nitrogen, the 0–0 band has the greatest intensity (Fig. 6.16)

The mass difference between isotopic molecules manifests itself mainly in the vibrational and rotational energies (as moments of inertia are involved) and the corresponding wavelengths of emitted light. The isotope shift for an 0–0 vibrational transition is zero or small (in the case of $^{28}N_2$ and $^{29}N_2$ it is only 0.011 nm) but the difference becomes greater with increasing order of

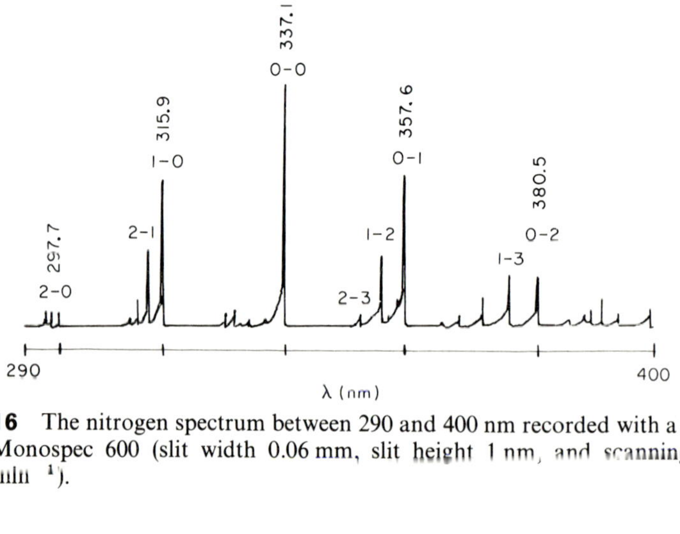

Fig. 6.16 The nitrogen spectrum between 290 and 400 nm recorded with a Hilger–Watts Monospec 600 (slit width 0.06 mm, slit height 1 nm, and scanning speed 2.5 nm min^{-1}).

the vibrational transition. At the same time the intensities of the bands, decrease rapidly, so that a compromise must be made between increasing isotope effects and the decreasing intensity. In the case of nitrogen, the 2–0 transition at 297.7 nm for $^{28}N_2$ has been generally chosen (Clusius and Becker, 1947; Hoch and Weisser, 1950; Hürzeler and Hostetler, 1955; Obolenskaya and Zhadkova, 1964; Leichnam *et al.*, 1968). This transition has the additional advantage that there are no interfering bands from impurities except for carbon monoxide which has one at 297.6 nm (Meier and Müller, 1965).

If the same electronic transition probabilities are assumed for the isotopic molecules, the intensities of the band heads can be used as a measure of the isotopic molecular concentrations. Of course, the rotational bands also exhibit isotope effects so that the isotopic bands differ slightly, but these effects only influence very precise evaluations. The wavelengths of the band heads of the three isotopic molecules of nitrogen are 297.7 nm for $^{28}N_2$, 298.3 nm for $^{29}N_2$ and 298.9 nm for $^{30}N_2$. In an historical review, Middelboe (1979) indicates other bands which have been used by several authors. Burridge and Hewitt (1980) used the 1–0 transition (wavelength 315.9 nm) for the determination of ^{15}N percentage natural abundances and values below (down to 0.04 atom %).

C. Emission spectrometer characteristics

For the necessary excitation of nitrogen, the gas in a closed system was originally excited in Geissler tubes, but these tubes are inefficient, quite complicated and costly to prepare. The most convenient method is the electrodeless discharge where gases are excited by microwaves in the megahertz frequency range. Microwave discharge has found increasing application, as this discharge source has many attractive characteristics: (i) it can produce a high degree of ionization, disassociation and excitation without undue heating of any background gas; (ii) no electrodes are necessary, so that the reaction tubes are free from contamination and less subject to damage; (iii) the source has no dangerous high-voltage contacts.

High frequencies from 13 to 2450 MHz have been used and there are different opinions about the best frequency. Broida and Chapman (1958) stated that 150 MHz allows a wider pressure range for the nitrogen gas. Leichnam and Capitini (1964) found, when working with atomic spectra with some thousand megahertz, a high background, broader lines and self-absorption. By lowering the frequency to a few hundred megahertz the discharge operates at lower temperatures and pressures, and the disadvantages of the higher frequency decrease or even disappear. The advantage of using 2450 MHz for excitation is the longer life of the discharge by reducing "clean-up". This phenomenon is explained by the fact that the excited gas molecules strike the discharge tube walls with sufficient velocity to become fixed in or to the wall, but this is less pronounced at the higher frequencies, because of the shorter travel length of the molecule.

Microwave power generators especially designed for the excitation of nitrogen have a fixed frequency. In general, the frequencies of 13, 27 and 100 MHz are used in commercial instruments. This microwave generator needs a matching antenna or cavity which is used as an excitation stand. The sample is fixed into this excitation stand where ignition is effected.

The microwave generators for the excitation of nitrogen have fixed power outputs of 50–60 W. For a stable discharge a power of 10–20 W is adequate. The ideal device for the transfer of power is a cavity (also called a discharge, gas or resonant cavity or cavity resonator). The types recommended are the Everson cavity with a quarter wave and the Broida cavity with a three-quarter wave both at 2450 MHz. The latter should yield a more stable discharge (Marshall and West, 1970; Woodward, 1970).

Experience has shown that cooling of the excited sample container prolongs the life-time of the sample discharge. The cooling is done with compressed air or with a fan. Since heating during excitation also broadens the lines or bands and causes poor resolution, the cooling of the sample tubes is important. In the case of nitrogen, the walls of the tube at the height of the plasma should not exceed 50–60 °C (Hürzeler and Hostetler, 1955; Ferraris and Proksch, 1972).

The light produced by the sample is focused by a quartz lens on the entrance slit of the monochromator. To avoid the influence of other light sources, a chopper can be placed next to the excitation stand. The light enters the monochromator and is diffracted there by means of a NaCl prism, quartz prism or quartz grating. A NaCl monochromator is cheaper than one with a quartz prism; however, the crystal inside the monochromator must, of course, be kept dry. Grating monochromators have the advantage that dispersion is linear, that the luminosity is about 20 times higher (because of the shorter path length of the light beam and lower absorption by the ruled surface than by the prism), and that the optical system is simpler. This usually results in a lower price for a grating with the same reciprocal dispersion as a quartz prism monochromator. The monochromator should have an automatic scanning device with a speed of 0.5–1.5 nm min^{-1}.

The light beam then is deflected by a mirror system and is focused at the exit slit of the monochromator, where a photomultiplier is placed. The signal of the photomultiplier is amplified and the output of the amplifier is displayed on a recorder or a digital integrator with a printout system. If a chopper has been used in front of the excitation stand, the amplifier needs to incorporate a demodulator. One commercial instrument uses a prism which

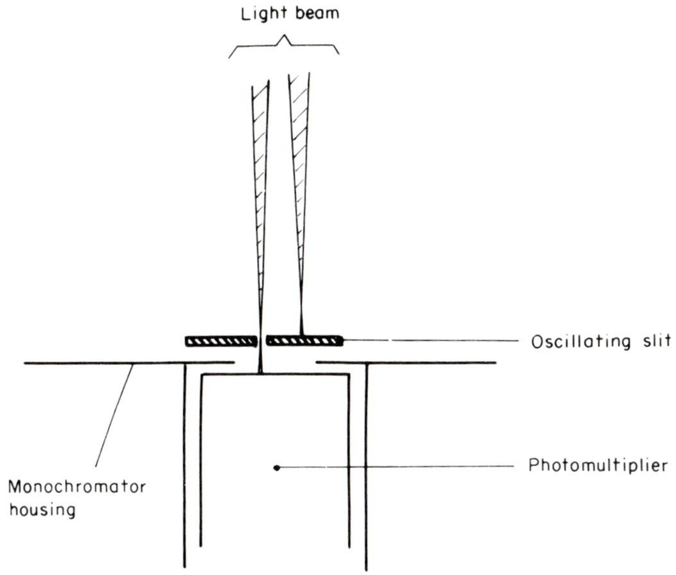

Fig. 6.17 Illustration of an oscillating slit for an emission spectrometer.

is adjusted so that the light beam produced by nitrogen samples is focused precisely at the exit slit of the monochromator. Instead of a scanning device, an oscillating slit (Fig. 6.17) is placed in front of the photomultiplier (Gerstenberger *et al.*, 1974; Gerstenberger and Meier, 1980). The amplitude of the oscillation is adjusted in such a way to cover the distance between the nitrogen bands involved in the measurement.

The photomultiplier used must be sensitive in the ultraviolet where the N_2 bands are emitted. The high voltage, up to about 1250 V supplied to the photomultiplier, must be highly stabilized.

The amplifier, whether it is a d.c. or an a.c. amplifier (in the case when a chopper is used), must have a dynamic range of at least 1000 to allow measurements of samples which are below natural abundance, as the ratio $^{28}N_2$:$^{29}N_2$ in the case of natural nitrogen is 135.6. The advantage of an a.c. amplifier which amplifies modulated light signals only is that light produced by other sources is rejected and results in a background which is about 100 times lower than that of a d.c. amplifier.

The recorder used should have a chart width of 250 mm and a response time of 0.5 s, and the chart speed should be selectable from 5 to 50 mm min^{-1}.

D. Commercial instruments

Most commercial instruments of modern design use a monochromator with a grating. The older NIA-1 model and the new N-150 model of JASCO (Japan) employ a recorder for the registration of the band intensities. The older NOI-4 model and the new NOI-5 model as well as the Isonitromat series of VEB Statron (German Democratic Republic) are still equipped with a NaCl prism. The NOI-5 uses a recorder like the JASCO models. The Isonitromat 5200 and 5201 are equipped with an oscillating slit previously described. The Isonitromat is equipped for the automated analysis of clear Kjeldahl extracts. The sample amount needed for a measurement is 120 μg of nitrogen. The range for measurements is from natural abundances to 30% ^{15}N. The function of this instrument is similar to that of the NOI-6 developed at the Institut für Stabile Isotope, Leipzig, GDR. This prototype has a monochromator with a grating. The NOI-6 has been improved in many details (mainly in the electronics) compared to the Isonitromat, but the description of the function stays the same.

The conversion of sample NH_3 to nitrogen gas is normally done in a unit which is an integral part of the instrument. The sample (Kjeldahl extract) is pipetted into the reagent container. The reagent container is linked to a helium tank, a reservoir filled with hypobromite solution and an exit port with drying reagent which is in turn connected via the discharge tube to a vacuum pump. The helium is used as a carrier gas. To enable an adequate flow of sample to the system, capillary leaks are used before and after the gas has passed the discharge tube part of the instrument. When the hypobromite is added to the Kjeldahl extract, nitrogen is produced and flows through the discharge tube where it is excited and can be measured. The detection of the nitrogen bands is done by an oscillating slit described before. The oscillator for driving the slit permits the use of an oscilloscope to show the different emission bands needed for the measurement. A four-channel analyser is used to integrate the bands of interest. Therefore, a microprocessor can calculate the results based on the data applied by the digital integrators of the four-channel analyser to produce printouts of band intensities. For automated analysis, 10 μg of nitrogen are required. With sample tubes, which can be analysed on the NOI-6 model only, 1 μg of nitrogen is necessary for a measurement. The precision of the results by either method is ±1.5% relative deviation at a confidence level of 95%. The range for measurements with the NOI-6 is from 0.365 to 80% ^{15}N abundance.

E. Data evaluation

The measurement of the band intensities alone does not always result in the calculation of the true ^{15}N abundance. When measurements are carried

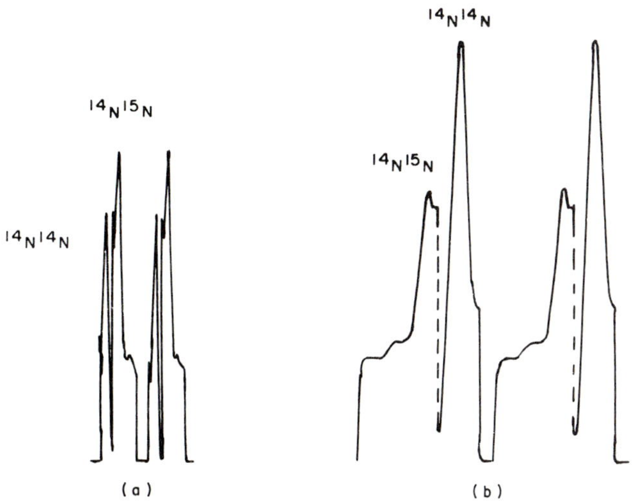

Fig. 6.18 Emission spectra of natural abundance nitrogen recorded with (a) NOI-4 and (b) NIA-1 spectrometers.

out on the NOI-4 or NOI-5 the use of calibration curves is necessary to end up with the true band intensities. A set of calibration tubes must be analysed and the values obtained have to be compared to those stated on the calibration tubes. It should be pointed out that at least two different curves must be used for the range from 0.366 to 90% ^{15}N abundance. Instruments employing a grating, in general, do not need calibration curves. Yamamuro (1981) gives curves for the 2–0 and the 3–1 bands showing the actual versus apparent percentage of ^{15}N for the NIA-1 spectrometer of JASCO.

Figure 6.18 shows spectra of natural nitrogen measured with the NOI-4 and NIA-1 models. The measurement of the band intensities is done in a similar fashion to that discussed under mass spectrometry. As the three isotopic molecules in the discharge are in equilibrium, only the two-peak formulae are necessary for calculation of the ^{15}N percentage abundance.

Due to the poorer resolution of the light beam as compared to the resolution of an ion beam, care of the background lines for $^{28}N_2$ and $^{29}N_2$ has to be taken.

In the range from 0.366 to 14–15% ^{15}N the percentage abundance of ^{15}N can be calculated according to equation (6.3), but above these values the formula

$$\text{Percentage } ^{15}\text{N} = \frac{100}{R' + 1}, \quad \text{where} \quad R' = \frac{\text{Intensity of } ^{29}\text{N}_2}{2 \times \text{Intensity of } ^{30}\text{N}_2}, \tag{6.16}$$

should be used. Here, as in the case of mass spectrometry, the peak height is a measure of the intensity, and the intensity is a measure of the relative concentration of the corresponding isotopic molecules.

The total intensity is obtained by multiplying the peak height by the sensitivity used; e.g.

$$R = \frac{18 \times 100}{14 \times 1},$$

where 18 is the peak height of the $^{28}N_2$, 14 is the peak height of the $^{29}N_2$, 100 is the sensitivity for $^{28}N_2$ and 1 is the sensitivity for $^{29}N_2$.

VI. The isotope dilution method

The isotope dilution method can be used to calculate the nitrogen concentration of a sample with known isotopic abundance. The sample is mixed with a known amount of nitrogen with different isotopic abundance than the sample. A measurement of the sample, the mix and the tracer is necessary for the calculation of the concentration. The following formulae can be used when the measurements have been performed on equilibrated samples, mix and tracer (de Bievre, 1971):

$$c = c_t\left(\frac{R_t - R_{mix}}{R_{mix} - R_s}\right)\left(\frac{1 + R_s}{1 + R_t}\right), \tag{6.17}$$

where c is the concentration of the sample, c_t is the concentration of the tracer, R_t is the $^{15}N:^{14}N$ ratio of the tracer, R_{mix} is the $^{15}N:^{14}N$ ratio of the mixture, and R_s is the $^{15}N:^{14}N$ ratio of the sample, or

$$c = c_t \frac{\left(\frac{R'_t}{2}\right) - \left(\frac{R'_{mix}}{2}\right)}{\left(\frac{R'_{mix}}{2}\right) - \left(\frac{R'_s}{2}\right)} \; \frac{1 + \left(\frac{R'_s}{2}\right)}{1 + \left(\frac{R'_t}{2}\right)}, \tag{6.18}$$

where R' is the $^{29}N_2:^{28}N_2$ ratio.

A different formula which can also be used in emission spectrometry was reported by Faust *et al.* (1976):

$$c = c_t\left(\frac{A_t - A_{mix}}{A_{mix} - A_s}\right), \tag{6.19}$$

where A is the atomic abundance.

The following example illustrates that equations (6.17), (6.18) and (6.19) give the same results, where c_t, the concentration of the tracer, is 10 mmol of nitrogen:

	$^{15}N:^{14}N$ ratio	$^{29}N:^{28}N$ ratio	^{15}N percentage abundance (%)
Sample	0.003 67	0.007 34	0.366
Mixture	0.199 04	0.398 07	16.6
Tracer	19.0	38.0	95.0

Using the $^{15}N:^{14}N$ ratio,

$$c = 10\left(\frac{19 - 0.19904}{0.19904 - 0.00367}\right)\left(\frac{1 + 0.00367}{1 + 19}\right) = 48.29 \text{ mmol of nitrogen;}$$

using the $^{29}N:^{28}N$ ratio,

$$c = 10\left(\frac{(38 - 0.39807)/2}{(0.39807 - 0.00734)/2}\right)\left(\frac{1 + (0.00734/2)}{1 + (38/2)}\right)$$

$$= 48.29 \text{ mmol of nitrogen;}$$

using the ^{15}N percentage abundance,

$$c = 10\left(\frac{95 - 16.6}{16.6 - 0.366}\right) = 48.29 \text{ mmol of nitrogen.}$$

The isotope dilution method also enables the calculation of isotope abundances when the concentrations of nitrogen in the sample and in the tracer solution are known. An interesting and important use of the addition of tracer nitrogen is in the case where the amount of nitrogen in the sample is so small that manipulation is difficult. The added volume of the tracer makes the collection of the gas and its subsequent analysis possible. The only information required is the isotopic composition of the tracer and of the mixture, and the concentration of the nitrogen in the tracer and the sample. The isotope abundance in atom % is then calculated by the equation

$$A_s = A_{mix} - (A_t - A_{mix})(c_t/c). \tag{6.20}$$

Using the values from the previous example we obtain

$$A_s = 16.6 - (95 - 16.6)(10/48.29) = 0.365 \text{ atom \%}.$$

The result varies slightly from the true value. The estimation of errors is given by Faust *et al.* (1976). The best results are obtained when the ^{15}N percentage abundance in the mixture is around 50%.

VII. Summary and conclusion

Mass and emission spectrometry are analytical tools for the determination of ^{15}N. Mass spectrometry has the advantage that more accurate results compared to emission spectrometry are possible. Emission spectrometry, on the other hand, is less expensive and only requires samples at least 50 times smaller for the analysis. The sample preparation method is similar for both techniques. Whether the "wet" Kjeldahl–Rittenberg method or the "dry" modified Dumas method for sample preparation is chosen, the method has to be adapted to the special needs of the sample material. For the reader interested in delving deeper into the field of ^{15}N analysis, a vast compilation of references and abstracts is given by Faust (1981).

Future trends show that minicomputers or microprocessors will be used more and more to automate mass and emission spectrometers. This will allow measurements with more precision in less time. Integrators and printers will be used instead of recorders. Graphical displays and the possibility to dump the graphics will allow more control and the calculated final results will be printed. Permanent control together with calibration measurements will correct for any change in the measurement system. Disadvantages will be that maintenance of these instruments will become more complicated, and highly qualified specialists rather than the usual operators will be required.

References

Baker, P. R. W. (1961). *Talanta* **8**, 57–63.

Balestieri, C. (1968). *Life Sci.* **7**, 269–274.

Beynon, J. H. and Brenton, A. G. (1982). "An Introduction to Mass Spectrometry", University of Wales Press, Cardiff.

Bremner, J. M. (1965). In "Methods of Soil Analysis" (C. A. Black, ed.), pp. 1149–1178 and 1256–1286, American Society of Agronomy, Madison, Wisc.

Broida, H. P. and Chapman, M. W. (1958). *Anal. Chem.* **30**, 2049–2055.

Brüel, D., Holter, H., Lindstrøm-Lang, K. and Rözits, K. (1946). *Compt. Rend. Trav. Lab. Carlsberg. Ser. Chim.* **25**, 289–324.

Burridge, J. G. and Hewitt, E. W. (1980). *Anal. Chim. Acta.* **118**, 11–28.

Capindale, J. B. and Tomlin, D. H. (1957). *Nature (Lond.)* **180**, 701–702.

Clusius, K. and Becker, E. W. (1947). *Z. Naturforsch.* **A2**, 154–158.

Clusius, K. and Rechnitz, G. (1953). *Helv. Chim. Acta* **36**, 59–62.

Conway, E. J. (1947). "Microdiffusion Analysis and Volumetric Error", 2nd edn, Crosby-Lockwood, London.

Cook, G. B., Goleb, J. A. and Middelboe, V. (1967). *Nature (Lond.)* **216**, 475–476.

Cook, G. B., Goleb, J. A. and Middelboe, V. (1968). *Anal. Chim. Acta* **43**, 229–234.

Cook, G. B., Goleb, J. A. and Middelboe, V. (1970). *Appl. Spectr.* **24**, 608–610.

de Bievre, P. (1971). *In* "Stable Isotope Dilution: A Possible Tool for the Analytical Chemistry of Noble Metals. Symposium on Nuclear Analysis Techniques in the Production and Industrial Use of Noble Metals", Report no. 71, Brussels, Commission of the European Communities.

Dumas, J. (1834). *J. Pharm. Chim.* **29**, 144–151.

Faust, H. (1965). *Isotopenpraxis* **1**, 62–65.

Faust, H. (1967). *Isotopenpraxis* **3**, 100–103.

Faust, H. (1969). Paper presented at the IAEA Research Coordination Meeting on Recent Developments in the Use of ^{15}N in Soil–Plant Studies, Sofia, Bulgaria, December, 1969.

Faust, H. (1981). *Zentralinstitut für Isotopen- und Strahlenforschung, Leipzig Mitt.* **49**, 1–66.

Faust, H., Müller, G. and Störl, H-J. (1976). *Isotopenpraxis* **5**, 205–210.

Faust, H., Bornhak, H., Hirschberg, K. and Birkenfeld, H. (1980). *Zentralinstitut für Isotopen- und Strahlenforschung, Leipzig Mitt.* **29**, 217–225.

Ferraris, M. M. and Proksch, G. (1972). *Anal. Chim. Acta* **59**, 177–185.

Fiedler, R. and Proksch, G. (1972). *Anal. Chim. Acta* **60**, 277–285.

Fiedler, R. and Proksch, G. (1975). *Anal. Chim. Acta* **78**, 1–62.

Fiedler, R., Proksch, G. and Köpf, A. (1973). *Anal. Chim. Acta* **65**, 435–443.

Gerstenberger, H. and Meier, G. (1980). *Isotopenpraxis* **17**(4), 150–155.

Gerstenberger, H., Meier, G., Mauersberger, K., Müller, G., Papst, D., Riedel, D., Rose, W. and Schmidt, G. (1974). *Fachz. Labor.* **18**, 452–455, 547–549 and 612–615.

Hoch, M. and Weisser, H. R. (1950). *Helv. Chim. Acta* **38**, 2128–2134.

Holt, P. F. and Hughes, B. P. (1955). *J. Chem. Soc.* **1955**, 95–98.

Hoskins, J. L. (1944). *Analyst* (*Lond.*) **69**, 271–277.

Hürzeler, H. and Hostetler, H. V. (1955). *Helv. Chim. Acta* **38**, 1825–1827.

Hüser, R. (1963). *Z. Anal. Chem.* **197**, 16–20.

Hüser, R., Habfast, K. and Bradke, M. (1960). *Z. Anal. Chem.* **176**, 429–433.

Kienitz, H. (1968). "Massenspektrometrie", Verlag Chemie, Weinheim.

Kirk, P. L. (1934). *Mikrochem. Mikrochim. Acta* **16**, 13–19.

Kirk, P. L. (1950). *Anal. Chem.* **22**, 354–361.

Kjeldahl, J. (1883). *Z. Anal. Chem.* **22**, 366–371.

Leichnam, J. P. and Capitini, R. (1964). Rapport CEA no. R-2428.

Leichnam, J. P., Middelboe, V. and Proksch, G. (1968). *Anal. Chim. Acta* **40**, 487–502.

Lloyd-Jones, C. P., Adam, J. S., Hudd, G. A. and Hill-Cottingham, D. G. (1977). *Analyst* (*Lond.*) **102**, 473–476.

Mariotti, A., Letolle, R. and Guillemot, J. (1978). *Analusis* **6** (10), 421–425.

Marshall, G. B. and West, T. L. (1970). *Anal. Chim. Acta* **51**, 179–189.

McDowell, C. A. (1963). "Mass Spectrometry", McGraw-Hill, New York.

McKenzie, H. A. and Wallace, H. S. (1954). *Aust. J. Chem.* **7**, 55–59.

Medina, R., Hoppe, W. and Schmidt, H. L. (1978a). *Fresenius Z. Anal. Chem.* **292**, 403–407.

Medina, R., Schmidt, H. L., Hoppe, W. and Traut, L. (1978b). In "Stable Isotopes. Application in Pharmacology, Toxicology and Clinical Research" (T. O. Baillie, ed.), pp. 303–309, Macmillan Press, London.

Medina, R., Hoppe, W. and Schmidt, H-L. (1979). In "Proceedings of the Third International Conference on Stable Isotopes" (E. R. Klein and P. D. Klein, eds), pp. 153–156, Academic Press, New York and London.

Meier, G. and Müller, G. (1965). *Isotopenpraxis* **1**, 53–62.
Middelboe, V. (1970). *Appl. Spectr.* **24**, 608–614.
Middelboe, V. (1979). *Roy. Vet. Agr. Univ. Yearbook* **1980**, 1–22.
Middleton, G. and Stuckey, R. E. (1951). *J. Pharm. Pharmacol.* **3**, 829–835.
Obolenskaya, L. I. and Zhadkova, N. T. (1964). *Khim. Sel. Khoz.* **1**, 13–21.
Ogg, C. L. and Willits, C. O. (1950). *J. Ass. Off. Agr. Chem.* **33**, 179–186.
Perrin, C. H. (1953). *Anal. Chem.* **25**, 968–971.
Perschke, H. and Proksch, G. (1971). Report no. IAEA-PL-341-16, IAEA, Vienna.
Perschke, H., Proksch, G., Kerö, E. and Mühl, A. (1971). *Anal. Chim. Acta.* **53**, 459–463.
Proksch, G. (1969). *Plant Soil* **31**, 380–384.
Reid, A. S. J., Cumming, G. L. and Webster, G. R. (1967). *Plant Soil* **26**, 196–198.
Rittenberg, D. (1948). *In* "Preparation and Measurement of Isotopic Tracers" (D. W. Wilson, A. O. C. Nier and S. P. Riechman, eds), pp. 31–42, Edwards, Ann Arbor, Mich.
Roboz, J. (1968). "Introduction to Mass Spectrometry, Instrumentation and Techniques", John Wiley & Sons, New York.
Rolle, W. (1962). *Kernenergie* **5**, 403–405.
Ross, J. and Martin, E. (1970). *Analyst* (*Lond.*) **95**, 817–822.
Sims, A. P. and Cocking, E. C. (1958). *Nature* (*Lond.*) **181**, 474.
Smith, J. H., Legg, J. O. and Carter, J. N. (1963). *Soil Sci.* **96**, 313–319.
Sommer, K. and Kick, H. (1965). *Z. Anal. Chem.* **220**, 21–29.
Tompkins, E. R. and Kirk, P. I. (1942). *J. Biol. Chem.* **142**, 477.
Wingo, W. J., Davies, O. L. and Anderson, L. (1950). *Anal. Chem.* **22**, 1340–1346.
Woodward, C. (1970). *Anal Chim. Acta* **51**, 548–553.
Yamamuro, S. (1981). *Soil Sci. Plant Nutr.* **27**, 405–419.

Index

D

E

L

M

N

S